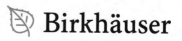

Reinhard Haberfellner • Olivier de Weck
Ernst Fricke • Siegfried Vössner

# Systems Engineering

Fundamentals and Applications

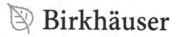 Birkhäuser

Reinhard Haberfellner
Institute of General Management
and Organization
Graz University of Technology
Graz, Austria

Ernst Fricke
BMW AG
Munich, Germany

Olivier de Weck
Engineering Systems Division
MIT
Cambridge, MA, USA

Siegfried Vössner
Engineering and Business Informatics
Graz University of Technology
Graz, Austria

ISBN 978-3-030-13433-4     ISBN 978-3-030-13431-0   (eBook)
https://doi.org/10.1007/978-3-030-13431-0

Library of Congress Control Number: 2019934964

This book is published under the imprint Birkhäuser, www.birkhauser-science.com by the registered company Springer Nature Switzerland AG
The registered company address is: Gewerbestrasse 11, 6330 Cham, Switzerland

# Preface to the English Version

This book is a translation of the successful German textbook "Systems Engineering – Grundlagen und Anwendung," which is in its 14th edition and focuses on the basics of problem-solving and the necessary ambidexterity to turn problems into solutions – systems design and project management. In recent years, we have witnessed increasing requests by European, American, and Asian colleagues, along with German universities, as more and more lectures are being given in English, to provide an English translation of our book.

What are the success factors of the book we are presenting here? Is there really a need for an additional book on systems engineering in English? What advantages does it offer in comparison with other offerings? In our view, there are several reasons for this publication.

The term "systems engineering" was originally coined at Bell Labs in the 1940s. Best practices and formal methods of systems engineering have emerged since the 1950s and have been codified in a number of standards and handbooks. These standards are very helpful in giving structure and consistency to the systems engineering process. However, they often promote a specialist nomenclature and mindset and a rather rigid approach to systems engineering that is sometimes difficult to apply in practice. This is especially true for small and medium-sized enterprises in addition to those from the non-aerospace and defense industries. Often, these concepts and standards were the product of teams consisting of several hundreds or even thousands of members from various sectors and companies, which makes these concepts more extensive, more complex, and more difficult to understand. Looking at projects in industry, the basics are often forgotten or poorly implemented.

Hence, what is needed is a structured, common-sense, and transparent approach to solving problems that can be understood by a large range of individuals and not only by highly skilled technical professionals. This is what this book offers.

Our concept is originally based on the work of A.D. Hall,[1] which somewhat inexplicably has lost its prominence over the years. A *small group of engineers*[2] in the industrial engineering department of ETH-Zürich (BWI) revived his ideas in the 1970s and developed a similar concept, which has been continuously refined over the last few decades and may therefore be called the Hall/BWI approach.

We believe that our approach has *essential strengths*, which are made up of a number of facets:

- The idea of the systems engineering concept can be presented in a *single graphic view* for easier clarification (see Fig. 1)[3].
- *Systems thinking* is well integrated into the concept
- The *process model* as an essential part of the concept is divided into four modules, which may be combined according to the characteristics and needs of a particular project (Fig. 2.7, later).

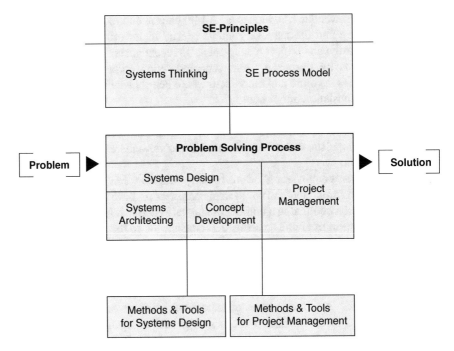

**Fig. 1** The systems engineering concept shown as a "systems engineering manakin"

---

[1] Hall, A.D. (1962): A Methodology of Systems Engineering.

[2] Büchel, A. (1969): Systems Engineering. Eine Einführung.

[3] The notion of "problem" in the figure, as we define it here is the difference between a situation as it currently exists (= present state), which includes problems and/or opportunities, and a future imagined state (= solution as the target state) that is hopefully "better" than the current situation, even if state of this future target at the beginning of the project is quite vague, uncertain or even controversial.

- The *"agile" methodologies*, originating from software engineering, are covered, and are differentiated against and where possible integrated into our approach in a well-disposed manner, avoiding unnecessary ideology.
- *Special methods, techniques, and tools* are *not part of our core model*, but may be individually added according to the needs of the project (size, subject area, branch, corporate rules, etc.). They may come from multiple disciplines.
- The concept is of *general relevance* and may be used in many branches and tasks using industry-specific terms, and the principles may be applied to large projects and to rather small ones. This is demonstrated in three case studies.
- One of the objectives of this book is to make *good engineering easier to understand and teach.*
- In contrast to other approaches, our book provides an approach for general *problem solving* rather than a set of hard-wired recipes.

In other words: We do not offer rigid processes or procedures but rather a *mental framework that gives direction* but has to be *interpreted intelligently* by project teams and their leaders. The success of a project is in our opinion not mainly based on the use of *cutting-edge methods and tools*, which often are not known or understood by other team members. Many experienced experts are convinced that the reason why most projects fail in practice is that *basic systems engineering wisdom* was not observed by the project teams and/or the decision-making authorities. This is what we want to focus on.

The *target audiences* for our book are widespread: the way of thinking and approach presented here are suitable for:

- Engineers and engineering students from all faculties (mechanical, electrical, industrial, mechatronics, informatics, civil, architecture, etc.)
- Economists, MBA students, etc.
- Planners and designers of all types of systems: hardware, software, processes, organizations, logistics, etc.
- Practitioners in industry, government, and nonprofit organizations

We hope you enjoy this book. A classic approach for modern times.

| | |
|---|---|
| Graz, Austria | Reinhard Haberfellner |
| Cambridge, MA, USA | Olivier de Weck |
| Munich, Germany | Ernst Fricke |
| Graz, Austria | Siegfried Vössner |
| April, 2019 | |

---

The original version of this book was revised. The correction is available at
https://doi.org/10.1007/978-3-030-13431-0_17

# Systems Engineering Concept and Structure of the Book

Under this section you can find a description of the basic ideas of the systems engineering concept and the structure of the book

## The Basic Ideas of the Systems Engineering Concept

The reader will find a thorough description of a proven methodology that helps in dealing with problems, no matter what their nature.

A problem here is understood as the difference between what exists (= *actual state*) and the idea of a *desired state*, however vague, uncertain, and even controversial this may be at the beginning. This situation occurs in pronounced form in practically every project and is depicted with question marks in Fig. 2:

- In many cases, the *actual state* and its assessment are not (fully) known and it is necessary to carry out investigations and surveys to find out.
- The appraisal of the *actual state* cannot be seen in isolation from the knowledge of the ideas, expectations for the *desired state*.
- The *desired state* of course does not exist at the beginning and has to be elaborated.
- If both the *actual state* and the *desired state* are uncertain, the difference (delta) may not be judged uniformly, but rather large or rather small. Related to that, the way, the urgency, the allowable expenses for reducing the gap, etc., may be assessed differently and give rise to discussions.

If you want to find a solution to a problem, many factors are decisive. These range from knowledge about the situation, experience, methodology, a behavioral component, professional ethics, etc.

This is expressed in Fig. 3: systems engineering represents the methodological component in problem-solving and should help to coordinate the different factors with each other. However, we do not think that methodology alone solves any

**Fig. 2** A problem is the difference between an actual and a desired state

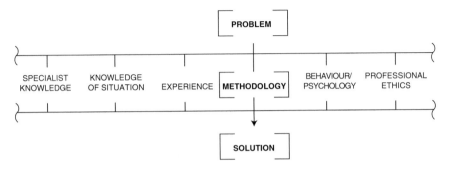

**Fig. 3** Systems engineering as a methodological component in problem-solving

problems. Although methodology is located in the center of Fig. 3 it is not the only and not always the most important factor in problem-solving: the success or failure of a project may be caused by any other component, as shown in Fig. 3.

These considerations have led to the system engineering concept as shown in Fig. 1 and briefly described below.

## Structure of the Book

This book consists of seven parts:

*In Part I, Systems Engineering principles*, the intellectual framework of our systems engineering approach is described. These are *Systems Thinking* (Chap. 1) as a tool to structure situations and circumstances, to represent it in their contexts, and thus to help to a better understanding, clear demarcation, and problem-adequate shaping of solutions. In Chap. 2 our *systems engineering process model* is outlined. This consists of various basic principles and four components that help to subdivide the development and implementation of a solution into manageable substeps. In addition, the process model can – like building blocks – be assembled from its components, and thus be aligned to the size and complexity of a project.

The systems engineering principles are first described in a simplified form, without regard to special and exceptional cases, which are described in later parts.

In *Part II, Problem-Solving Process*, the application aspects of the systems engineering principles and especially those of the process model are dealt with. Also, limitations or enhancements that were stated in Part I are explained here.

The problem-solving process is conceptually structured in *Systems Design* (= content structure of the solution, Chap. 3) and *Project Management* (= organizational aspects – Chap. 4, such as agreement on a project contract, establishing the project group, their organizational integration into the parent organization, establishing a steering committee, agreement of dates of reporting, etc.).

*Part III, System Design*, is structured into *Systems Architecting* – in terms of giving a basic structure to the solution, finding the solution principle – and into *Concept Development*, which deals with the more concrete manifestation of the architecture.

Chapter 5, *Systems Architecting*, gives a brief description of the meaning, task, options and impacts of the choice of a certain system architecture. In Chap. 6, *Concept Development*, the single steps of the problem-solving cycle are deepened: situations analysis, formulation of objectives, concept synthesis, concept analysis, evaluation, and decision. These steps offer the same logic and basic structure to be repeated at the various levels of concretization.

*Part IV, Case Studies* summarizes the *Systems Engineering Basics* of the Hall/BWI approach as represented here (Chap. 7). In Chap. 8, *Private House Building*, a simple project – to build a single-family house – is described, from the definition of the basic intent to the detailed planning. As an example of a very complex system, *Airport Planning* is described in Chap. 9, showing the systems engineering methodology in script-like dialogs. In Chap. 10 we added an additional case study: planning a *smart city* with a very interesting and attractive *science tower*.

*Part V, Systems Engineering for Practice*, gives practical assistance for projects. In Chap. 11 we give 7 basic recommendations for the application of systems engineering in your project. Chap. 12 describes *Typical Weak Areas in Projects (Stumbling Blocks)*. Chap. 13 contains *Activities Checklists*, which are intended to support the quality of project execution, and in Chap. 14, the *Characteristics of Successful Project Management* as the results of an empirical study are outlined.

*Part VI, Methods, Techniques, and Tools* (MTTs), first provides an overview of MTTs, which can be used in systems engineering (Chap. 15), without becoming an integral part of the methodology. In Chap. 16 about 100 MTTs are briefly described, which can support the system design and project management and which come from a variety of disciplines.

The back matter is the collection pot for different directories: answers to the questions to knowledge and comprehension, which are given after each of the former chapters; list of figures; bibliography; and alphabetic subject index.

This book is intended both as a textbook that will assist the user when accessing the topic of systems engineering. On the other hand, it is designed as a reference guide for later use. This purpose is supported by an index, which facilitates access to keywords and terms.

## Set of Questions[4]: Systems Engineering Concept in General

1. Describe the basic ideas of the systems engineering concept.
2. What significance do methodology and process have in problem-solving? Is methodology the only component? If not, what are the others?
3. What is the role of project management within the systems engineering concept?

---

[4] Please try to answer these questions on your own first. In the back matter you will find answers; however, we would admit other reasonable answers as well.

# Acknowedgments/Special Mentions

We would like to thank all contributors and authors of earlier versions of the book whose contributions we were free to use: Prof. Alfred Büchel, Dr. Peter Nagel, Dipl.-Ing. Heinrich von Massow, Dr. Klaus Rutz, and Dr. Mario Becker

Thanks to Eric Bye and Hermann Schibli in addition to Phil Herrick, Peter Moertl, and Donatella Turrini for translating the English version of the book.

# Contents

# List of Figures

# Part I
# SE-Principles

# Chapter 1
# Systems Thinking

Systems thinking and the systems engineering process model are important principles of systems engineering (Fig. 1.1).

## 1.1 Purpose and Terminology

### 1.1.1 Systems Thinking as a Part of the Systems Engineering Concept

Systems thinking is understood here as a way of thinking that enables better understanding and designing of complex phenomena (= systems).

Systems thinking includes notably:

- Terms to describe complex entities and their relations
- Model-based approaches to illustrating real *complex phenomena* without having to simplify them unduly
- Approaches that support *holistic thinking*

### 1.1.2 Basic Terms and Characteristics of Systems

Certain *basic terms* are used to describe systems. These terms must first be defined and characterized.

---

The original version of this chapter was revised. The correction to this chapter is available at https://doi.org/10.1007/978-3-030-13431-0_17

© Springer Nature Switzerland AG 2019, Corrected Publication 2021
R. Haberfellner et al., *Systems Engineering*,
https://doi.org/10.1007/978-3-030-13431-0_1

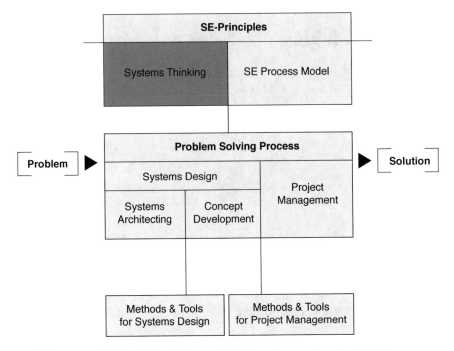

**Fig. 1.1** Systems thinking within the framework of the systems engineering concept

### 1.1.2.1  Systems/Elements/Relationships

In normal usage, many phenomena are described as systems, e.g., IT system, communications system, transport system, planning system, solar system, economic system, and educational system. There may also be added supplements such as "the corporation, a socio-technical-economic system" or "the pond, a biological system."

All these examples share common features that S. Beer formulated as follows:

"The word system stands for connectivity. By that we mean every accumulation of parts that stand in relation with each other. What we define as a system is therefore a system because it comprises parts that stand in relation with each other and in a certain respect form a whole" (Fig. 1.2):

- Systems therefore consist of *elements* (parts/components), meaning in a very general sense the building blocks of the system.
- Elements have *properties* and *functions*: for physical elements the properties could be, for example, dimensions or color. The function corresponds mostly to the intended purpose of an element within the framework of a certain system context.
- Elements themselves can be seen as systems.
- Elements are linked to one another through *relations*. This term may also be understood in a very general sense. It could concern material flow relations, information flow relations, situational relations, cause–effect relations etc.

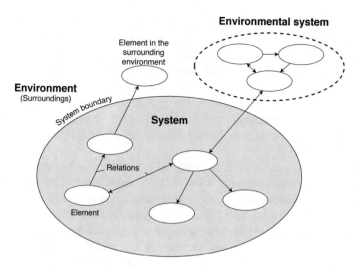

**Fig. 1.2** Basic terms of systems thinking

### 1.1.2.2 System Boundary/Environment

A system boundary is understood to be a more or less arbitrary border between the system and its surroundings or the environment in which it is embedded. The systems that are of interest in systems engineering are generally open. Their elements display relations not only among themselves but also with their surroundings.

Located within the *environment* or in the surroundings of a system are systems or elements that lie outside the system boundary, but that nonetheless affect the system or are influenced by the system. To emphasize the character of these systems, one speaks of environmental systems.

It is typical that a greater (stronger, more important) degree of relations exists or is recognized within the system boundaries than between the system and its environment. This "preponderance of internal connection" (N. Hartmann) creates an entity that can be considered as a system. Thereby, the system boundary does not have to be physically visible. It can be purely conceptual in nature and can even differ according to the point of view –more on this later.

### 1.1.2.3 Structure of a System

Elements and relations form a construct and thereby demonstrate an order. This is called the *structure* of a system. In it we can recognize arrangement patterns or ordering principles of the elements and the relations in a system. Examples of structural forms of this type are: hierarchical structures, stellar structures, network structures, layered structures, and structures with feedback.

*An Application of the Terminology Using the Example of an Industrial Enterprise*
When an industrial enterprise is understood as a system, it is thereby characterized
as consisting of many different *elements* or components. These elements may
include employees, machines, organizational rules, products, raw materials, intermediate products, departments, and much more. Many *relations* in an industrial
enterprise are operative and therefore of great significance for its functioning. These
relations connect the elements among themselves, for example, as relationships of
material flow, information flow, energy flow, social relationships, command paths,
and work sequences.

Because it is an open system, an industrial enterprise also has surroundings that
are needed for survival and that is relevant to understanding its way of functioning.
Examples include: customers, market requirements, competitors, suppliers, employment market, state of technical knowledge, associations, cooperating partners,
agencies, laws, ecological surroundings, natural resources, proprietors, and banks.
Many different relations exist between the system and its environment, such as
material, information and energy flows, in addition to value flows, among others.

Although the boundary between system and environment is arbitrary, in many
cases – taking into consideration the purpose of the investigation or the purpose of
design – there are indications that can guide the drawing of boundaries. The following criteria would be appropriate for a boundary identification: a system comprises
those elements that have relations with one another and have a certain form, figure,
function, and property of the whole. Thus, the elements that are less important to the
characteristic of the whole are relegated to the surroundings (environmental system), even though they may have an effect on or be influenced by the system or
single elements. Taking the example of an industrial enterprise, everything that
belongs to it, spatially and/or legally and/or organizationally, lies within the system
boundary. Everything else belongs to the environment.

### 1.1.2.4  Subsystems

Taking one element of a system as a system itself by breaking it down, that is, by
constructing elements at a lower level and connecting them by means of relations
(Fig. 1.3), one then speaks of a *subsystem*. A department that is subdivided into
several work places, for example, would thus be a subsystem of an industrial
enterprise.

### 1.1.2.5  Suprasystems

When several systems are combined into a more comprehensive system, one uses
the term *suprasystem*. A suprasystem of an industrial enterprise would be, for
example, a multicorporate enterprise.

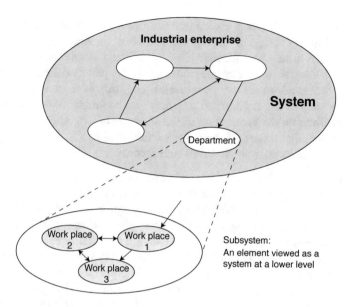

**Fig. 1.3** System and subsystem

### 1.1.2.6   System of Systems

The term "system of systems" or SoS is found repeatedly in the scientific literature. It is usually not clearly separated from the subsystems and suprasystems described above.

Here, the term is to be interpreted as follows: an SoS is a system consisting of several individual systems and further distinguished by two characteristics (Maier 1998):

- Each single system of this SoS is not exclusively dependent on its superordinate system, but can be operated independently of it, alone, or together with other systems. Therefore, it has its own purpose or (customer) benefit, which it can fulfill independently of other systems.
- Each single system of this SoS is primarily developed independently of the other systems and it can also be acquired separately.

Together, these systems produce an additional or greater benefit than a system by itself could enable.

In this context, a corporate group, consisting of several industrial enterprises, would not necessarily be a SoS:

- If the corporate group coordinates individual industrial enterprises through industry-wide management and leads them in a certain strategic direction (strategic holding), it would be a suprasystem, according to the traditional perspective. The term SoS would be dispensable.

– But if the corporate group is seen primarily as a finance holding that manages single subsidiary companies merely on the basis of financial demands, one could understand it to be a SoS: the individual firms would merely have to meet profitability goals within the group and would have no explicitly defined task/function with respect to other subsidiary companies. They may even have been purchased independently of one another and they may also be sold again without any noteworthy effects on the functioning of the other subsidiary companies. The additional benefit that a subsidiary company contributes to a finance holding would be its financial contribution to the overall success.

In this way an automobile, consisting of different subsystems such as chassis, car body, and engine, would not be an SoS: the chassis could not be operated alone (without body or engine) and fulfill its purpose. A typical example of an SoS, on the other hand, would be the integration of a mobile phone or an app into a car. Mobile phones, apps, and vehicles are to a large extent developed independently of each other and can fulfill their purpose/benefit without the other system. However, it is an additional benefit for the customer if the mobile phone can be operated from the steering wheel and uses the car's microphone and speakers. Thus, with an SoS, the design of interfaces to the systems with which it will be connected is particularly important (see also the following consideration regarding the black box).

### 1.1.2.7  Systems Hierarchy

When a system is subdivided into several levels (Fig. 1.4), the result is a hierarchical system structure, a *systems hierarchy*. This shows how the terms *system*, *subsystem*, and *element* relate to one another. The level of resolution, which is no longer

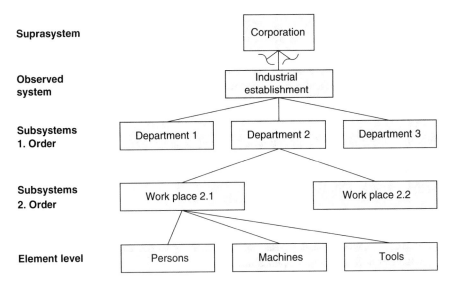

**Fig. 1.4**  Systems hierarchy

**Fig. 1.5**  Black box with
input/output relations

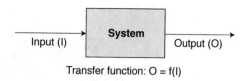

Transfer function: O = f(I)

subdivided – even if only provisionally – represents the *element level*. The elements can be seen as black boxes. The different subsystem levels lie in between the element level and the system level.

### 1.1.2.8  Black Box, Gray Box, and White Box

One speaks of a *black box* approach when the internal construction of a phenomenon is still without significance or unknown (Fig. 1.5). Only the function (the purpose) in addition to the present or desired inputs and outputs (results) are of interest. This is an important tool for reducing complexity.

One speaks of a *white box* when the exact connection between output and input is of great significance (for example, for a simulation) or when internal connections were analytically deduced and have to be observed in detail. This applies, for example, to finite element-models.

A *gray box* is roughly or partially structured. Roughly structured means that the degree of detailing is low and partially structured means that the degree of detailing of the structure varies: some areas of the system represented in the box would be structured in a rudimentary way or not at all, and others would be structured in more detail.

The gradual transition from a black box to a gray box to a white box can be seen in Fig. 2.2, Sect. 2.1.1 (basic idea of the top–down process).

### 1.1.2.9  Aspects of a System/System Types

Every system that consists of elements and relations can be observed and described from different viewpoints, through a type of filter. Therefore, certain qualities or properties of elements or their relations come to the fore. Each description should be called an *aspect* of the system.

An industrial enterprise could be considered under the aspects of material flow, information flow, value flow, energy flow, command paths, or authority. The corresponding aspects result in different structures, segments, properties of elements, relationships, and the like coming to light or gaining in significance (Fig. 1.6; see also Sect. 1.2.6).

Systems are often referred to synonymously as *complex* or *complicated*. But these two terms should be clearly distinguished. For this, the following distinction (see Fig. 1.7), is useful (Ulrich and Probst 1995).

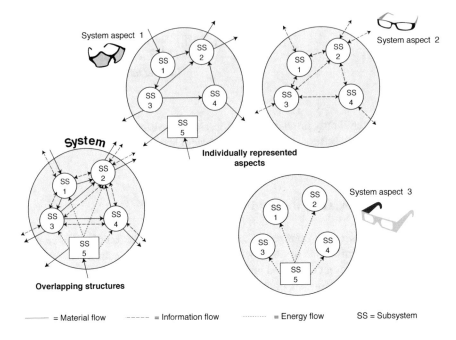

**Fig. 1.6** Aspects of a system

### Simple System

A simple system consists of a few elements that are firmly and permanently connected to each other and that display a low intensity in their relations. Owing to their simplicity, these systems can be explicitly described in their entirety – in special cases even by means of mathematical–analytical methods.

This simple basic model can then be expanded into two dimensions, as shown in Fig. 1.7: on the one hand toward an increasing number and diversity of components or connections and, on the other, toward more dynamic interactions (i.e., variable over time) or interconnections between the elements. From the combination of these dimensions, three further types of system emerge:

### Massively Interconnected, Complicated System

Massively interconnected, complicated systems are characterized by a great number of elements and by the great variety of those elements (as illustrated in the Fig. 1.7 by squares and circles). The elements here are connected statically, which is illustrated in Fig. 1.7 by solid lines. Because of the size of the system, it is often very difficult to describe such systems explicitly. This also applies to describing the behavior of the system, which generally can only be achieved through computer simulation (with all the associated limitations).

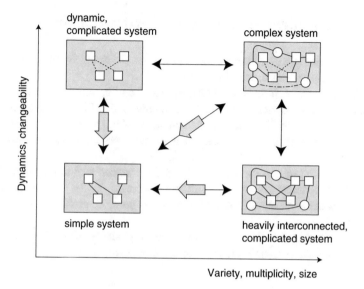

**Fig. 1.7** System types. (Adapted from Ulrich and Probst 1995, Bandte 2007)

**Dynamic, Complicated System**

Dynamic, complicated systems are distinguished by the temporal, often also nonlinear variability of the connection between elements with regard to their interaction, strength, and structure (as illustrated in the Fig. 1.7 by broken lines). Although the number and strength of the interactions are comparatively low, it is difficult to describe such systems quantitatively or to predict their behavior because of their dynamic character.

**Complex System**

When a system also exhibits a great number of diverse elements and connections, and these connections are dynamic, one speaks of a complex system. In such systems there are often system-wide interactions between single elements. Here, it is even more difficult to describe or understand systems than with dynamic, complicated systems.

In systems engineering, with the methods introduced in this book, we are attempting as far as possible, to model systems as simple systems, and to avoid or reduce complexity.

## 1.2 Approaches to System View

In the following, different approaches to viewing systems are described, that are of significance in the context of applying the systems approach.

### 1.2.1   System Models as a Basis for Systems Thinking

An essential principle of systems thinking involves clarifying systems and complex relations through model-based illustrations.

Models are abstractions and simplifications of reality and therefore they reveal only partial aspects. Hence, it is important that the models be sufficiently meaningful with respect to the situation and the statement of the problem. This means that the question of usefulness and relevance to the problem should be part of all reflections.

The following sections offer initial reflections that can lead to meaningful statements about a concrete system. In doing this, we create simple, graphic models that are meant to demonstrate the actual relations, to promote problem awareness and discussion.

These basic reflections would also be the prerequisite for a possible quantitative treatment, for example, in the form of input/output calculations.

### 1.2.2   The Environment-Oriented View

With the environment- or *surroundings-oriented view*, we first ignore the system and concentrate on the relations between the system and its surroundings. The system itself is seen as a black box. A good starting point for this approach is to consider the type and extent of external factors that influence the way in which the system functions. Here, it is useful to differentiate between the surrounding systems and the relations with the observed system.

With regard to the example of a company in Fig. 1.8, one would consider, for example:

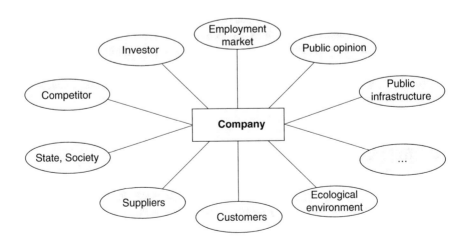

**Fig. 1.8** Environment-oriented approach

- Who are the customers? What kind of customer relations and needs are there? How well do we satisfy their needs?
- Who is the competitor? What influence does the competitor have on our company? How good or bad are we in comparison?
- Which regulations/laws apply to the company? How do they influence our business?
- What is our specific ecological environment? Which output variables of our company influence it? How strongly? In what way?

### 1.2.3   Effect-Oriented View (Input/Output View)

The effect-oriented systems view, or *black box view*, starts with the question: which important influences or inputs from the environment, together with the system's possible behavior, result in effects on or outputs to the environment (Fig. 1.5)?

Provided that a (mathematical) function can be specified to describe the regularities of converting inputs to outputs, one speaks of a so-called *transfer function*. The intrinsic structural cause-and-effect relations within the system are not of interest in this approach. To that extent, the system is a black box. But because internal relations often cannot be completely ignored when applying the transfer function, the black box concept is sometimes relativized by the term *gray box*. Concerning the *white box*, the internal relations of a transfer function are provided in detail.

Examples:

- Energy balances for companies: what goes into what condition of aggregation, what comes out? How much of it is used effectively, how big are the losses?
- Material- or pollutants balances: what goes into the system? In which quantities? In what condition and in which quantities does it come out again?
- Productivity indices or efficiency calculations of every kind are based on these considerations.

The effect-oriented approach is thus a good aid for roughly evaluating the condition and the quality of a system. It is accordingly also a good aid for roughly characterizing problem areas and solutions. Before beginning a detailed examination or design, one defines the scope of rough function blocks (as black boxes), defines their assumed or desired functions and their interactions (input/output), and only then enters into structure-oriented and thus more detailed reflections.

### 1.2.4   Structure-Oriented View

With this view, one considers the elements of a system and their relations; the dynamic effect mechanisms are of special interest here. This perspective is suitable for explaining how an output is created from an input or – when a solution concept is presented – how an input is to be transformed into the desired output.

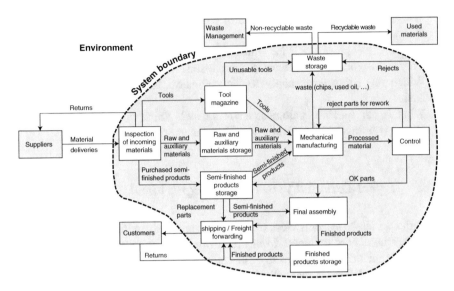

**Fig. 1.9**  Manufacturing company, graphic illustration of the system aspect "material flow"

The material flow of a manufacturing company is outlined in Fig. 1.9. The illustration could serve as an entry model for analyzing questions of transport, problems of processing times, etc. Should information problems be treated, the information flows and the information processing facilities would have to be analyzed and illustrated.

With a structure-oriented view, the structural make-up and the structural relations within the system come to the fore. Intra-system elements and relations are defined and illustrated. Flow structures, process structures, effect mechanisms, etc., are of special interest.

### 1.2.5  Aids for Illustrating Relations or Structures

*Graphs*
Often one uses *graphs* to illustrate a structure, as in Fig. 1.9. Elements of the system are the so-called *nodes* of the graph, which are drawn as circles or squares. Relations (so-called *edges*) are expressed through lines. According to the type of assertion made, the relations can be represented as directional (with arrows) or nondirectional (without arrows). This allows, for example, information or material flows, cause–effect relations, etc., to be expressed. When graphs are drawn freehand, such illustrations are also called bubble charts.

*Matrices*
Another form for illustrating structures and system relations is the matrix (Fig. 1.10). In this case, the components are ordered in a table, with the use of lines and columns. Existing relationships are expressed, either through markings at the intersecting

| Output element □→ from  \  Input element →□ to | Supplier | incoming goods | Tool magazine | Raw materials storage | Semi-finished products storage | Material waste | Workshop | Quality control | Final assembly | Finished products storage | customer |
|---|---|---|---|---|---|---|---|---|---|---|---|
| Supplier | | 100 | | | | | | | | | |
| incoming goods | 5 | | 10 | 70 | 20 | | | | | | |
| Tool magazine | | | | | | | 10 | 5 | | | |
| Raw materials storage | | | | | | | 70 | | | | |
| Semi-finished products storage | | | | | | | 100 | | 65 | | 10 |
| Material waste | | | | | | | | | | | |
| Workshop | | | 5 | | | 10 | | 170 | | | |
| Quality control | | | | | 155 | 5 | 10 | | | | |
| Final assembly | | | | | | | | | | 65 | |
| Finished products storage | | | | | | | | | | | 65 |
| Customer | | | | | | | | | | | |

**Fig. 1.10**  Manufacturing company, matrix illustration of the system aspect "material flow"

points of the columns and lines, or through a specification of the relationship intensity (for example, numerical values). On this basis, with the help of matrix technical algorithms, structures can be optimized with respect to the targets to be determined. This is used, for example, in the → Design Structure Matrix.[1]

## 1.2.6   Aspects of the Systems View

It was already pointed out earlier that systems and their elements and relations can be described as "filtered" through different aspects (points of view). Examples of such system aspects are:

(a)  System "Corporation"

- Elements (subsystems): for example, functional areas such as sales, production, development, purchasing
- System aspects: information flow, ordering process, cause of cost, material flow, etc.

(b)  System "Europe"

- Elements (subsystems): for example, states, political entities such as Germany, France, Switzerland, and Italy
- System aspects: trade, flow of goods, transport networks, currency exchange rates, etc.

---

[1] → meaning: this keyword will be treated separately in Part VI, Methods, Techniques, Tools.

(c) System "Human Being"

- Elements (subsystems): for example, body parts such as head, torso, arms, or legs
- System aspects: nervous system, blood circulation, etc.

The structure-oriented view of a system under different aspects (points of view) resembles the observation of the system through different filters. This idea has already been illustrated in Fig. 1.6. This may lead to the following considerations: elements of a system can be relevant to several system aspects and thus appear in different representations. Thus, the production department, for example, is undoubtedly significant with regard to the system aspects of material flow, cause of cost, and information flow. Likewise, the head and torso of a human being are components of the system aspects of the nervous system and blood circulation.

Different system aspects serve to reduce complexity only temporarily, as they are inextricably linked to one another. Thus, information flow drives material flow, that is, it provides the material flow with different types of information (processing stage, current place of storage, volume of incoming material, outgoing goods, etc.).

The statements and insights that can be deduced from a system illustration are crucially influenced by the respective system aspect (= the spectacles). The concept of moving individual aspects to the foreground while deliberately neglecting others – at least for the time being – arises from the orderly handling of complexity.

## 1.2.7  Application of System-Hierarchical Thinking

Systems thinking is meant to counteract the danger of delimiting circumstances or problems too narrowly. In particular, the surroundings-oriented view and the model concept of the open system point in this direction. Expanding the horizon of observation is, however, also associated with the risk of an increasing inability to act, because the number of elements and relations is no longer manageable. Here, the concept of system-hierarchical thinking, in connection with the black box principle, facilitates the orderly handling of this complexity.

Thus, a system (problem area or solution) is at first only structured in a rough manner by forming a manageable and deliberately limited number of subsystems and by showing the relations that appear to be essential.

To be able to ignore detailed aspects of the view for the present, subsystems are considered to be black boxes. Being aware of this, one deliberately distances oneself from a structure-centered view. Only when no adequate statements are possible at the level of the rough overall view is the black box approach abandoned in favor of the structure-oriented approach. In this way, depending on the actual problem, one can move around at one time at the level of the overall system, and at another time at the subsystem level, without losing the overall context.

Thus, the concept of systems hierarchy represents a principle that is comparable with the view through a "zoom lens" As needed, one adjusts the lens for the close-up

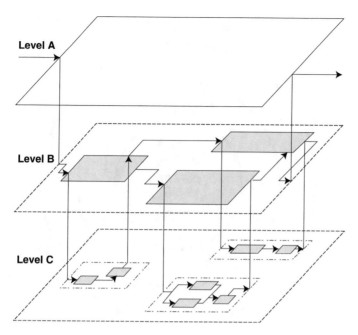

**Fig. 1.11** Stepwise resolution of a system

or for the wide-angle shot (Fig. 1.11). Thereby, in connection with the hierarchical approach, two lines of thought become possible:

- The *subsystems* view is aimed downward by prompting the question: what elements can a system, a subsystem, etc., be composed of? Each component that is hierarchically resolved in this way can be separately considered as a system with a surroundings-, impact-, and structure-oriented view.
- The *suprasystems* view examines to which superior system(s) a system belongs (Fig. 1.12). These types of reflections focus on observing a larger context and can help with the search for the proper entry level.

## *1.2.8 Final Comments*

In the preceding sections we have explained important characteristics of systems and appropriate system views. This was intended to show how the understanding of complex systems can be facilitated. The sequence in which the methods of approach were treated also corresponds, almost, to the sequence of mental steps one should follow in the practical application of systems thinking.

However, it should have also become clear that not all surrounding influences, not all aspects and structures, can be investigated in a concrete application. How extensively observations are to be carried out is a question that should always be answered considering fitness for purpose and the relevance of the problem.

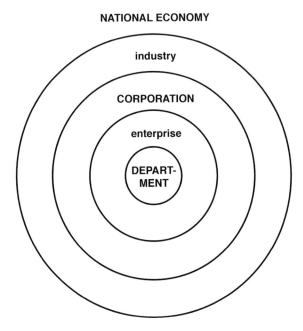

**Fig. 1.12** Layer illustration of suprasystems

## 1.3  Agility of Systems

The term "agility" has been increasingly used in connection with systems engineering. It means active, mobile, versatile, and lively.

### 1.3.1  The Concept

The English term "agile" can now be interpreted in different ways in connection with systems engineering, whereby two entirely different issues are included[2]:

"*Agile systems* engineering": here, the term "agile" refers to the result, to the developed system. Once the system has been delivered and implemented, it should be agile, mobile, etc., in the sense of being variable and modifiable later on. This can be a very important and meaningful concern in a rapidly changing environment.

"Agile *systems engineering*" means something completely different. The process of engineering should be designed to be agile, i.e., flexible, so that it can accept and implement new or modified requirements during development.

_____

[2] Haberfellner, R. and de Weck, O.: Agile *Systems Engineering* versus *Agile Systems* engineering.

We address agile processes in Sect. 2.2.2 (Agile Process Models). The basic idea of *agile systems* is explained below by means of examples. In Part III, Chap. 5, we look again at agility and further aspects of changeable systems, in addition to appropriate architectural principles.

In this context, we refer particularly to the explanations in Sect. 2.2.4 (Keeping Options Open as an Approach in Support of Agility) and Sect. 2.2.5 (Real Options Approach for the purpose of gaining maneuvering room), which facilitate and expand these considerations.


## 1.3.2  Three Examples of Agile Systems

(a) *The Soccer Stadium in Klagenfurt/Austria: European Championship (EC) 2008*

"The … EC-Stadium in Klagenfurt will be able to accommodate 30,000 spectators right from the start. This was the required size for the three Klagenfurt EC preliminary matches. After that, however, it would have been much too large and would therefore have to be reduced to 12,000 seats. According to the architect Wimmer, it would have been disappointing not only that the stadium were half as big after the EC, but also if no further use for dismantled parts could be found. Mastering the process of a controlled reduction is not only a creative but also a logistical challenge.

The stadium has therefore been designed in such a way that a second stadium can be built from the parts that are needed solely for the EC matches. Wimmer cannot yet say anything about a possible location; for the moment he is only concerned that it is possible. Reusability pertains above all to the stands. Because they are constructed from steel, they can be erected again elsewhere. Whether for the European Championship or thereafter – the stadium is under a high exploitation pressure."[3]

We are not concerned here with the method of construction; it is only the fundamental concept that is of interest.

A general conclusion can already be drawn: systems need *flexibility,* especially when:

- They are expensive and require considerable initial investment.
- They are long-lasting, for example, longer than 10 years. The probability that requirements change naturally increases with the duration of a system's lifecycle.
- The costs of subsequent rebuilding would be extremely high.

If these conditions are given, one should think about how much flexibility should be implemented in the system from the start – as a type of insurance both against underused overcapacity and as an opportunity to avoid an unexpected lack of capacity.

---

[3] Source: O. Elsner in the Vienna daily press *Der Standard*, print edition. 3.9.2005, page 16.

(b)  *The Automobile industry in general*

One way of implementing flexibility in a production concept is to establish the *product concept* intelligently and, for example, in the sense of an architecture concept that enables a large number of product variants by using a small number of architectures.

Another way is to implement flexibility in the *production facilities*: car manufacturers must –to be cost-effective – strive for a full capacity of their plants, close to the maximum limit. For example, BMW in Europe and Toyota in Japan reach such utilization levels. An important and successful path toward this goal lies in combining several derivatives in a production line, whereby the respective products should find themselves at different stages of their product life-cycle.[4] To the extent that the new product takes hold, the quantities of the old product can be successively reduced. This process can be supported through appropriate marketing measures; for example, through special models with full optional extras, if the old product cannot yet be discontinued, or through a reduction of sales efforts for the old product if the capacities are already being used for the new product. Similarly, there can be a much better response to the different seasonal demands of the various models.

(c)  *Automobile industry – Magna Steyr Fahrzeugtechnik (Automotive Engineering) MSF, Graz/Austria*

The MSF plant in Graz was intentionally designed for flexibility. It can simultaneously produce a series of completely different vehicles on the same assembly line. In addition, the concept has two further subtleties: the sequence of the different vehicles is discretionary and the relative manufacturing data was coming from the USA (Chrysler) and from Germany (Mercedes). The systems of production planning and production steering of the MSF in Graz was therefore capable of communicating with several differently organized systems. As of 2005,[5] the following vehicle types were simultaneously assembled on the *assembly line*:

- Chrysler Voyager + PT Cruiser
- Jeep Grand Cherokee + Mercedes M Class
- Chrysler Voyager + Jeep Grand Cherokee + Chrysler 300C

The bodyshell work always took place separately owing to different work cycles. Painting was possible in the same facility for the Chrysler Voyager, Jeep Grand Cherokee, Chrysler 300C, Mercedes G, Saab Cabrio, and Mercedes E-Class.

A general implication is that such flexible facilities are tied up with *higher investments* (equipment, tools, space requirements, storage for parts, etc.). The *demands* made on the *assembly workers* are also higher when the ability to assemble different vehicle types in a random order is required. Such additional costs, however, should

---

[4] Neue Züricher Zeitung, 3.9.2004

[5] Chrysler and Daimler were still merged in 2005. The current clients of MAGNA (OEMs) do not permit production facilities to be shared with other OEMs. If needed and accepted, the conditions required to re-establish this type of agile production systems still exist.

be compared with the costs associated with unused resources or nonsaturation, if extra assembly lines would have had to be provided for all vehicle types. This form of flexibility is more likely to be implemented when the production facilities cannot be saturated with a single product or a single platform and/or if the additional investments, overhead costs, etc., associated with a new plant did not pay off.

The company BMW is following a similar concept, e.g., in their plant in Dingolfing, where models of the 3-, 4-, 5-, 6-, and 7-series are assembled.

The agility of systems described here does not as a rule emerge by itself, but must be aimed for deliberately and implemented in the concepts.

## 1.4   System Dynamics

Up to this point, we have considered systems primarily from static points of view and we have not taken into consideration that it obviously makes sense to model the processes in the systems and thereby derive insights about the functioning or the optimal design of a system. The methods of "system dynamics" make this type of endeavor possible. Here, we sketch out only the basic considerations. For a deeper treatment of the subject, we refer the reader to the pertinent literature.[6]

Short description (according to J. Sterman):

- Our languages, English, German, French …, are linear languages. Their sentence structures transform everything we want to express into a world outlook that expresses: "x causes y" or x ➔ y.
- This linearity misleads us into focusing on one-way relations instead of on circular or reciprocal causes, such as x ➔ y and y ➔ x.
- Unfortunately, the most irritating problems and those most difficult to penetrate, with which corporations and their managers are confronted today, are caused by a network of tightly inter-connected circular relationships.
- *Systems thinking* is – like our colloquial languages – also a *language*. Fortunately, it directs our attention to relations and dependencies that are tightly interwoven, and thus it serves as a type of visual language.
- A *graphic illustration* of a system is thus an effective aid to communication because it can express the essence, the kernel of a problem in a format that is easily remembered, even though it abounds in logical deductions and insights: it brings precision to our insights, it forces us to explicitly enunciate *conceptual models*, it allows these to be discussed and examined, and thus it gives expression to our view of the situation.
- Of course it requires work before we can fluently master this language (Figs. 1.13 and 1.14).

---

[6] The classic: Forrester (1977): Industrial Dynamics

  More recent literature: Sterman (2006): Business Dynamics. Senge (1990): The Fifth Discipline.

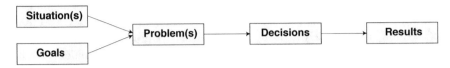

**Fig. 1.13** Linear thinking – normally, we concentrate on finding solutions to our problems (according to Sterman 2006)

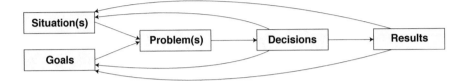

**Fig. 1.14** Should we not also deal with problems that are newly created through our solutions (circular thinking; according to Sterman 2006)

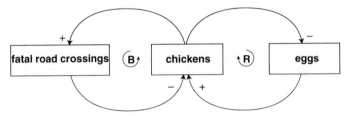

+ = effect "same direction"
− = effect "opposite direction"
B = Balancing Loop (compensating, generally stabilizing loop)
R = Reinforcing Loop

**Fig. 1.15** It is difficult to make circular relationships intelligible if we do not have the proper language (according to Sterman 2006)

How does one speak circularly?

Interpretation of Fig. 1.15:

- Right loop: chickens lay eggs, the more chickens, the more eggs (+ = same direction). Eggs turn into chickens, the more eggs, the more chickens (+). An R-loop (= reinforcing loop) emerges.

- Left loop: chickens cross the street and are run over; the more chickens, the more often it occurs (+ = same direction). The more chickens are killed, the fewer remain (− = opposite direction). Thus, a B-loop results (= balancing loop).

Interpretation of Fig. 1.16:

- There is a certain amount of work (*unfilled orders*). Getting it done *requires* an appropriate *work force*. The greater the amount of work, the greater the *required work force* (+ = same direction). It is assumed that the *manpower currently available* will be adapted to the required work force (+ = same direction).

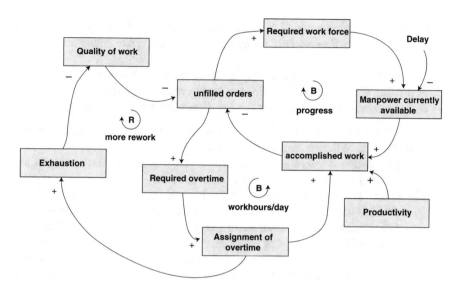

**Fig. 1.16** Work-to-do project model (according to Sterman 2006)

If *delays* should occur, the effect "−" (= opposite direction) is to be expected. The greater the existing work force, the greater the extent of the *accomplished work* (+). Increased *productivity* also has this effect (+). The greater the extent of the accomplished work, the fewer the number of *unfilled orders* (−). Overall, a *balancing loop (B)* is produced.

- The *assignment of overtime* can accomplish an effect in the same direction + (B). Overtime, however, also has a negative effect: the more there is, the higher the degree of *exhaustion* + (= same direction). But increasing exhaustion lowers the *quality of work* (− = opposite direction), which − because of the need for *reworking* − leads to an increase in the amount of *unfilled orders* (− = opposite direction). *A reinforcing loop (R)* is produced.

The method of system dynamics, in a simple manner and with capable software, allows the way in which measures or elements affect one another to be comprehended or described − through the formulation of functional relations (formulas), in tabular fashion (different discreet values), and much more.

Although mapping a system's feedbacks in causal diagrams is valuable in its own right, simulation is essential in understanding how such systems behave. Our ability to mentally simulate even the simplest systems is extremely poor (even systems with no feedback and only a single-state variable).

## 1.5  Summary

- Systems thinking should be applied when the task is to analyze or structure complex phenomena that can be designated and explained as a system. The approaches shown here represent, on the one hand, the lenses through which systems are to be observed, and, on the other hand, they are the basis for illustrating system models. Thereby, the danger of forgetting or disregarding something important is lessened.

- In the systems engineering concept, the systems approach is used in the analysis of existing systems (problem field) and in the creation of solutions.
- Systems thinking can be characterized particularly by the following views:

The *surroundings-oriented view*, which serves to identify the external factors that influence the system and vice versa.

The *effect-oriented (black box) approach*, which is at first interested only in the inputs and outputs, but not in the internal structure.

The *structure-oriented view*, by which the internal structure of a system is to be identified, understood, or determined. Here, the dynamic aspects such as flows, processes, and internal effect mechanisms are of special significance.

The notion of the existence of different *system aspects* refers to the fact that every situation can be observed and analyzed from different viewpoints (putting on different spectacles), which move various attributes of elements and/or relations into the foreground or deliberately neglect them. To avoid undue oversimplification, a system, as a rule, therefore, cannot be described by a single structural model.

The *hierarchical view*, *looking upward*: this approach leads to system models of suprasystem(s). Thereby, other, more comprehensive system boundaries, become apparent. This view is especially suitable for promoting a more comprehensive and holistic way of thinking.

The observation of the system as a *hierarchically* built structure (*looking downward*): every system can be described in ever-increasing detail over a series of levels. This point of view is the basis for models of subsystems ranked as first, second, third, etc.

We speak of so-called *systems of systems (SoS)* when systems that have been developed independently of each other and fulfill their function independently of a certain system context, are assembled into a system or incorporated into a system.

The idea of *network thinking* is meant to facilitate the identification of cause–effect chains or defects and intervention opportunities.

- The difficulty of *drawing the borders of the problem field* is that a more comprehensive view increases the effort required for analysis, but also increases the chance of a good solution.
- Reflections based on systems thinking are significant in the context of both *teamwork* and *project management*.
- *Agile systems* are systems that after completion or delivery can still be changed without the need for great additional expenditure, that is, they can be quickly adapted for new or changed requirements. This agility, however, does not emerge by itself; rather, it should be recognized as a necessity and must be specifically designed, that is, it must be considered in the system architecture.

- *System dynamics* is a way of thinking and a graphic language that allows us to represent, quantify, and simulate the dynamic behavior of systems, in particular the feedback functions, in the form of stabilizing or reinforcing loops, and to be represented, determined quantitatively, and simulated.

## 1.6 Self-Check of Knowledge and Understanding: Systems Thinking

1. What is meant by systems engineering systems thinking?
2. What defines a "system"?
3. In which cases is it not appropriate to use the term "System"?
4. Sketch an example of a system with its components and their relationships that is imbedded in an environment and interacting with other systems.
5. What is a "system of systems" (SoS)? What is the main difference between a subsystem and an SoS?
6. When using a hierarchical, level-based approach to describe a system (Fig. 1.4), what defines the lowest level of detail?
7. What is meant by "black, grey and white box" models for describing real-world phenomena?
8. Give an example where different system aspects (looking through different spectacles) can help to model the whole system with its components, properties, and relationships.
9. Choose a system based on an arbitrary example for further investigation: structure the system hierarchically. Then think which system aspects would be interesting.
10. Explain the difference between a complex and a complicated system?
11. What are the three main points of view that can be used to create a system model:
12. What is the tradeoff when using structure-based models for subsystems and components?
    Use this trade-off to explain why hierarchical system models often use black or grey box subsystem/component models at higher levels and the aforementioned structure-based models mainly at lower levels.
13. Give an example of "agile systems" and discuss the challenges for systems engineering in that context.
14. In what way can system dynamics help with modeling systems?

## Literature

Ackoff, R. L. (1971): Towards a System of Systems Concepts
Bandte, H. (2007): Komplexität in Organisationen

Beer Stafford (1959): Cybernetics and Management.

Checkland, P. (1999): Systems Thinking, Systems Practice

Chestnut, H. (1965): System Engineering Tools.

Chestnut, H. (1967): System Engineering Methods.

Churchman, C. W. (1979): Systems Approach.

Crawley, E., Cameron, B., Selva, D. (2015): System Architecture: Strategy and Product Development for Complex Systems

Department of Defense (2008). Office of the Deputy Under Secretary of Defense for Acquisition and Technology: Systems Engineering Guide for Systems of Systems. Version 1.0. August 2008

Dörner, D. (1997): Logic of failure

Forrester, J. W. (1977): Industrial Dynamics

Gomez, P. und Probst, G. J. (2007): Die Praxis des ganzheitlichen Problemlösens

Haberfellner, R. and de Weck, O. (2005): Agile SYSTEMS ENGINEERING versus AGILE SYSTEMS engineering

Hall, A. D. (1962): A Methodology for Systems Engineering

Maier, M. (1998): Architecting Principles for Systems-of-Systems

Meadows D. et al. (2004): Limits to Growth: The 30-Year Update

Negele, H. et al. (1997): ZOPH — A Systemic Approach to the Modeling of Product Development Systems

Popper, K. (1959): The Logic of Scientific Discovery

Popper, K. et al. (1998): The World of Parmenides

Senge, P. M. (1990): The fifth discipline

Senge, P. M.; u. a. (1994): Fifth Discipline Fieldbook

Sterman, J. D. (2006): Business Dynamics

Steward, D. V. (1981): The Design Structure System

Ulrich, H.; Probst, G. (1995): Anleitung zum ganzheitlichen Denken und Handeln

For details of the bibliographic references, see "Bibliography" at the end of this book.

# Chapter 2
# Process Models: Systems Engineering and Others

The systems engineering process model described below contains a series of recommended actions and guidelines that have proven their worth in practice and constitute an essential component of the systems engineering methodology. Its integration into the systems engineering concept can be seen in Fig. 2.1.

Other process models – such as the Waterfall Model, the Vee Model, INCOSE methodology, simultaneous (concurrent) engineering and agile models – are described and discussed comparatively following a description of the systems engineering process models.

## 2.1 Components of the Systems Engineering Process Model

The process model is based on *four basic principles*, which should be regarded as combined into usable components. These are the concepts in which it is appropriate:

- To proceed *from the general to the detailed*, and not vice versa
- To follow the principle of *thinking in variants*, i.e., essentially not being content with a single variant (generally the first available one), but rather consistently looking for alternatives
- To structure the process of systems development and implementation according to temporal perspectives (*phased approach*)
- To apply a kind of work logic as a formal procedural guideline in solving problems, regardless of their type and of the phase in which they occur (*problem-solving cycle, PSC*)

These four components form a meaningful whole, for they can be linked together. We first describe them individually and independently of one another to make their

---

The original version of this chapter was revised. The correction to this chapter is available at https://doi.org/10.1007/978-3-030-13431-0_17

© Springer Nature Switzerland AG 2019, Corrected Publication 2021
R. Haberfellner et al., *Systems Engineering*,
https://doi.org/10.1007/978-3-030-13431-0_2

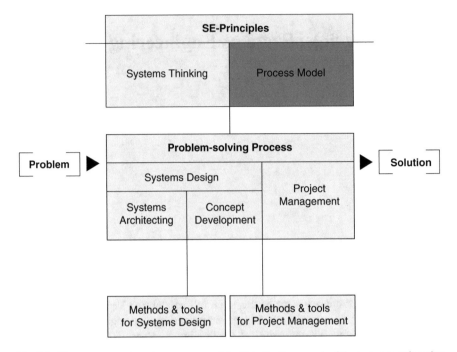

**Fig. 2.1** The systems engineering process model within the framework of the systems engineering concept

basic idea easier to understand. Then, for didactic reasons, we later address divergent ideas, modifications, extensions, and restrictions. The observations in Parts II through V of this book contain more concrete statements and recommendations for action, along with explanations and examples from case studies.

### 2.1.1   The Principle "From the General to the Detail" (Top–Down)

#### 2.1.1.1   Basic Idea

Frequently, planners who spend little time discussing questions of a basic nature, and instead quickly propose concrete solutions, are considered to be especially competent. They thereby create the specific conditions for addressing the universally beloved detailed questions in which, as we well know, the devil resides. Here, no case should be made for any major discussion on a general, less concrete level. But we do not think there is much sense in trying to deal with the devil in the detail right away – and ignoring his grandparents, who may be hidden in an inappropriate, mistaken, or non-existent overall concept.

Dealing immediately with detailed matters may be appropriate and allowable in cases of small problems or detailed improvements of a functioning solution or with

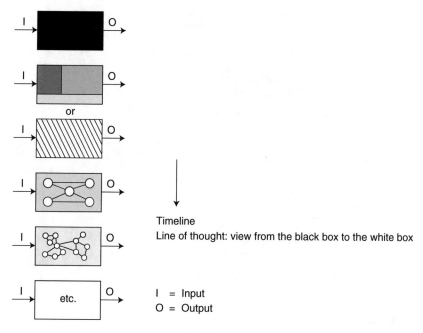

**Fig. 2.2**  Principle "from the general to the detail" (top–down)

routine problem-solving that came up earlier in a similar form and was already dealt with. For this, the methodology of systems engineering is not absolutely essential. The methodology of systems engineering should be used for solving problems that are difficult to comprehend, complex in nature, and/or substantially intertwined with the environment, and not for solving problems in which the mutual impact is evident and thus requires no closer scrutiny or deliberation.

The basic idea of the process presented in Fig. 2.2, *from the general to the detail*, has already been broached with an explanation of systems-hierarchical thinking (Figs. 1.4 and 1.11). It originates from the black box principle and expresses the gradual resolution of a black box into gray and white boxes with different shades of gray. The second box from the top in the Fig. 2.2 is intended to indicate that systems components can be conceptualized or represented in various degrees of resolution. The shaded gray box underneath it shows a uniform degree of detail.

### 2.1.1.2   Application to the Structuring of the Initial Situation (Problem-Structuring) and to the Draft Solutions

At the start of a project, areas that are to be developed and delineated may need closer investigation or changes should or may be made. Important components or areas of the problem field and the factors that influence them should be identified and presented in their dependencies. In terms of *systems thinking* (see Part I, Chap. 1) these are the elements of a system and their relations, in addition to

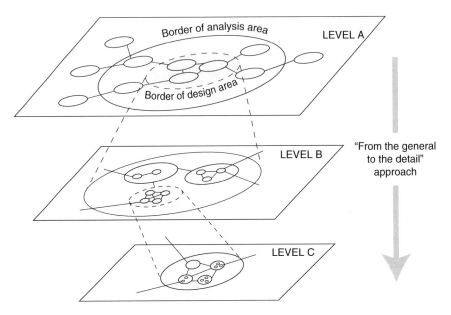

**Fig. 2.3** Narrowing the field of observation

important environmental elements and their relationships. Only when the problem is structured and defined clearly enough for the planners and their customers does it make sense to address the qualitative and quantitative investigation of the details to define the design area and to systematically develop draft solutions for them.

The basic idea of this consideration is presented in Fig. 2.3 and should indicate the narrowing of the frame of reference as the project moves forward.

The areas of analysis (situation) and design areas (solutions) do not need to be identical. The outer circle on Level A marks the *borders of the analysis area*; the inner dotted one, the *design* area, in other words, the area within which changes can and should be made. The process of increasing concretization and detailing is indicated on Level B.

With this consideration, the underestimated risk involved in the application of systems thinking should be faced: the definitely desirable and recommended thinking in effect relationships should not mislead anyone into needlessly turning small problems into large problems. The recommendation to expand the observation horizon, especially at the beginning of a project, is thus consistently connected to the demand for narrowing, that is, a skillful and conscious delimitation.

### 2.1.1.3  Alternatives to the "From the General to the Detail" Approach

One fundamental alternative to the top–down approach would be the reverse, the bottom–up approach, which would mean that we begin with the detail and that the whole is a product of the sum of the individual steps. This approach may be appropriate under special conditions, for example, when dealing with improvements to an existing and functioning solution (see Sect. 6.3.3.2 or Improvement (Melioration) Projects, Sect. 6.5.2). In this case, timely knowledge of the details is important, for rapid measures with deliberately limited effectiveness are usually the main goal. But with new constructions and reconfigurations on a larger scale, we consider it essential to develop a general concept that can represent the indispensable orientation frames for the planning and execution of partial steps.

The so-called agile process models occupy a special position and offer advantages in very dynamic, continually changing situations. The logic tends to be bottom–up, with relatively small increments of development, which, considering their priorities, can continually be reconsidered, along with rather frequent iteration steps (Stelzmann 2011).

### 2.1.1.4  Summary

The principle *from the general to the details* is intended to express the following:

- The field of view is *first to be grasped broadly* and subsequently to be *narrowed down* gradually. This applies to the analysis of the problem field, the starting situation, and the drafting of solutions.
- The analysis of the starting situation (of the problem field) should not begin with detailed inquiries before the *problem field* is *structured* roughly, is *embedded* in its *environment* or *isolated* from it, and the *interfaces* are defined (often only in the sense of a working hypothesis).
- In structuring the solution, the first *general objectives* and a *general solution framework* should be established, whose degree of detailing and concretization is gradually increased as the rest of the work on the project advances (this does not exclude later modification, correction, and perhaps even discarding of a selected framework). To some extent, concepts on higher levels serve as *guidance* for the detailed design of the solution.

Supplementary considerations, modifications of the top–down principle, and application instructions can be found in Parts II and III.

## 2.1.2  The Principle of Variant Creation

### 2.1.2.1  Basic Idea

For practically every task or every problem, there are several possible solutions. One important principle of systems engineering is thus not to be satisfied with the very first solution idea, but rather taking as broad an overview as possible of the solution variants that are conceivable at a specific level of consideration. These *solution principles* can also be understood to be different variants of a *systems architecture* that always have very distinctive characteristics. One should therefore consistently attempt to be aware of the basic idea on which a solution is based, and seek alternatives before beginning to develop it in greater detail. This helps us to remember the correlation with the starting situation and to direct attention to other conceivable solution principles.

Of course, the principle of variant creation also applies to the next lower system levels. As a rule, different variants of a basic structure or even architecture are also conceivable at a subsystem level. But the multiplicity of variants would grow so quickly, with consistent variant creation at all levels, that it could scarcely be managed if the *variant creation* did not involve a simultaneous, stepwise *reduction* in the resulting multiplicity of variants.

To be able to make a selection, one must of course be able to get a general picture of the *characteristics*[1] and the *consequences* – whether desirable or undesirable – that can be expected with the choice of a specific solution. One should thus have rather specific notions of what the individual solutions *look like*, how they *operate*, how high the expected *costs* are, what the expected *benefit* is, what *advantages* and *disadvantages* they entail in terms of *desirable* or *undesirable* outputs or side effects, etc. This is indicated by the increasingly detailed structuring at the various levels in Fig. 2.4:

The following would be a generally valid approach. Starting with a specific task, one should formulate several basically conceivable solution principles in terms of systems architecting, structure them to the point where it is possible to form a picture of their effects, requirements, and consequences, and then choose the variant that offers the greatest promise of success. For a specific solution principle, various configuration variants are possible at the next lower concretization level (variants of overall concepts), among which a choice must again be made.

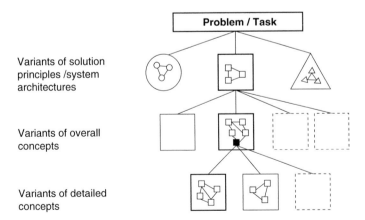

**Fig. 2.4** Staged variant creation and elimination, connected to the "from the general to the particular" action principle

---

[1] Engl. "-ilities", such as *stability, flexibility, expandability, scalability*.

In this stepwise top–down approach, the effect-oriented view and the structure-oriented view come into play alternately: at one specific level, there is above all the question of what *effect* the system or the various elements of the system should produce (= *effect-oriented view* or *black box principle*). At the next lower level, this view must be abandoned and one must consider how elements are to be structured, in other words, how the solution is to be configured so that the desired effect is produced (*structure-oriented view*).

However, the necessity of this stepwise setting of the course entails the fact that decisions have to be made based on an incomplete insight into detailed problems and on insufficient information. It may thus be that the right way is not immediately found, or even that a point is reached when the selected way proves to be impracticable. Then, the only recourse is to return in the planning process to a higher step and search for the solution in another direction –the excellent qualifications of the people involved in the planning notwithstanding.

This risk could be circumvented by basically deferring the choice and deciding only when all the possible solutions at all concretization levels have been worked out in all variants. However, because of the workload and time required, this method has no practical meaning. Commissioning parties and planners should keep this in mind whenever a selected solution subsequently proves to be impracticable.

However, under special conditions, within a restricted time frame, it may indeed be appropriate, or even necessary, to pursue several solution principles simultaneously. This is especially true when the risk associated with the commitment to a specific solution principle is relatively high and the additional effort for the simultaneous pursuit of multiple variants appears in relation justifiable.

It is particularly important for the commissioning party to realize the significance of the above-mentioned *interim decisions* to avoid surrendering the illusion that he or she could intervene in the decision to go ahead with implementation without consequences and influence the solution with his or her ideas, as the multiplicity of variants would already have been so restricted at the transitions between the individual hierarchical steps that the solution could scarcely be influenced any longer at this time, but rather only accepted or rejected.

### 2.1.2.2   Principle Variants Versus Detail Variants

*Principle variants* are those clearly differentiated from others by their basic idea, and to which – consciously or unconsciously – a particular architecture has been given that can be distinguished from other possibilities. *Detail variants*, on the other hand, are those that are based on the same fundamental idea, but shape it differently in detail.

In practice, planning and decision situations are sometimes characterized in that variants that are not different in principle, but only in the details, are submitted for a decision. This indicates that the course has either already been set consciously, or that soon it will have to be set unconsciously. The former obviously occurs when the previously explained approach principles are followed consistently, and is quite harmless. However, the latter case should give rise to the following cautions: careful, we are about to put time and money into the selection of variants that essentially differ very little from each other! Are there really no fundamentally different alternatives? Have we done our homework on variant creation?

The so-called *improvement plans* (in the sense of improvements to existing solutions) constitute a special case. These often have no clearly distinguishable alternatives, as the improvements may result from a number of individual measures that can be combined in many ways or broken down into individual steps. There may be a whole series of improvement measures for every variant, and they may have little in common with one another and may even be contained in different variants, e.g., in the sense that a greater melioration may include all the medium-sized measures, and that these include all the small measures. For example, consider the renovation or the remodeling of a building, the variants for improving the logistics of an enterprise, the reworking of a product concept, the improvement of a legislative text, etc.

### 2.1.2.3   Alternatives to the Principle of Variant Creation

With respect to methodology, we see no alternative to the principle of variant creation. Even though this is frequently violated in practice – often due to a lack of time – we consider this principle to be a special feature of good planning. As far as a lack of time is concerned, lots of energy is often invested – also under time pressure – in the variation of details, whereas it could be invested more effectively in developing or considering a variety of basic variants. This is not only a question of methodology, but also one of mindset and mental agility. If, despite all efforts, an alternative to a specific solution is not even conceivable, which is rarely the case in practice, the futile search for alternatives had entailed not only disadvantages, but also one advantage: greater certainty concerning the truth of the selected way.

One phenomenon observed in practice also involves the fact that sometimes basically different variants crop up at a point in time when, because of the work already done, it would be time for realization: This may be a sign of changed value judgment that later gives rise to possible solution ideas that were not conceivable earlier, e.g., planning in the public sector under strong political pressure. But it can be a sign of methodological weakness if, for reasons of convenience, lack of inspiration, or for other reasons, a particular solution direction is settled prematurely and unnecessarily.

### 2.1.2.4   Summary

We consider the principle of variant creation, of thinking in alternatives, to be an indispensable component of good planning. This is a methodically basic attitude, and – when the principle of *from the general to the detail* is followed – it should not lead to any significant increase in planning effort.

If this principle is not followed, there is a higher risk that fundamentally different solution approaches are introduced into the discussion in the late phases of development. Possible consequences include the discussion stalling or the planning stopping and the return to a higher level. Both results are unsatisfactory: the waiver to a better solution, due to the advanced development phase, which usually is connected to time pressure, in addition to the waiving of some detailed planning done needlessly.

## 2.1.3   The Principle of Structuring into Project Phases as a Macro-logic

### 2.1.3.1   Basic Idea

The idea of subdividing the development and realization of a system into individual phases that can be separated logically and temporally from one another represents a concretization and an expansion of the principles "from the general to the detail" and "variant creation" explained above. The purpose is to structure the development of a solution into manageable partial stages, thereby facilitating a stepwise process of planning, deciding, and concretizing with predefined milestones or correction points. A distinction must be made between the *life cycle of a system* or a solution on the one hand and the *project phases* that, on the other hand, serve the development and realization of the solution. This is presented in Fig. 2.5, in which the characteristic results of the individual project phases are provided in the center of the illustration.

The number of project phases, and even the formalism with which they are processed, are surely dependent on the type, scope, and significance of a project. Smaller projects can generally be developed satisfactorily by using a smaller number of phases and less formalism. The designation of the individual phases is also of secondary importance, as it is influenced by the business sector, the task, the terms used in the company, and many other factors. What is important is that the complexity of a problem statement and the risk of a wrong decision can be reduced through purposeful structuring into individual steps of planning, decision-making, and implementation.

The phase model is first described in its simplest form. The purpose and the contents of the individual phases are explained. Subsequently, we move into *other process models* (Sect. 2.2) and into *supplements* (Sect. 3.2.3).

### 2.1.3.2   The Individual Project Phases

In the following, the early phases (the preliminary and main studies) are treated more thoroughly than the later phases. This is mainly because the methodical aspects are of particular importance in the early phases, whereas special expertise and methods dominate in the phases approaching implementation, which are not

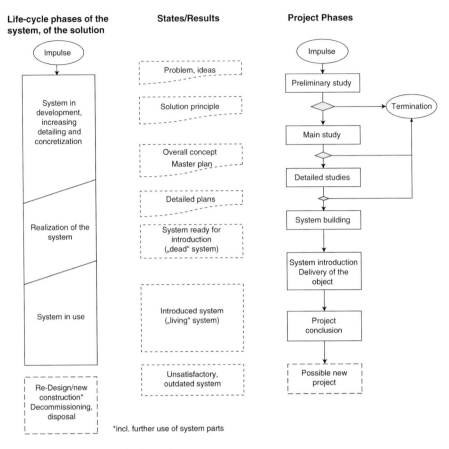

**Fig. 2.5**  Phase concept – basic version

universally valid and thus cannot be described in general terms. On the other hand, as the project progresses, the effort of the practical execution increases noticeably; the early phases generally require less effort in this regard.

**Impulse**

This rather unstructured phase encompasses the time between the perception of a problem, the unease with the current situation, the detection of an opportunity, the emergence of more or less vague solution ideas, etc., and the resolve to undertake something concrete, e.g., to start an organized investigation in the form of a preliminary study. The problem statement can then either be formulated relatively clearly already, or it may consist merely of vague suspicions concerning problems and their causes. It does not matter where the impulse comes from. It is far more decisive that it be accepted by a person or an entity with the competence to initiate concrete actions, issue a project mandate – initially for a preliminary study only – and of course to grant the necessary personnel, finances, informatics, infrastructure, and other resources.

The impulse phase is short if adequate awareness of the problem (psychological stress), awareness of the opportunity, readiness to act, appropriate personnel, and the means for execution at least are available in a preliminary study. It takes a long time if these factors are not very clearly defined, especially on the side of the commissioning party or the entity authorized to issue the mandate. Unfavorable prerequisites for the execution of a project exist in particular if, just to get started, major promises must be made to put into use a possible solution that of course at this early stage cannot be particularly legitimate. Promises of this type reduce the intentionally delayed planned possibilities for correction, modification, or even termination of an unnecessary or an unjustified development on the basis of emotional or prestige barriers.

It is also important that in the impulse phase, no decision should be made concerning the realization of a solution, but only concerning whether or not a preliminary study should be initiated. With respect to the preliminary work required, e.g., agreeing on a project mandate, building up a project group, etc., we refer the reader to Part II, Sect. 4.2.1, "Start-Up Work."

**The Preliminary Study**
The purpose of the preliminary study is to clarify or to ascertain with reasonable effort:

- How broad the *scope of analysis contest of the study* shall be (the boundaries of the problem field)
- Which *mechanisms* have an effect in the problem field
- What the *problem* or the *opportunity* consists of
- Whether the *right problem/ right opportunity* is being addressed
- In what way and to what *extent* is there a *need for a* new or modified *solution*
- What area a new or modified solution should encompass (i.e., where the *boundaries of the design area* should lie)
- What *requirements* the solution should satisfy (system or design goals)
- Whether the *solution principles* are *basically conceivable* (variants) and whether they appear to be feasible for technical, economic, political, social, psychological, temporal, ecological, and other reasons
- Which *solution principle* offers the *greatest promise of success*, for which the relevant evaluation criteria are to be worked out
- Whether by replacing an existing solution it is possible/desirable to reconstruct on an existing system architecture or whether a new architecture is required.

If we establish a relationship between the project phases and the principle of a staged variant creation (Fig. 2.4), the preliminary study encompasses the uppermost level, that is, the working out of the problem and the variants of solution principles that come into question, including the decision basis for choosing the variants that promise the greatest success. In our view, the preliminary study is explicitly not limited to picturing the actual state (although at least a general knowledge of it is obviously a component of a preliminary study), but rather it clearly goes beyond that, because it must also offer possible solutions.

In the preliminary study, one should first deliberately take into consideration the investigation area (see Level A in Fig. 2.3) and also adequately consider the environment in which the solution is to function subsequently, and with which it is interrelated.

As the nature and extent of the needs are not always clear at the beginning of a project, preliminary study is of particular importance. Often at the start, there are only a few known symptoms of an unsatisfactory situation, a possible danger, or clues for an opportunity, and there exist only vague objectives. These symptoms must be investigated to determine their causes and possible ways of eliminating them; the opportunities and risks must be worked out clearly before it is possible to set the course for developing solutions.

Closely connected to the question of needs is the question of delimiting the analysis area and the subsequent design area. In the systems approach, which can be very helpful with this delimitation, there is a certain inherent tendency – because of the challenge of considering external factors and their influences – to expand the original task. However, it is by no means stated that this is always appropriate. Although in most cases a better mutual balance of the various subsystems or system aspects can be reached through a more comprehensive systems concept, the fact remains that the necessary expenditure of time and money can increase considerably, and yet the available resources and the available time are generally limited.

Thus, preliminary study can be understood as a *clarification process* that should be followed by a fundamental decision by some decision-making body. The systems designer should work out the possible problem solutions, the consequences that arise from them, and the requisite preconditions; generally, he or she has no authority to decide on the way forward for a project.

The termination of a project at the end of a preliminary study should thus not be scored as a stigma or an admission of incorrect deliberation. This is far more likely an intentional course setting, which can also lead to termination for sure. Thus, the effort of a preliminary study can be compared to an insurance premium that is intended to reduce the risk of an undesirable outcome. Heeding this reflection increases the probability of terminating in a timely and orderly fashion less promising projects, before great expense has been incurred in the planning (Fig. 2.5).

If it is decided to proceed with the project – possibly with modified objectives – the probability of a later termination is thus reduced, even if not to zero. Also, during the main study, and even after completion of various detailed studies, it may prove necessary to terminate the development and renounce the implementation because of better insights into the relationships among the problems and the possible solutions (Fig. 2.5). But with further progress, the likelihood of this should decrease.

The following may serve as checks for evaluating the *quality of a preliminary study:*

– Is the *task* or the *problem* defined clearly enough?

   Do we know which problem or which task we want to solve and why?
   Is it adequately defined?
   Is its connection to the *environment* clear?

Is there clarity on this within the project team and with the commissioning party?

- Is the *design area* adequately *defined*, known, and delimited sensibly as the area in which changes should or may be made?
- Are the *objectives* in terms of the demands on the solution clear (which functions should be implemented, economic objectives, human resources/social objectives, temporal, ecological objectives, etc.)?
- Is there an adequate *comprehensive view* over basically conceivable *variants* (solution principles)?
- Can these variants be *evaluated* with respect to their suitability (including requirements and consequences)?
- Is it possible to *make a decision* on a specific solution principle? Can this be justified logically and comprehensibly?
- Are the *critical assumptions* or *components* known?

Note: This check list should not be used to deliberately lengthen preliminary studies. It involves basic and not detailed evaluations.

### The Main Study

The purpose of the main study is to concretize and refine the structure of the overall system based on the solution principles (architecture, framework) selected in the preliminary study. This results in overall concepts (variants) that should facilitate a valid assessment of the functionality, usefulness, and profitability, plus any negative consequence of the possible solutions. Now, the focus is on the specific creation of the solution itself. The environment is important, especially to the extent that it affects the further elaboration of the concept designs or is influenced by them positively or negatively. Particular attention must be paid to interfaces.

Critical system components, in other words, those that are particularly important and for which there is reason to suspect that later on they could pose problems when worked out in detail, should be brought to the fore. Detailed investigations and concepts can thus be worked out in the form of well-defined detailed studies within the framework of the main study (or, in extreme cases, as early as the preliminary study). We address this later in greater detail when we treat the case study applications (Part II, Sect. 3.2.3). In extreme cases they can lead to a termination of the development, and this offers the advantage that little or no unnecessary planning is done.

The degree of detailing and the basic observation levels (overall system versus sub- or aspect systems) are thus sometimes *difficult to distinguish from one another.* We return to this in greater detail later on, in the treatment of the application aspects (Part II, Sect. 3.2).

The result of the main study is the decision on an *overall concept* (framework, master plan) that should make it possible to organize further development and realization within an orderly framework (Note: the term *concept* is to be understood broadly here. Depending on the state of the development phase, it can involve a graphic plan, a design drawing, supported by a verbal description, tables, etc., plus combinations of these types of presentations.).

An overall concept as developed in the main study should:

– Present a master plan for the next phases
– Make investment decisions
– Facilitate the definition of partial projects
– Set the priorities for carrying out detailed studies and system building

The following can serve as *checklist of questions* for the final evaluation of the *quality of a main study*:

– Is the suggested *overall concept convincing* and *feasible* (functionally, economically, with regard to personnel, organizationally)?
– Is there a *comprehensive view* of *conceivable alternatives*?
– Are the *critical components* known?
– Is the situation *ready for the decision*? Is the decision generally considered worth supporting? Is it inwardly and outwardly reasonable or manageable?
– Especially on the side of the project team, are there *clear ideas* about how one should move forward?
– Are any necessary or foreseen *priorities* for further detailing or realization clear? For example, has it been determined in what logical and/or temporal sequence the detailing or realization should take place?

**The Detailed Studies**
In this phase, the objects treated are individual *subsystems* or *system aspects* that are picked out from the system concept for temporary, segregated treatment. The demarcation of the problem fields or design areas becomes increasingly easier, as the requirements for the partial solutions are generally deducible from the overall concept. The observation field is now narrowed down drastically.

*The purpose of detailed studies* is:

– To work out detailed solution concepts and make decisions on appropriate design variants
– To concretize the individual partial solutions to the point that they can subsequently be built and introduced as smoothly as possible. Thus, the clarification of important partial problems, which, because of a particular detail concept, are to be expected in the following realization phases, belongs in the task area of the corresponding detailed study.

With respect to the interaction and the integration of partial solutions, see Part II, Sect. 3.2.3.3.

The following can serve as *checklist questions* for the final evaluation of the *quality of each detailed study*:

– Are the *requirements* derived from the *overall concept* met by the detailed concepts? Do the detailed concepts meet the goals?
– Can the detailed concept be *embedded* in the framework of the *overall concept*; can it be integrated? Does it perform its intended functions? Does it exhibit characteristics that are undesirable from the viewpoint of the overall concept?

– Is it *concretized* in such a way that it can be subsequently built?

Note: here, the analysis criteria explained in Part III, Sect. 6.3.4 can be used.

## Systems Building and Tests

The purpose of systems building is the building of solutions in the broadest sense, e.g., the construction of buildings and facilities, the making of products (machines, devices, perhaps prototypes, pilot runs), the creation of IT software including documentation, connected with the detailed preparation of organizational measures and the creation of user-oriented documentation or operating instructions, the organization of lines of communication, the defining of organizational regulations that should apply in cases of a malfunction or breakdown, the appropriate training of users, service personnel, and others (perhaps overlapping with the next phase), the determination of maintenance or service procedures and intervals, etc.

The objects treated here are partial or overall solutions that should be prepared for introduction. Tests or trials before introduction can be particularly significant. It is necessary to distinguish between *individual tests* that deal with trials for individual components, and *system tests* (integration tests), in which the proper functioning of the overall system is tested. Sometimes, it is even usual to provide specific project phases for performing systems tests, because defining the acceptance and testing procedure can be of particular significance.

## System Implementation

Only relatively small, simple solutions, after appropriate preparation, can be implemented in their entirety without great risk. By and large, with complex systems, an abrupt implementation can be very risky because of the variety of unforeseeable side effects; thus, this should be realized gradually. In such cases, one begin with an overall concept, but make the detailed implementation of further steps dependent on the initial implementation results. This is possible in particular with organization projects, and is naturally more difficult with construction, machine, and installation projects.

For the purposes of know-how transfer, adequate training of the handler, operator, and end user is especially important. Before acceptance of the system, the goals, specifications, or guarantees should be reviewed by the commissioning party, customer, developer, operator, and others. This phase ends with formal delivery by the producer, possibly in conjunction with a closing ceremony.

## Project Conclusion

The project also concludes with the proper acceptance of a solution by the commissioning party. Now, there are a number of final tasks that must be carried out, such as billing, debriefing (lessons learned as a learning opportunity for carrying out similar projects in the future), dismissing the project group, etc.

## Usage and Maintenance

Now that the project is over, the utilization lifecycle begins. Operation experiences should be collected to improve the existing solution or to design similar systems. The solution is to be consolidated, maintained, and perhaps improved.

**Reconfiguration, Redesign, Decommissioning**

If, during the use of the system, it becomes apparent that a reconfiguration of fairly significant magnitude, or even a redesign of the solution, is necessary, there will be an impulse for a new preliminary study, and the entire procedure begins anew at this point (Fig. 2.5). This may be the case when further use is no longer allowed, justified, or (economically) sensible.

Changes of smaller magnitude normally require no formal development procedure. They are made during the usage phase, likewise without distinguishing among the various project phases. A formal treatment of requests for changes may help in the organized collection of change requests and thus facilitate an orderly, coordinated reconfiguration.

The redesign of a system commonly involves the decommissioning of an existing system. To complete removal smoothly, the decommissioning must also be the subject of thoughtful planning. With systems whose decommissioning may be difficult, this issue (removal) must be given greater consideration as early as possible in the design process. The removal may even turn into a separate project.

### 2.1.3.3   Other Phase Models

The phased approach presented here is to be understood to be representative of a large number of *phase-oriented* process models. Even if the number and the designation of the phases are different, all retain the same basic idea of subdividing a project into phases, distinguished temporally from one another. In addition, a number of *modifications* and *variations* of the phase concept are possible, which change it in partial areas or interpret it differently, without essential questioning.

1. Of course, depending on the scope and the complexity of a project, the planner is free to include more or fewer phases:

   – A *combination (merging)* of the preliminary, main, and perhaps even detailed study phases into a single development phase in small and manageable projects is just as conceivable as
   – An *expansion*, e.g., by introducing a pre-feasibility study (before a preliminary study) or a separate testing and acceptance phase

2. Later on, we address the possibilities of (partially) doing without a clear temporal border between phases, in terms of an *overlapping approach* – see Part II, Sects. 3.2.3.5 (Dynamics of the Overall Conception) and 3.2.3.6 (Temporal Overlapping Process).

### 2.1.3.4   Alternatives to the Principle of Dividing Into Sequential Phases

One alternative would be to set aside the phased approach. This may make sense with small projects, or with complex projects, where requirements are developed and understood in parallel to developing concepts. We address approaches that are critical with respect to the phase concept in Sects. 2.2.1.6, 2.2.1.7 and 2.2.1.8, for example, prototyping, versions concept, and simultaneous engineering, because we view them as alternatives or complements to the entire systems engineering process model and not just to the phase model.

Alternatives that should be taken seriously are *agile methods* such as Scrum, originating in software development, and spreading visibly to hardware projects. Causes are the cumbersome nature of the phase model – see Sects. 2.2.2, 2.3, and 4.7. If one analyzes carefully, following the Scrum approach, there are phases of a kind, but with a different logic and no pre-defined manner.

### 2.1.3.5    Summary

The phase concept represents the logical expansion of the principles "from the general to the detail" and "variant creation."

- It offers a time-structured pattern that helps in dividing the development of a solution into manageable partial stages.
- A stepwise planning, decision-making, and realization process, with predefined stops or adjustment points, reduces the complexity of project handling, and provides mutual learning opportunities for the planners, the implementers, and the commissioning party.
- In the sense of a macro-logic, the concept of the project phases can be seen as a *management-oriented module* of the process model. It demands contact between designers and commissioning party/management at previously defined points, a common decision-making process, and a decision.
- Further instructions for application are in Part II, Sect. 3.2.3.

## 2.1.4    The Problem-Solving Cycle as a Micro-logic

### 2.1.4.1    Basic Idea

The following explanation of the PSC is based on the Dewey problem-solving logic (Hall 1962). Within the framework of the systems engineering concept presented here, it represents a kind of *micro-logic* (in contrast to the macro-logic of the phase model), which should be used *in every project phase* whenever any type of problem comes up. The following simple partial steps (Fig. 2.6) are the *focus* of such a *micro-logic*:

- *Search for objectives or goal-setting:* where are we? What do we want/need? Why?
- *Search for a solution:* what are the possibilities, what are the ways of getting there?
- *Selection:* which one is the best/most appropriate?

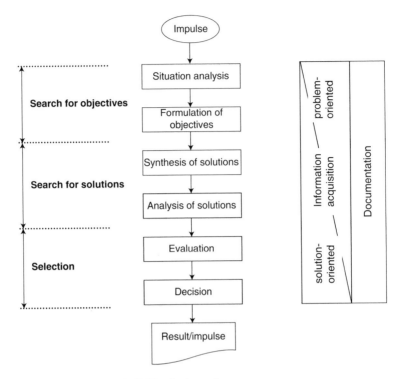

**Fig. 2.6** Problem-solving cycle (PSC) – basic version

The status of the development or implementation work (project phase) has a crucial impact on the content and the degree of detailing for these steps. The PSC can also be applied in agile approaches, namely in each "sprint."

The PSC is of initial interest from the viewpoint of the logic of the development. Figure 2.6 presents mainly the content, the "what," of the individual procedural steps and their logical consequences. In Part II, Sect. 3.3.2, these guidelines, the "how," is emphasized further; there are also reference points that deal with the concrete procedure and application-related expansion or narrowing.

### 2.1.4.2   The Individual Steps in the Problem-Solving Cycle

**Impulse**
The impulse is to a certain extent to be understood as a trigger that sets the work logic in motion. At the beginning of a preliminary study, the impulse is to be understood as the initial spark and is identical to the impulse that also sets the preliminary study in motion (Fig. 2.5).

However, the impulse can also have the character of a concrete task: if a *result* has already been reached in earlier planning steps (e.g., a decision on a particular solution principle), it is then a question of concretizing this result in the next-lowest planning step. Generally, this means that the parts of a concept are singled out and detailed.

The logic of the PSC would be in both cases an appropriate guideline for going forward.

**Situation Analysis**

The *purpose* of the situation analysis (assessment of the situation) consists in familiarization with the initial situation and the task, or generally clarifying it, understanding it, and creating a basis for the setting of concrete goals.

At an early stage (e.g., the preliminary study), this may involve taking a closer look at the symptoms of an unsatisfactory situation, for the purpose of achieving a better understanding of the problem, at possible opportunities, and dangers or their causes.

At a later stage in the project (e.g., detailed studies) the emphasis is on dealing with the concrete initial situation at that time. The following are relevant questions: which overall concept was agreed upon? Which special conditions or restrictions must be considered for further design, including among others the interfaces of adjacent elements?

In the situation analysis *four characteristic views* can be distinguished; they are closely related to one another and can be used alternately or simultaneously:

1. The *system-oriented view* originates from systems thinking and should help in structuring the starting situation, especially with respect to functionality.

   - The problem field can be made transparent, that is, processed and delimited, with the help of the terms *element, relation, black box, suprasystem, subsystem, environment, system boundaries,* etc.
   - Various observation aspects of the same situation can be both highlighted and distinguished from one another with the help of the term *system aspect (filter or glasses).*
   - An analysis of the influencing variables makes it possible, for example, to work out the source, type, and extent of a possible influence on a particular project.
   - Dynamic, process-oriented views allow process operations and behavioral characteristics in the problem field to come to the fore.

   The system-oriented view provides deeper insights into the situation and gives rise to the following observations:
2. The *cause-oriented* (diagnostic) view is aimed at describing the *symptoms* as concrete manifestations of an unsatisfactory situation or a possible opportunity or danger, at ascribing them to the appropriate elements of a situation description, and ultimately work out *possible causes.*
3. The *solution-oriented* (therapeutic) view directs attention at solution ideas and possible interventions (catalog of resources, state of the art, models, etc.) and is very helpful in understanding the problem more effectively, defining the intervention area, and working up realistic goals. With an idea of what a better *target condition* could look like, it naturally becomes easier to scrutinize the

*actual state*. Solution-oriented concepts should, however, not get out of hand in the situation analysis, as the real solution development comes later (synthesis/analysis).

4. The *future-* or *time-oriented view* approach is overlaid on these three perspectives, which can consist, for example, of the following questions:

   - How will the situation develop in the future (short-, medium-, and long-term) if no action is taken today (development of the problem field)?
   - Which important developments must be identified in the solution field?
   - Which effects can possible interventions produce; in what direction can or will the changes lead?

In addition, *boundary conditions* for the solution search are worked out and secured in the situation analysis; they should be understood as determinants from:

   - The system environment (natural, economic, technical, human resources, social, legal, and perhaps even emotional and other types)
   - Earlier decisions that for the moment cannot or should not be influenced
   - The commissioning party's ideas, such as cost limits, deadlines, etc., or
   - The actual state in which constituent elements are seen as (perhaps only temporarily) unchangeable.

The *result* of the situation analysis should produce both *qualitative* and *quantitative* information that provides a better view of the problem. All the considerations used in this context contain not only objective facts, but single out the facts, connect them with opinions, mindsets, and individual evaluations. One characteristic of a good situation analysis is that the compilers take pains to distinguish between facts and their interpretation.

Based on the situation analysis it may become necessary to revise some objectives expressed in the problem statement. This applies especially at the early stages of a system development (preliminary study). In the advanced phases (development of detailed concepts), the impulse rather takes the character of a concrete task and the situation analysis may become correspondingly shorter, for the content and the framework of the further activities have already been narrowed down greatly through earlier decisions. For more on situation analysis, see Part III, Sect. 6.1.

**Formulation of Objectives**

Generally ideas and expectations are expressed as early as the impulse or the project mandate: what is to be accomplished or avoided through the redesign or the re-structuring of a system, e.g., requirements with respect to performance and scope of services, costs, and availability. However, especially during the early stages of a project, there is the difficulty that these ideas or expectations are built on a relatively insecure information basis. Frequently, neither the problem nor the solution field (e.g., the state of the art) is particularly well-known. Thus, a concrete and realistic formulation of objectives in the very early stages is of course difficult. Such goals should rather be taken as preliminary working hypotheses.

In accordance with the problem solution cycle presented here, the *formulation of objectives* is placed at the *end of the search for objectives*, and not at the beginning. The results of the situation analysis should be used as a source of information for the setting or adjusting the specific goals.

These goals, as oriented to a particular project, should be adequately tuned to higher-level goals, e.g., the objectives of the company, and they should support them as far as possible.

The *purpose* of the formulation of objectives is thus a systematic summarizing of the purposes that should underlie the search for solutions. It thus makes sense to use certain *ground rules* as reference points. In particular, the objectives should be:

- *Solution-neutral*, in other words, the functions or effects (the "what") of solutions should be described, and not the solutions themselves (also the "how")
- *Complete* and thus should contain all the important requirements for the desired solution
- As *operational* as possible in the sense of precise and understandable
- *Realistic*, i.e., it should consider the objective circumstances of the situation, plus the social circumstances and the subjective values, especially of the decision-makers, the opinion=makers, and the affected parties

*Mandatory, Desired, and Recommended Objectives*
The diverse demands on a formulation of objectives often cannot be met completely, and of course there may also be contradictions and inconsistencies among conflicting demands.

To establish priorities concerning the importance of goals, the distinction between mandatory and desired objectives has proven useful. *Mandatory objectives* are those that *must* be met; *desired objectives* are those that should be achieved as thoroughly as possible, but not imperatively (nice to have). Some authors also recommend an additional category of *recommended objectives* whose importance falls between the mandatory and desired objectives. We agree with this recommendation and adopt the term.

The recommended or desired objectives form the starting point for a set of criteria for subsequent evaluation. Included hereinafter is a list of operational subgoals used to measure the quality of the solution concepts as developed later on. This list can and will be expanded during the search for a solution, when the specific requirements for particular solutions and their alleged consequences are known.

The decision (or approval) for objectives is the conclusion of the step of formulating objectives; at this point, the goals agreed upon by the commissioning party and the project group should be set down. The goals worked up should also be stated as the binding basis for further planning work. However, it has to be taken into consideration that legitimate change requests may possibly occur later on, which may lead to subsequent adjustments. These should be made as clear and transparent as possible.

The degree of concretization and detailing of objectives is of course dependent on the current project phase. During the early stages, the goals are more global and only partially qualitative and oriented toward the overall solution; in later phases they are more detailed, primarily quantitatively substantiated, and concentrated on the partial solution.

For the process of formulating objectives see Part III, Sect. 6.2.5. Regarding cooperation between the commissioning party and the project group in formulating objectives, we refer the reader to Part II, Sect. 3.4, Expanded Problem-solving Cycle. Concerning the increasing concretization of objectives in the phase process we refer the reader to Part III, Sect. 6.2.3.5, Thinking in Terms of Objectives and Means, and Part IV, Chap. 8, Case Study 1: Private House Building: Additional Domicile.

**Synthesis of Solutions**
The synthesis of solutions is the constructive, creative step in the PSC. The *purpose* of the synthesis is to develop solution variants appropriate to the level of concretization of each phase, from the results of the situational analysis, based particularly on knowledge of the situation, on understanding of the problem, the formulation of objectives, and related challenges. This may involve drafts, concepts, constructions, detailed guidelines for implementation, etc. The level of concretization of the variants should be adequate to allow comparison of the individual variants and choice of the most appropriate one.

In this step, creativity techniques are of particular importance. For further information on synthesis, see Part III, Sect. 6.3.

**Analysis of Solutions**
Although the synthesis can be designated as a *synthesizing–constructive* step in the PSC, the solution analysis is the critical, *analytical–destructive* step. The *purpose* of the analysis is to assess (validate) whether a solution, a concept corresponds to the stated requirements, or whether it exhibits fundamental flaws, which of course are easier to repair as long as the solution exists only on paper. With increasing concretization of a solution, the analysis becomes more elaborate, more solid, and more detailed.

In particular, it is a question of determining whether:

- *Formal* aspects, such as the agreed-upon mandatory objectives, can be achieved
- The individual draft solutions are at the proper *concretization level* for the corresponding phase, or whether unimportant parts are too detailed and essential parts have not yet been addressed
- A solution is *capable of integration*, "outward looking"
- The *functionality* of a solution is discernable and it can be evaluated with it ("inward look")
- Questions concerning the *operational efficiency* (e.g., safety, reliability, operability, maintainability, etc.) can be answered
- The *requirements* and *consequences* of selecting the solution just analyzed can be evaluated in economic, technical, personnel, social, emotional, ecological, and other terms

With the increasing concretization of a solution in the phase process, this analysis step becomes more elaborate, more concrete, and more detailed. Thus, the analysis creates the basis for the subsequent evaluation, from which it must, however, be separated conceptually.

In the *analysis*, it is a question of assessing every individual solution for its usefulness and suitability. On the one hand, this serves as pre-selection, in which unsuitable or clearly less desirable solutions can be discarded early, and on the other hand, as an impulse for a targeted improvement of solution drafts by subjecting them to revision and synthesis.

The *evaluation* involves systematically comparing the remaining variants considered to be basically suitable.

In the course of the analysis, it is entirely possible to encounter essential features of a solution that previously were neither sought nor expected, but which nonetheless come up and may be either desirable or undesirable. These should be used as an opportunity to augment the criteria plan already begun in the step formulation of objectives.

Often, it is not possible to neatly separate synthesis and analysis from one another temporally, because at the moment when a solution idea comes up, the critical controversy also immediately begins. This gives expression to a rather *intuitive analysis* that has little predictability and that should be avoided by most → creativity techniques ("the principle of deferred judgment," not prematurely criticizing ideas). On the contrary, in the problem-solving process, a *systematic analysis* in particular is called for, which should be used formally and consciously when there are essential planning results, and perhaps should even be implemented by other persons or with their support.

With reference to the hierarchical presentation of the procedural principle *from the general to the particular* (Fig. 2.2), a particular sequence of synthesis–analysis can also contain several detailing and concretization steps, whereby the number of variants with increasing detailing is expanded and reduced several times. One proceeds to the next step and conducts a formal evaluation and selection only after achieving concrete results. This reduction in variants in the analysis framework can be considered an intermediate decision that, once the situation is clear, can also be made with reduced formal effort. For more information on synthesis/analysis, see Part III, Sect. 6.3.

**Evaluation of Solutions**

The *purpose* of the evaluation consists of systematically comparing *suitable variants* with one another to determine which is the most appropriate. The criteria for the evaluation are, of course, the objectives that have steered the development and the pre-selection. Thus, only variants that meet all mandatory objectives are accepted for evaluation. The relevant elimination is made in the solution analysis.

A formal evaluation of variants thus makes sense if the presumed best variant is not immediately apparent. The difficulty lies in the fact that sometimes solutions with very different characteristics and manifestations must be compared.

The required criteria for evaluation are ultimately determined from the mandatory or desired objectives as developed in the framework of the formulation of objectives, and from any additional characteristics, conditions, and consequences of the individual solutions. This consists of a list of partial aspects that are considered essential for evaluating the quality of the individual draft solutions.

There are many methods and techniques that can be used in the evaluation phase, e.g., → plus-and-minus balance sheet, → value-benefit analysis, → cost-benefit calculation, → cost-effectiveness analysis, → economic feasibility calculation, and → real options. Such methods should not be regarded as instruments that take the place of decision-making, however. They merely make the decision-making situation transparent, for on the one hand, they force the decision-maker to think about his/her criteria and structure them. On the other hand, a clear view is required of the extent to which the individual criteria are considered to have been met.

By reducing irrationality and arbitrariness or intuition as the only criteria, evaluation methods can help to improve the quality of decisions. For further information on evaluation see Part III, Sect. 6.4.

**Decision/Selection**
The purpose of this step is to appoint the solution variant for further processing, based on the evaluation results. With respect to the "how" of conducting the evaluation and decision, refer to Part III, Sect. 6.4. For the division of duties between commissioning party and project team, see Part II, Sect. 3.2.3.1.

**Result**
The result of the planning activities may consist of finding a satisfactory solution that either can serve as an *impulse* for the next project phase (e.g., main study or detailed studies) or can now be realized, built, and implemented.

However, it may also be that no satisfactory solution is found, or the problem cannot be solved with the currently available personnel, financial, or material resources.

In this case, the following courses of action are conceivable:

- The project is terminated; the existing status is not changed, or is changed only insignificantly.
- The cooperation with the systems developers up to this point is ended (revocation of trust).
- The demands on the solution are trimmed back (goal reduction).
- One returns to a higher system level, perhaps to deal with the existing problem based on a different concept.
- The problem is rewritten or defined in a totally different way.

**Information Procurement**
Information is needed during all steps of the PSC: in the search for objectives, for solutions, and in the selection. The type and extent of the information procurement should take into account, the varying requirements of the individual steps (Fig. 2.6).

In the *search for objectives* information procurement should be mainly problem-oriented; in other words, it should serve the identification and delimitation of the problem, the development of realistic objectives so that the solution can be worked out, in addition to the basic clarification of possibilities for intervention and solution.

In the *search for a solution* and *the selection*, information gathering must become increasingly solution-oriented, that is, oriented toward the development and evaluation of particularly functional and instrumental solution concepts.

An overview of the various procedures of information procurement or processing is given in Part VI.

**Documentation**

The results and interim results of the individual steps should be documented in a comprehensible manner. This increases credibility and makes subsequent detailing and modifications easier or possible.

### 2.1.4.3 Alternatives to the Problem-Solving Cycle

Many of the process models, just like the systems engineering PSC, are based on the Dewey problem-solving model; thus, there are some unmistakable similarities. This especially includes models that describe management decisions and that are promoted by various consulting firms.

In addition, there are also process models that place clear emphasis on the sequence of actions – especially involving the sequence of steps in the PSC. We want to divide them into actual-state-oriented and desired-state-oriented models and subsequently analyze them briefly.

**Actual-State-Oriented Process Models**

These include in particular the *classical process model*, which is characterized by the following process steps:

- *Assessment* of the *current condition*
- *Critical review* of the current condition
- *Development* of a solution for the *desired state*

The following critical arguments refer less to what the method claims than to what it does not claim, and to an often uncritical practice:

1. The term *assessment of the current condition* almost imperatively directs the gaze to the present or the past; the methodology makes no statement concerning the assessment of development trends in the problem and solution areas.
2. For an inexperienced planner, an actual assessment can turn into occupational therapy in a state of perplexity. For example, extensive surveys are conducted and information is gathered, about which nobody really knows whether it is needed at all, or for what, and whether the degree of detail in the inquiry and the evaluation of the questioning are even appropriate. The more time these surveys require, the greater the probability that the area of investigation will change or be

modified by measures introduced in the meantime, and that the results obtained so far will be largely invalidated.

3. In addition, an excessively detailed work with the current circumstances, especially at the beginning, can significantly blur the view toward the future or, for that matter, the concept of a fundamental improvement.

4. Moreover, the fundamental question arises whether the logic of this procedure is even correct: is it meaningful to criticize an actual state at all, without having defined a reference basis for this criticism? Or stated differently: what is the actual state compared to in this criticism? It is not and cannot be the desired state, for that is not developed until the next step. But at least the requirements of the desired state should be defined in conjunction with this criticism, or should be worked out simultaneously. If this does not happen, an evaluation of the current condition is possible only on the basis of preconceptions. This is not fundamentally negative; we all work with preconceptions, and without them we are incapable of making judgments or acting. The only negative aspect is that the methodology recommends no structuring of the preconceptions and thus of the *collective* mindset (e.g., analogous to the "formulating objectives" step in the PSC).

On the basis of these considerations, we deem this methodology to be a usable general outline for writing a report. A developed solution can certainly be argued effectively on the basis of this outline with the actual state: what is it like today? What is unsatisfactory? (= critique). Desired state: what should it be like in the future?

However, in our opinion, it is too sketchy and imprecise for a process recommendation.

### Desired-State-Oriented Process Models

These methodologies[2] present one radical deviation from the procedural method described above: the actual state for developing a solution is initially of secondary importance. An *ideal concept* is conceived on the basis of a short functional allocation, and only afterward are the specific conditions, the so-called actual state, worked out. Then trade-offs on the ideal concept or alternatives to it are elaborated until a useful compromise between the requirements of the as-is state and the possibilities for implementing the ideal solution is found.

The ideal solution thus provides the structure for the investigation of the actual state. This is an *advantage*, because then there exists a reference basis for the actual assessment. One know which questions must be followed up in the as-is assessment: those that help in evaluating the usefulness, suitability, or practicality of the ideal concept. As a result, the actual state assessment necessarily turns out to be structured and target-oriented.

However, it can be a *disadvantage* to probe the needs and problems of the actual state and its causes primarily from the viewpoint of a rashly selected solution. In that case, fundamental problems may remain undiscovered.

In addition, thinking in terms of ideal concepts and the subsequent discovery of those factors that hinder or preclude their implementation can involve a significant potential for frustration.

---

[2] e.g., Nadler, G. (1967): Work Systems Design. The Ideals Concept.

**Position of Our Concept**

The logic of the PSC lies between the two extreme forms outlined above. The preferred one depends on the given problem, and to a large extent, it is a matter of taste, that is, based on a personal style of thinking and working:

- In routine situations that are adequately structured because of the experience of the developer/designer/project engineer and that have a predetermined development process, the classical process methodology may be adequate.
- In situations where actual or presumed difficulties and limitations to the actual state predominate so that there is a tendency to eventually accept the latter as the best compromise, a radical avoidance of the actual state in the sense of the "IDEAL concept" may be appropriate.
- For most cases, though, the logic of the PSC is suitable, which, with respect to information search, can be characterized as follows:

  The definition of the requirements of a solution is necessary and not possible without knowledge of the existing unsatisfactory situation and its environment.

  The familiarity with possible solutions – in terms of the "state of the art" – supports the formulation of the requirements. But these do not have the character of IDEAL concepts, but are at first still neutral.

  A situation analysis is necessary, even with new developments, to clarify the needs or the opportunities that led to its impulse.

  The information search should correspond to the respective detailing phase and proceed in measured steps (from the general to the particular).

  At first, it should be problem-oriented; in other words, it should serve the structuring and the delimiting of the problem, the definition of the requirements, and the assessment of possible developments and interventions.

  The targeted investigation of quantitative information should be carried out only after qualitative problem structuring.

  The information gathering is increasingly solution-oriented as the process goes on, that is, it is oriented toward the development and evaluation of concrete draft solutions.

### 2.1.4.4 Summary

As the fourth building block of the systems engineering process model, the *PSC* constitutes a type of micro-logic that can be used as a guideline for dealing with problems or tasks in every phase of a project. This logic can be simply summarized as follows:

1. *Searching for objectives* or *concretizing objectives,* consisting of the steps situation analysis and formulation of objectives, which should direct attention to the following questions:

   - Where are we?
   - What do we want/where do we want to go?
   - Why?

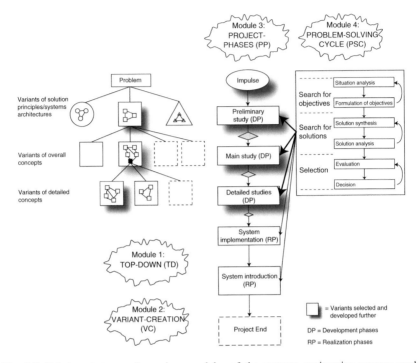

**Fig. 2.7** Relations between the various modules of the systems engineering process model (arrangement by tendency)

2. *Searching for a solution*, consisting of solution synthesis and analysis:

   – What possibilities exist for getting there?

3. *Selection*, consisting of evaluation and decision:

   – Which possibility is the best or the most appropriate?

With an absolutely acceptable reduction to these basic steps, we believe that the PSC is applicable as a universal thought pattern to simple and highly complex problems.

### 2.1.5   Relations Between the Individual Components of the Process Model

The four components of the systems engineering process model constitute building blocks of an overall methodology between which meaningful relations exist or can be established. The arrangement pictured in Fig. 2.7 should be taken only as a basic trend. We consider this modular setup to be particularly characteristic and a special strength of the systems engineering concept:

**Interaction of the components of the process model:**

1. The principle *from the general to the detail* (top–down) and the *principle of variant creation* are presented in the left part of Fig. 2.7.
2. The *project phases* module concretizes these principles in terms of management orientation where the phases are presented with a temporal pattern on which the various levels of the top–down procedure can be arranged:

   - The *preliminary study* involves working out the *problem* and developing various *solution principles.*
   - The *main study,* developing various variants of *overall concepts.*
   - The *detailed studies,* developing *detailed concepts.*

3. The principle of *variant creation* is also represented in the graphic representation of the principle *from the general to the detail*, and of course it is also an essential component of the search for a solution (synthesis/analysis) in the PSC.
4. The *PSC* is the micro-logic that makes it possible to attack any problem.

**The Significance of the Process Steps in the Problem-Solving Cycle**

The PSC as a whole undoubtedly takes on the greatest significance in the development phases (the preliminary, main, and detailed studies) because most of the problems that occur here should be solved methodically and appropriately. In the realization phases (systems building and systems introduction), routine processes and a situation-oriented improvisation tend to take on additional significance. But in principle the PSC can be used by any problem occurring during implementation.

However, it is not only the significance of the PSC as a whole that changes during the course of the project phases, but also the significance of the individual process steps:

- In the PSC of the preliminary study, because of the fundamental soft settings, the *search for objectives* (situation analysis and formulation of objectives) should be given particular attention; then, it may diminish in the main study and the detailed study.
- The *search for a solution* (synthesis and analysis) is important in the preliminary study, the main study, and the detailed study. The preliminary study deals with basic approaches (architecture decisions); the others deal with their design.
- The same applies to the *selection* (evaluation and decision).

It should be further noted that the focus of the individual steps changes successively throughout the development: as solutions are developed it becomes increasingly narrower, but in steps that integrate the developed parts into superordinate concepts, the focus should become increasingly broad.

## 2.2   Other Process Models

Other process models are described below. They can be regarded as alternatives to the systems engineering process model (Hall/BWI) explained above. In this context it is helpful, inspired by B. Boehm, to distinguish between "plan-driven" methods and "agile methods".[3]

*Plan-driven methods* specify a high degree of structure – in the sense of a clear sequence of steps – and combine this with the expectation that qualitatively highly advanced solutions can be worked out in an efficient manner. The systems engineering process model described predominantly belongs in this category.[4] Other representatives of this group include the waterfall model, the Vee model, the International Council on Systems Engineering (INCOSE) methodology, IEEE-15288, and simultaneous (concurrent) engineering.

Despite the undisputed advantage of being able to contribute a logical procedural structure to projects, the *plan-driven methods* are sometimes vehemently criticized, especially in combination with software projects. In particular, they are accused of making the (software) development process unnecessarily cumbersome and requiring long development times, but without producing satisfactory results, because they impose very restrictive behavior regarding subsequent, valid specification changes. From this need – beginning in software development – the so-called *agile methods* have developed.

*Agile methods* are procedures that are differently structured, with a view to reacting flexibly and adaptively to new results, insights, needs, and desires.

### 2.2.1   Plan-Driven Models

The systems engineering process model with its four components, as mentioned, predominantly belongs to the plan-driven models category – even though it certainly has agile characteristics (see Sect. 2.2.3.2).

In the following, we present nine more plan-drive models. We think we can show that the systems engineering process model is a more comprehensive approach that helps in interpreting its alternatives in addition to also being open to the situation-dependent adoption of good ideas. In addition, the confrontation with other approaches also helps to clarify one's own viewpoint, emphasize it, or put it into perspective.

---

[3] Boehm, B. and Turner, R. (2004): Balancing Agility and Discipline.

[4] In later sections, we loosen the procedures that have been outlined so far in simplified terms for didactical reasons and show that a certain degree of agility is also possible with the systems engineering process model.

### 2.2.1.1 Waterfall Model

The waterfall model is the oldest and best-known process model for systems development, and it is a traditional top–down development approach. In its original version, it was a linear (non-iterative) process model in software development.[5]

A project is subdivided into individual phases in which the results (outputs) of one phase constitute the input for the following one – thus the term *waterfall*. This is a result- or milestone-oriented model, and it therefore corresponds in its basic idea to the phases module in the systems engineering process model. However, because the latter entails four modules in total, the two concepts are not directly comparable with one another.

To confront the rigidity of the theoretical model, the model was expanded, iterations were allowed and represented by arrows pointing backward (Fig. 2.8). This is also often referred to as the "splashing" waterfall model.

The waterfall model is generally used to advantage where requirements, achievements, and procedures can be described relatively precisely in the planning phase and seem to stay stable.

The *advantages* of the waterfall model are as follows:

– A clear delimitation of the phases is at least theoretically possible: activities are to be carried out completely in a given sequence. Each activity must be concluded before the next one begins.
– Simple possibilities for planning and control: at the end of every activity there is a completed result in the form of documents, models, or software/hardware, i.e., the waterfall model is a results-driven model.

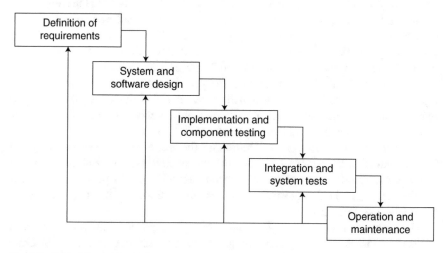

**Fig. 2.8** Waterfall model with iterations

---

[5] First described and criticized in Royce (1970).

- The process is top–down.
- It is a simple, comprehensible model that is easy to manage in the linear form presented. With stable requirements and a clear assessment of costs and scope, this can be an effective model.

The following are *disadvantages* and problems:

- Phases clearly delineated from one another are often unrealistic – in reality the transition between them is fluid: parts of one system can still be in the planning stage, whereas others are already in implementation or in use.
- In practice it is often unavoidable to return to previous phases.
- As user participation is provided only in the startup phase and subsequently the design and the implementation are done without stipulating any participation by the user or the commissioning party, the model is rather inflexible with respect to new knowledge, insights, and needs for change.
- To avoid subsequent changes to the requirements, an early commitment is pursued (this continues until the "signing" of the lists of requirements). The consequences of this kind of bureaucratic behavior can be the commissioning party's dissatisfaction with the system delivered when subsequent changes are refused, or expensive changes result from the multiple iterations of the process.

### 2.2.1.2   Vee Model

The Vee model is a combination of a top–down and a bottom–up approach. Top–down customer goals are converted into technical requirements and specifications for the overall system, and later into subsystems and concepts, then the subsystems are created and integrated bottom–up, and finally the overall system is delivered with respect to the original goals. In particular, an effort was made to remedy the disadvantages of the waterfall model and to link together the iterations and integration steps that are possible in the spiral model (subsequently, see Part I, Sect. 2.2.2.1). The following description is limited to the principles that are contained in most representations (Fig. 2.9). The designation *Vee model* characterizes the V-shaped representation of the individual process elements, structured according to their coarse temporal position and their depth of detail.

The *descending left side* of the V represents the breakdown of the customer goals into technical specifications, initially for the overall system, and then for the subsystems. The downward iterations encompass all steps for understanding user requirements, and the demonstrations of feasibility, all the way down to the level of the smallest system elements.

The *right side* represents the *integration* of the components into subsystems and of the latter into the overall system, along with the simultaneous *verification* in terms of a comparison with the requirements defined on the other side of the V. The upward iterations support the technical basis, making especially certain that the solution is

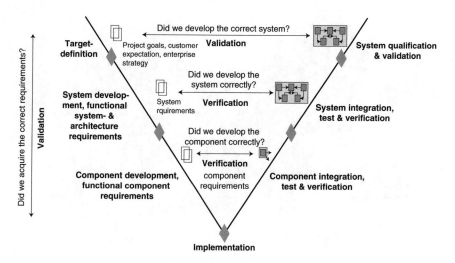

**Fig. 2.9**   Vee model. Dependencies between requirements, verification, and validation. (Authorized source: 3DSE Management Consultants GmbH, Munich)

also capable of ensuring the user's requirements in reality. Finally, the system is validated to be sure that it satisfies the client's needs. The result of the integration on the right side of the V is thus always verified against the requirements on the left side and then validated. As shown in Fig. 2.9, in some Vee model representations, any iterations are possible in all directions.

The first ideas involving the Vee model as a process approach appeared at the end of the 1970s.[6] Today, there are multiple interpretations of this model, both in software development and in overall systems development.[7] The model was expanded in succession into the Vee Model 97 and later into the V Model XT (XT = extreme tailoring). This makes the system adaptable to particular needs (tailoring), and it facilitates a stronger orientation toward agile and incremental approaches.

The underlying principles of the Vee model are applicable not only to IT projects, but more and more to other development projects too, e.g., in mechanical engineering/mechatronics (Fig. 2.10).

---

[6] Jensen, Randall W.; Tonies, Charles C. (1979): Software Engineering.

[7] Forsberg, K.; Mooz, H. (1991): The relationship of Systems Engineering to the Project Cycle.

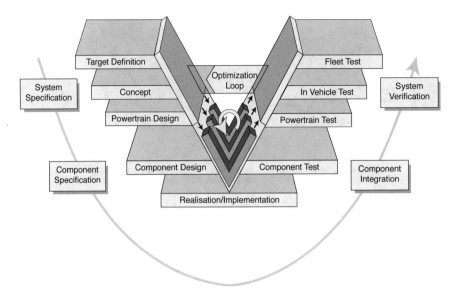

**Fig. 2.10** Use of the Vee model in the development of automobile power train systems. (Authorized source: AVL List GmbH, Graz)

### 2.2.1.3   The SIMILAR Process

This partial model is intended to illustrate the individual steps, starting with the customer needs and their exploration and going all the way up to the successful delivery of a solution (Fig. 2.11).[8]

The SIMILAR model consists of a combination of steps that correspond to steps of the PSC in the Hall/BWI process model on the one hand, and on the other, to various project phases. The individual steps are as follows:

– State the problem, including formulation of the requirements of the solution
– Investigate alternatives, but only once (not several concretization steps)
– Then, follow the steps that we would assign to the project phases: model the system, integrate (modules), launch the system, and assess performance

Bahill and Gissing describe the individual steps in the following.[9]

**State the Problem**
The problem statement starts with a description of the top-level functions that the system must perform: this might be in the form of a mission statement, a

---

[8] After Bahill, A. T.; Gissing, B. (1998); *Systems Engineering Process.*

INCOSE (International Council on Systems Engineering) has incorporated the SIMILAR process into its manual to illustrate the systems engineer's procedure.

http://www.incose.org/practice/fellowsconsensus.aspx

[9] Bahill and Gissing (1998).

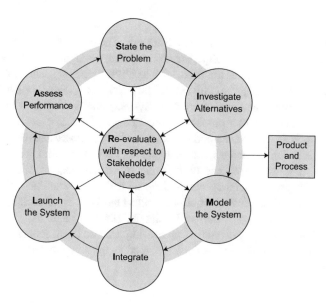

**Fig. 2.11** The SIMILAR process. (Authorized copy: Bahill and Madni 2017. Springer Nature)

concept of operations or a description of the deficiency that must be ameliorated. Most mandatory and preference requirements should be traceable to this problem statement. Acceptable systems must satisfy all the mandatory requirements. The preference requirements are traded off to find the preferred alternatives. The problem statement should be in terms of *what* must be done, not *how* to do it. The problem statement should express the customer requirements in functional or behavioral terms. It may be composed of words or as a model. Inputs come from end users, operators, maintainers, suppliers, acquirers, owners, regulatory agencies, victims, sponsors, manufacturers, and other stakeholders.

**Investigate Alternatives**
Alternative designs are created and are evaluated based on performance, schedule, cost, and risk figures of merit. No design is likely to be best on all figures of merit; thus, multicriteria decision-aiding techniques should be used to reveal the preferred alternatives. This analysis should be redone whenever more data are available. For example, figures of merit should be computed, initially based on estimates by the design engineers. Then, concurrently, models should be constructed and evaluated; simulation data should be derived; and prototypes should be built and measured. Finally, tests should be run on the real system. Alternatives should be judged for compliance of capability against requirements. For the design of complex systems, alternative designs reduce project risk. Investigating innovative alternatives helps to clarify the problem statement.

(Note: this step would correspond to the principles "from the general to the particular" and "variant creation.")

## Model the System

Models are developed for most alternative designs. The model for the preferred alternative is expanded and used to help manage the system throughout its entire life cycle. Many types of system models are used, such as physical analogs, analytic equations, state machines, block diagrams, functional flow diagrams, object-oriented models, computer simulations, and mental models. Systems engineering is responsible for creating a product and also a process for producing it. Therefore, models should be constructed for both the product and the process. *Process models* allow us, for example, to study scheduling changes, create dynamic PERT charts, and perform sensitivity analyses to show the effects of delaying or accelerating certain subprojects. Running the process models reveals bottlenecks and fragmented activities, reduces cost, and exposes the duplication of effort. *Product models* help to explain the system. These models are also used in tradeoff studies and risk management.

As previously stated, the systems engineering process is not sequential: it is parallel and iterative. This is another example: models must be created before alternatives can be investigated.

## Integrate

No man is an island. Thus, systems, businesses, and people must be integrated so that they interact with one another. Integration means bringing things together so that they work as a whole. Interfaces between subsystems must be designed and subsystems should be defined along natural boundaries. Subsystems should be defined to minimize the amount of information to be exchanged between the subsystems and well-designed subsystems send finished products to other subsystems. Feedback loops around individual subsystems are easier to manage than feedback loops around interconnected subsystems. Processes of co-evolving systems also need to be integrated. The consequence of integration is a system that is built and operated using efficient processes.

## Launch the System

Launching the system means running the system and producing outputs. In a manufacturing environment this may mean buying commercial off-the-shelf (COTS) hardware or software, or it may mean actually making things. Launching the system means allowing the system do what it was intended to do. This also includes the system engineering of deploying multi-site, multi-cultural systems.

This is the phase where the preferred alternative is designed in detail; the parts are built or bought (COTS ), the parts are integrated and tested at various levels leading to the certified product. In parallel, the processes necessary for this are developed – where necessary – and applied so that the product can be produced. In designing and producing the product, due consideration is given to its interfaces with operators (humans, who need to be trained) and other systems with which the product interfaces. In some instances, this causes interfaced systems to co-evolve. The process of designing and producing the system is iterative, as new knowledge developed along the way can cause a re-consideration and modification of earlier steps.

The systems engineers' products are a mission statement, a requirements document including verification and validation, a description of functions and objects, figures of merit, a test plan, drawing of system boundaries, an interface control document, a list of deliverables, models, a sensitivity analysis, a tradeoff study, a risk analysis, a life cycle analysis, and a description of the physical architecture. The requirements should be validated (Are we building the right system?) and verified (Are we building the system right?). The system functions should be mapped to the physical components. The mapping of functions to physical components can be one to one or many to one. But if one function were assigned to two or more physical components, then a mistake might have been made and it should be investigated. One valid reason for assigning a function to more than one component would be that the function is performed by one component in a certain mode and by another component in another mode. Another reason would be deliberate redundancy to enhance reliability, allowing one portion of the system to take on a function if another portion fails to do so.

**Assess Performance**
Figures of merit, technical performance measures, and metrics are all used to assess performance. Figures of merit are used to quantify requirements in the tradeoff studies. They usually focus on the product. Technical performance measures are used to mitigate risk during design and manufacturing. Metrics (including customer satisfaction comments, productivity, number of problem reports, or whatever you feel is critical to your business) are used to help manage a company's processes. Measurement is the key. If you cannot measure it, you cannot control it. If you cannot control it, you cannot improve it. Important resources such as weight, volume, price, communications bandwidth, and power consumption should be managed. Each subsystem is allocated a portion of the total budget and the project manager is allocated a reserve. These resource budgets are managed throughout the system life cycle.

**Re-evaluate**
Re-evaluation is arguably the most important of these functions. For a century, engineers have used feedback to help control systems and improve performance. It is one of the most fundamental engineering tools. Re-evaluation should be a continual process with many parallel loops. Re-evaluation means observing outputs and using this information to modify the system, the inputs, the product or the process.

Bahill and Gissing characterize their SIMILAR model as follows: the process of drafting, designing, and producing a system is iterative. If new knowledge and new experience are gained, they should be used to improve the design or the construction. Thus, a return to earlier steps in connection with the development of new solutions should be seen as normal, to the extent that it serves to improve the solution in terms of the project goals.

The results that a systems engineer delivers in the course of his work are a clear statement of task, a list of requirements (including verification and validation),[10] a

---

[10] According to IEEE norm *Verification* = Is the system built properly? and *Validation* = Is the right system built?

description of the functions and objects, a test plan, identification of the system boundaries and of the relevant environment, and definition of the interfaces. A list of the deliveries and performances (deliverables) agreed upon supports the verification that the requirements have been met. Further analyses to be conducted include sensitivity analyses of the evaluation results, risk analyses, life-cycle analyses, descriptions of the physical architecture, etc.

### 2.2.1.4  VDI Guideline 2221

The guideline was developed with a view toward closed systematics for the development/design of machine systems in collaboration with the VDI (association of German engineers). It contains a *step* and *phase* process from the *abstract to the concrete* and from the concept to the installation drawings, the *development and elimination of variants*, and a *microcycle* not presented here (determination of requirements, development of solutions, and evaluation).

The VDI guideline and the approximately corresponding parts of the systems engineering concept are compared in Fig. 2.12.

### 2.2.1.5  Engineering Design Methodology According to Pahl and Beitz

Pahl and Beitz divide up their product development methodology according to the following four steps (Fig. 2.13). As can be seen, it comprises the top–down and the phase approach.

1. *Task clarification*
   First, it is necessary to gather appropriate *information* about the customer needs, the market, the competitive situation, etc. This is the basis for preparing a *list of requirements* (specifications, functional specifications)[11] from which the product functions result.
2. *Conceptual design*
   The product functions are broken down into partial functions, for which various solution principles are sought. Connecting the various principles gives rise to solution variants. These are evaluated, and the preferred variant determines the working structure.
3. *Embodiment design*
   The individual functional components are structured to scale (= general assembly drawing) and analyzed, evaluated, and calculated. Then they are detailed into a full-scale detailed drawing. In this process, the technical flexibility must also be

---

[11] The terms *specification* and *functional specifications* are not used consistently in practice. A distinction could be as follows: a *specification* (requirement specification, client specification) describes the overall requirements of a "product" by the client/commissioning party or from that viewpoint. The *functional specification* (technical specifications, detailed technical concept, desired concept, overall system specifications, implementation specifications, or feature specifications) describes in concrete form how the contractor is meeting the requirements; this belongs to the contractor.

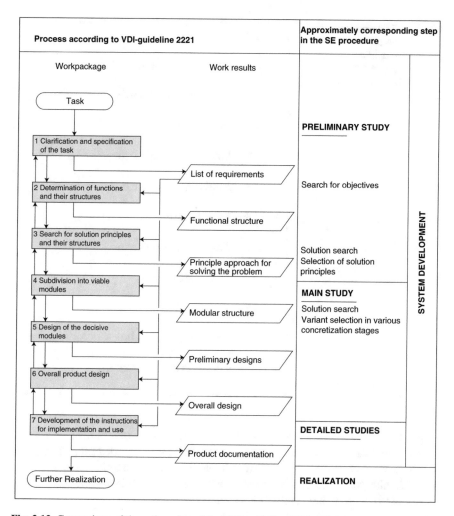

**Fig. 2.12** Comparison of the action plan of the VDI guideline 2221 and the systems engineering procedure (Jänsch, J. and Birkhofer, H. (2006): The Development of the Guideline VDI 2221)

taken into consideration. Styling considerations are addressed as needed and physical models are prepared. These sketches or designs are evaluated and a choice is made of the preferred variant(s). Functional models may be made, with which the function of the selected technical solutions must be verified. Often, a combination of several models is created before the detailed development of product function or construction groups occurs during a further work phase (transmission, motor, etc.).

4. *Detailed design*

The necessary manufacturing documentation is drawn up. If the individual component drawings are available, prototypes are manufactured and tested for early detection of weaknesses, flaws, or problems. Based on the existing error log, the

**Fig. 2.13** The main phases of the design engineering process. (Simplified representation according to Pahl and Beitz (Pahl et al. 2007)

**1. Task clarification**

  - Informative determination

  **→ List of requirements**

**2. Conceptual design**

  - Solution principle determination

  **→ Solution (concept)**

**3. Embodiment design**

  - Creative determination

  **→Ultimate design**

**4. Detail design**

  - Manufacturing-technical determination

  **→Product documentation**

product is reworked, and the specification may be adjusted. Further prototypes may be manufactured. In the so-called pilot run, it is determined whether resources such as tools and jigs are suitable for series production. With an initial production run/first series (often of reduced batch size), it is determined whether trouble-free production can be expected.

### 2.2.1.6  The Prototyping Approach

Prototyping as a procedural principle occurred in the mid-1970s in data processing. On the one hand, the trigger was the heavy phased process, with its relatively high number of formalisms (project mandates, decision reports, documentation, etc.), was also held partly responsible for the so-called user backlog; and, on the other hand, the trigger was the users' difficulties in having to specify concrete requirements at an early stage, without being able to imagine what ultimately would come from them. Often, the user only realizes after implementation of the solution what he should have wished for. Each deviation of the developed solutions from the expectations is seen as a disappointment on the part of both the developer and the user.

**Basic Idea**

The idea of designing a sort of prototype, initially at a relatively low cost, before developing the ultimate product, to visualize and evaluate the product's essential characteristics, was adopted long ago in machinery construction, engineering, and the building industry. There, it is very common to realize solution designs into physical models at various phases before the ultimate (mass-produced) product is manufactured, e.g., in the form of functional or laboratory prototypes, pilot models, etc. These do not involve the expectation that the prototype can be used by the commissioning party/user. It is rather intended to provide a better evaluation of the concept pursued so far, and may also serve for testing under operating conditions.

In the construction industry or in plant manufacturing, physical models are used that can provide a better idea of the subsequent solution. Modern design tools can often take the place of realizing physical models and thus save time and money (e.g., digital 3D models, computer simulations, etc.).

Basically four types of prototypes can be distinguished[12]:

- *Proof-of-principle prototype (model)*: this kind of prototype serves to investigate a particular design (layout, architecture) of a solution, without precisely simulating the outer form or the future materials of the finished product, etc. Among other things, this makes it possible to recognize which design options work and which ones do not, thereby identifying further development paths. For example: a new chassis is combined with an existing body.
- *Form study prototype (model)*: this should allow developers to assess the size, shape, look, and feel of a solution without making it in detail. It should reveal ergonomic factors or visual aspects of the end product. Such prototypes are often made from materials that are cheap and easy to work with (wood, clay, foam, etc.).
- *Visual prototype (model)*: as a mock-up of the subsequent solution, this kind of prototype should facilitate an evaluation of the esthetics, design, colors, and surface structure of the contemplated product. The functions that the subsequent product is to deliver are not yet represented, for example, a model or a photo-montage of a housing complex.
- *Functional prototype (model)*: this provides a fairly clear representation of the ultimate design, materials, functionalities, etc. But it can, for example, for reasons of cost, be reduced in size, for example, a production facility on a laboratory scale.

From various viewpoints, prototypes can be seen as:

- *Design aids* (= explorative prototyping), with the goal of concretizing solutions more quickly and thus achieving more efficient communication between developers and, for example, users. Prototyping should then help to close in on the user's needs. When these are clear, the prototype is perfected and made into a solution that is ready to be introduced.

---

[12] According to Wikipedia.

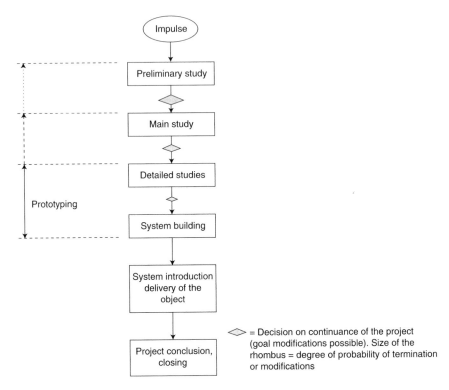

**Fig. 2.14** Prototyping as a design aid in the phase process

– However, prototyping can have different meanings, for example, when an especially *quick solution* is sought that does not need to be either complete or perfect. This (quick and dirty) solution is delivered to the user (e.g., pilot installation) and if necessary it is changed, adjusted, or perfected during the operational phase (evolutionary prototyping). The procedure is supported in the area of data processing by software tools that use powerful commands to facilitate a very quick and efficient procedure for the developer. In the early stages, this procedure may appeal to everyone concerned. But people should be aware of the danger of the addiction toward improvisation increasing and the solutions remaining quick and dirty.

**Comparison of Prototyping with Systems Engineering Concept**

Prototyping is to be unreservedly welcomed as a *design aid*. It is not a contradiction of the systems engineering action model, but can even support it effectively. However, prototyping should not replace conceptual phases (preliminary or main study), for these are required as orientation aids for the detailed development. But it could help in checking the desirability of a solution from the user's viewpoint, or the practicability of a concept. Thus, the main focus of the application should be seen mainly in the coming phases of detailed studies and system building, where a clear division between planning and implementation is often no longer necessary (Fig. 2.14).

A prototyping-oriented process that does not have the characteristic of a design aid, but that of working up and implementing a *quick solution*, may be applied with relatively small solutions that are needed only for a short time or for a special purpose. In addition to the undisputed advantages of quick availability, this type of process can still have some disadvantages for both the user and the designer:

- On the one hand, *speed* often comes at the expenses of *quality*, and inadequate quality can quickly dampen the pleasure of a quick solution. On the other hand, it is often difficult for developers to convince users, or decision-making bodies (management) of the necessary effort required to turn a functioning prototype into a well-engineered solution.
- Last but not least, we consider a general waiving of conceptual phases and a solely user-oriented development of application software to be shortsighted and even dangerous. How are solutions developed in this way (e.g., programs or programming systems) to be integrated subsequently if the interfaces were not designed with an overall view? How can they be maintained if the documentation is inadequate or non-existent because they were produced so quickly?
- Therefore, prototypes should, as in the original sense of the word, be disposable products that perform important functions temporarily, with the transition to what is known as the "versions concept" emerging, which is described in the next section.

### 2.2.1.7 The Versions Concept

The so-called versions concept exhibits similarities to the prototyping approach of the quick solution and is applicable in developments of all types (machines, devices, installations), and sometimes it may even be unavoidable.

**Basic Idea**
The basic idea is that a solution is not to be perfected at one blow, but rather that a first version should be designed and implemented, and made available to the user. On that basis, improvements become possible because of operational experience and take place from one version to the next (slow-growing systems). This involves a shift from the design orientation of the phase concept to an implementation orientation. The purpose and the similarities to the prototyping approach are thus unmistakable.

In addition to the definite appeal of such a process, especially with respect to the speed of development and the achievable visible progress, in our view, there are also a few attendant limitations. This process can lead to less careful design, because it is easier to shift problems or improvements to the next version. Furthermore, the versions concept places high demands on documentation and project administration, as at every point it *must* be clearly understood where each version is valid and how the individual components of a solution were implemented, or how they are interdependent. The configuration management briefly described in the glossary (Part VI, Chap. 15) is certainly is an essential requirement.

**Comparison of the Versions Concept with the Systems Engineering Concept**
Without taking a dogmatic view, we consider both concepts to be compatible, in particular if the versions concept is restricted to a weight shift inside the phases and is not intended to totally eliminate the phase concept. The development phases (preliminary, main, and detailed studies) are consciously streamlined. The planning horizon for the usage phase is quite short, because subsequent, modified versions must be reckoned with right from the start.

The phase model, the versions concept, and the prototyping approach can thus be combined very effectively: the first version is developed after a (shortened) phase model (with observance of the action principle "from the general to the particular" and the principle of variant creation). The prototyping approach is chosen for selected conceptual components (detailed studies, systems building). The creation of a second or third version no longer follows prescribed rules, unless comprehensive changes are involved, for which a (reduced) phase process could be used.

### 2.2.1.8  Simultaneous Engineering, Concurrent Engineering

Simultaneous engineering (also known as concurrent engineering) has its origin in product development. The trigger for this idea is the demand for shorter development times that are reflected in favorable, early market entry effects such as higher prices, greater market share, and thus savings through scale effects, positioning as market leader, higher accumulated profits, etc.[13]

**Basic Idea**
The concept seeks an acceleration of the development and implementation process – however, not through an intentional reduction of requirements, as with the versions concept, but rather through an extensive *parallelization* of processes (Fig. 2.15). Issues of production, procurement of production resources, suppliers, etc., should be included in product development as early as possible and in this way, they will be largely handled simultaneously.

**Basic idea of acceleration:**

1. Involves a *rethinking of the traditional process logic* for development projects with a view toward carrying out partial steps as far as possible in parallel rather than sequentially. In particular, the time-critical and independent partial steps come to the fore. The main maxim is critical in the sense of *concept-defining first*, so that the follow-up activities can begin without delay.
2. Requires a *holistic work approach*, in which the follow-up activities later encountered (production, materials management, logistics, cost accounting, integrated product teams, perhaps with external suppliers, etc.) are integrated into the

---

[13] Ribbens, J.A. (2000): Simultaneous Engineering.
Stjepandić Josip; Wognum Nel; Verhagen Wim J.C. (eds.) (2016): Concurrent Engineering in the 21st Century.

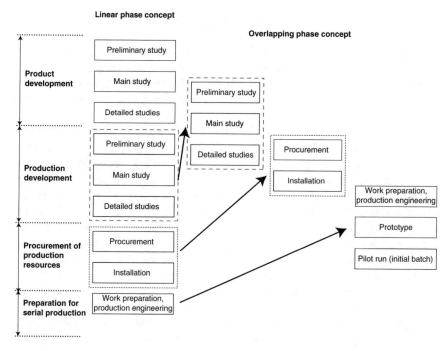

**Fig. 2.15** Simultaneous (concurrent) engineering as an overlapping phase concept

development process as early as possible. On the one hand, errors can be avoided early or additional possibilities can be used, and, on the other, the follow-up activities can be started earlier, that is, simultaneously. In this team-oriented approach many simultaneous engineering supporters see a significant potential for setting changes and innovations in motion.

3. Increasingly tends to promote the *inclusion of CAD, CAM, and CAx technologies* and thereby the acceleration or even the elimination of entire steps or step sequences. As an example, the use of CAD technologies in connection with design, accounting, parts list management, materials information, computational or pictorial simulations, etc., can make the time-consuming creation of physical prototypes or laboratory experiments and trials superfluous, or at least reduces it greatly.

## Comparison of Simultaneous Engineering with the Systems Engineering Concept

In our view, the concept of simultaneous engineering is not in contradiction to the systems engineering concept. It can be understood as an application-oriented interpretation of the phase concept. The phase concept is not seen as strictly linear, in the sense that the product development must be totally completed in detail before the development of the production resources can be addressed. We should not wait until even this development is completed to start procuring the resources – as shown in the left part of Fig. 2.15.

A partially simultaneous handling is made possible by an overlapping arrangement of the phases. This new structuring of the process logic can force us to fix central details during the very early phases so that the follow-up activities can be tackled quickly. This can have effects on the content of the early phases of product development.

However, there should be a minimum level of advancement in the sense of a temporal progressing. There is some danger, for example, *before* the selection of a solution principle, in conducting excessively intensive discussions with the "implementers," or letting them start too soon. The building and plant industry could serve as a warning example here. An opposing tendency can be observed: a careful and detailed design of the object is given clear preference before the start of realization. A quicker and especially more cost-effective construction process because of fewer subsequent changes is perceived there as an essential argument.

Although advantages are no doubt linked to the simultaneous handling of a detailed development of the product and to the preparation of production or even outside procurement, there are also *risks*, e.g., with subsequent, unforeseen changes in the product and the associated effects on production facilities that have already been ordered or procurement agreements on parts purchased from a third party. In the interest of speeding up the process, this risk may be acceptable. Good project organization and, in particular, good coordination and communication among the people involved in the project, are essential prerequisites.

There is another consideration here in the comparison with the systems engineering concept: the more subjects are handled in overlap, of course, the quicker the decrease in alternatives. This concerns both the detailed product design and eventually also the choice of production processes and external suppliers. This loss of options is less risky if one has already had some experience with a product, that is, if it involves the reworking of an existing product and/or production concept rather than a new development.

## 2.2.2  Agile Process Models

The *plan-driven methods* as described above are sometimes vehemently criticized, despite the unquestioned advantage that they can bring a logical process structure to projects, especially in connection with software projects. In particular, they are criticized for making the (software) development process unnecessarily difficult, requiring long development times, and usually producing unsatisfactory results because they may encourage a very restrictive behavior with respect to subsequent, justifiable specification changes. The so-called *agile methods* have grown out of this need.

This category includes mainly process models that were developed in connection with software development. Prototyping and the versions model are exceptions that can easily be transferred to any type of development, although here they have been assigned to the *plan-driven models*.

Agile process models are further developed and increasingly carried over to the development of systems comprising hardware and software components. We return to this matter later on.

### 2.2.2.1   Spiral Model

The spiral model is a process model in software development that was inspired by Barry W. Boehm as an alternative to the rigid waterfall model.[14] The division into phases is retained, but by combining with the prototyping idea, the phases occur in overlap. In addition, the model provides guidelines on the conditions under which individual phases should be repeated. The risk of specification flaws is reduced through the prototyping approach, and a better starting basis for follow-up activities is created through the experience gained in programming the prototypes. The specifications (design, user interface, etc.) are tested in the iterative prototyping process. As long as the prototype is not accepted, the specifications are changed or expanded. This is continued until a satisfactory result for the developer and the user is achieved.

In Fig. 2.16 the total effort is presented along the respective radius of the spiral and the project progress with the angular coordinate.

If the spiral model is compared with the systems engineering process model, one may say that both the project phases and the PSC modules are recognizable: the sequence of the quadrants essentially corresponds to the logic of the PSC (1. Goals, 2. Alternatives, etc.). With every *turn of the spiral*, the degree of detailing in the planning grows from the inside toward the outside (requirement, preliminary design, detailed design, etc.) or in the results (prototype 1, prototype 2, operational prototype, etc.), which would correspond to the *project phases* – but in an iterative and prototyping-oriented approach.

### 2.2.2.2   Agile Manifesto

What is known as the *Agile Manifesto* (Manifesto for Agile Software Development) plays an essential role in connection with the development of the agile methods; it is described in excerpts below[15]:

> *We are uncovering better ways of developing software by doing it and helping others do it. Through this work we have come to value:*
>
> – *Individuals and interactions over processes and tools*
> – *Working software over comprehensive documentation*
> – *Customer collaboration over contract negotiation*
> – *Responding to change over following a plan*

---

[14] Boehm (1988): A Spiral Model of Software Development and Enhancement".
[15] Beck, K. et al. (2001): Manifesto.

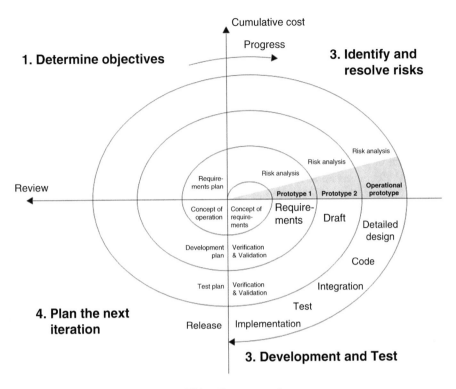

**Fig. 2.16** The spiral model. (Source: Wikimedia, commons)

*That is, while there is value in the items on the right, we value the items on the left more.*

And:

*We follow these principles behind the Agile Manifesto:*

- *Our highest priority is to satisfy the customer through early and continuous delivery of valuable software.*
- *Welcome changing requirements, even late in development. Agile processes harness change for the customer's competitive advantage.*
- *Deliver working software frequently, from a couple of weeks to a couple of months, with a preference to the shorter timescale.*
- *Business people and developers must work together daily throughout the project.*
- *Build projects around motivated individuals. Give them the environment and support they need, and trust them to get the job done.*
- *The most efficient and effective method of conveying information to and within a development team is face-to-face conversation.*
- *Working software is the primary measure of progress.*
- *Agile processes promote sustainable development. The sponsors, developers, and users should be able to maintain a constant pace indefinitely.*
- *Continuous attention to technical excellence and good design enhances agility.*
- *Simplicity – the art of maximizing the amount of work not done – is essential.*

- *The best architectures, requirements, and designs emerge from self-organizing teams.*
- *At regular intervals, the team reflects on how to become more effective, then tunes and adjusts its behavior accordingly.*

This Agile Manifesto arose in the climate of a general unease concerning the existing process concepts that were perceived as inflexible. At the same time, hope was pinned on the quicker availability of results, on the possibilities of an evolutionary adaptation of work and results to changed conditions, and not least, on satisfied clients.

In this field, a number of so-called agile methods have also been developed with a view to transferring the ideas contained in the Agile Manifesto to tangible project work. A few of the many agile methods are outlined briefly below.

### 2.2.2.3 Extreme Programming

Extreme programming arose from the needs and the insights of a small, top-quality software development team that was confronted with imprecise and continually changing requirements.[16] In comparison with other development methodologies, it is very lean and flexible. It is based on intuitively comprehensible insights that reach into upper extremes – hence the name *extreme programming*.

The basic constituent ideas of extreme programming are as follows:

- A distinction between decisions based on business interests and on project interests
- Test-driven development: writing test routines for program parts to be created and the continual use of these test routines throughout the entire program development
- Continuous integration of the individual components: integration of all system components and tests of the entire system several times a day
- Pair programming for the entire program development: two programmers with alternating roles in front of one display screen
- Simple, minimalist design: at the beginning, projects have a simple design, which continually develops further with the inclusion of the project customer(s) to add required flexibility or to eliminate unnecessary complexity (KISS,[17] YAGNI[18]);
- The quickest possible implementation of a minimal system and its further development in the direction that makes the best sense or is the most worthwhile.

Extreme programming was developed by Kent Beck, Ward Cunningham, and Ron Jeffries in the years 1995–2000, within a software development project for payroll accounting in the Chrysler Corporation. The Comprehensive Compensation System project itself was introduced in 2000 on the occasion of the takeover by Daimler.

---

[16] Beck, K.; Andres, C. (2008): Extreme Programming.

[17] KISS = Keep It Simple and Smart (variant: Keep It Simple, Stupid).

[18] YAGNI = *You Ain't Gonna Need It*, i.e., a constant rejection of functionalities that are currently not required, but may become necessary in the future.

In the following years extreme programming experienced great popularity and an equal measure of criticism. Among the main criticisms was the high demand for qualifications and team spirit among the project collaborators, a lack of scalability for medium to large projects, and a neglect of potential in far-sighted planning or the application of previous knowledge.

#### 2.2.2.4 Feature-Driven Development

Feature-driven development (FDD) is a collection of work techniques, structures, roles, and methods for agile software development.[19] FDD places the emphasis on the notion of feature: every feature represents an added value for the client. The development is organized using a feature plan. The chief architect plays an important role; he or she continually observes the overall architecture and the specialized core models.

Feature-driven development projects go through five process steps:

1. Developing an *overall model* with the goal of reaching a consensus on the content and the scope of the system to be developed and the technical core model.
2. Creating a *features list* and describing the features according to the pattern *action, result, objective*.
3. *Planning features* with respect to the sequence in which they are to be carried out. This is geared toward the mutual dependence of the features, their complexity, and the work load of the programming team. The features are assigned to individual development teams for further processing.
4. *Features design:* the development teams create sequence diagrams for the features, and the head programmer refines the class models on the basis of the sequence diagram. The developers then write the first class and method bodies. Finally, the results are inspected. If there are any technical ambiguities, the specialists are called in.
5. *Constructing features:* the developers program the prepared features. In the process component tests and code inspections are applied for quality assurance.

The first three steps are carried out in a few days. Steps 4 and 5 are conducted in continuous change, as every feature should be implemented in no more than 2 weeks.

#### 2.2.2.5 Scrum

Scrum is a collection of interrelated work techniques, structures, roles, and methods for project management within the framework of an agile product or software development.[20] The goal is to optimize the development environment, reduce the

---

[19] Coad, Peter; Lefebvre Eric; De Luca, Jeff (1999): Java Modeling in Color with UML.
     Palmer, S.R.; Felsing, J.M. (2002): A Practical Guide to Feature-Driven Development.
[20] Cf: Schwaber, K. (2008): The Enterprise and Scrum. Schwaber, K.; Beedle, M. (2001): Agile Software Development with Scrum. Also, Wikipedia.

organizational overhead, and come as close as possible to market demands by means of iterative prototypes.

Scrum contains few determinations a priori: teams or developers mostly organize themselves and select the methods used. The processes and the methods are continuously adapted to the current demands. Scrum builds on many basic assumptions of what is known as *lean production* and transfers experience from the automotive sector (Toyota) to software development. A central feature of both is the constant further development of the participants in the process, including the customer and partners, the manufacturing processes, and the tools and methods, with simultaneous, constant retention of the basic underlying assumptions: continually improving production to achieve the highest quality with the lowest expenditure (effort).

One central element of Scrum is the *sprint*, which designates the implementation of an iteration. Scrum allows for pre-defined iteration lengths for each sprint (e.g., 5–30 days). Before the *sprint*, the customer product requirements are collected in a *product backlog*, which states the features of the product to be developed. It includes all functionalities that the customer wishes, along with the technical dependencies.

There are three clearly defined roles for the staff of a project, who should pursue the same goals:

1. The *product owner* determines the overall goal that he and the team are to achieve, or he sanctions it. He makes the budget available and regularly sets the priorities for the individual product backlog elements. He also decides which features are the most important ones, among which the development team has a choice for the next sprint.
2. The *team* estimates the expenditure for developing the individual backlog elements and starts implementing practicable elements for the next sprint. The team works in a self-organized manner in the frame of a *time box* (the sprint) and has the right (and the obligation) to decide for themselves how many elements of the backlog must be achieved after the next sprint; these are known as *commitments*.
3. The *scrum master* has the task of directing the development and planning processes, plus overseeing the assigning of roles and rights. He maintains transparency during the entire development and encourages bringing to light the existing potential for improvement. He is not responsible for the communication between the *team* and the *product owner*, as they should communicate directly with one another. He stands by the team and makes sure that it can work productively.

Just like extreme programming, Scrum follows a hype cycle. Scrum is nowadays the most common agile process model.

After initial euphoria, critical voices multiply and report the unsatisfactory control possibilities of the development process, or relate a productive variant of Scrum whose use is restricted only to the lower development levels or defined scopes of development – see "Hybrid Forms of Project Organization" in Sect. 4.3.2.5.

#### 2.2.2.6   Crystal

This term encapsulates a whole family of software development methods that are differentiated with respect to the number of people involved and the risk level (the so-called *criticality*): depending on the number of people, various communication mechanisms are required. The effort for ensuring the correctness and the reliability of the system to be developed is determined by the risk level.[21]

The methods of the *crystal family* are labeled with color attributes. The spectrum runs from *crystal clear, crystal yellow, crystal orange, crystal red*, and *crystal magenta* to *crystal blue*. The simplest variant, *crystal clear*, is recommended for team sizes of two to six people – the largest variant, *crystal blue*, for 200–500.

In comparison to other agile methods, the methods of the *crystal family* are less dogmatic and formalized. Likewise, strict programming in teams of two (extreme programming's *pair programming*) is no more prescribed than involving the client in the program development on site.

According to the *crystal philosophy* there are also no static development methods for the development teams. New and appropriate methods are determined for each project. This means that with fairly simple projects, crystal becomes very similar to *extreme programming* and differs only in more complex projects, through more sophisticated procedures.

### 2.2.3   When Are Plan-Driven and Agile Methods Appropriate?

In the following, we attempt to define the appropriate areas for the methods that are clearly differentiated by their nature. This does not imply that one approach is necessarily right and the other wrong, but rather that it depends on the specific context.[22] The dominant application areas of the two method groups are described in Fig. 2.17 very briefly.[23]

#### 2.2.3.1   Is a Convergence Possible?

There are some major differences between the plan-driven and the agile methods, so that at first glance they appear to be incompatible with one another. The following questions are of interest in assessing the extent to which the systems engineering concept contradicts agile principles:

– Where does the systems engineering concept already show approaches to a certain agility, or support agile principles?

---

[21] Cf: Cockburn, A. (1998): Surviving Object-Oriented Projects.
    Cockburn, A. (2004): Crystal Clear

[22] Boehm, B; Turner, R. (2004): Balancing Agility and Discipline.

[23] Known as the situational approach in organization theory.

| Criterion | Suitability of agile methods | Suitability of plan-driven methods |
|---|---|---|
| Project size | Well suited for rather *small projects* and teams. Is based on tacit knowledge (knowledge stored in people's heads rather than explicitly outlined). Upward scaling thus limited. | *Large projects* and teams. Difficult to downsize to small projects. |
| Criticality (safety, etc.) | No experience with safety-critical products available. Difficulties to be expected with simple design and insufficient documentation. | Development of highly critical products |
| Dynamic of the environment | Dynamic environment and simple design, combined with continual structural adaptation (refactoring), are an excellent match for one another. | Stable environment allows large designs that then are implemented in detail. |
| Personnel | Requires continual presence of qualified personnel. Risky if not available, because of the dynamic indicated above | Qualified personnel required especially for the project definition. Elaboration? can also be carried out by less highly qualified personnel as long as the requirements remain stable. |
| Culture | Thrives in a culture in which people enjoy great latitude in design and empowerment. | Thrives in a culture in which people feel good, when their roles are defined with clear procedures and behaviors. |

**Fig. 2.17** Comparison of agile and plan-driven action methods. (Inspired by Boehm, G. et al.)

- With respect to which principles is the systems engineering concept neutral, or which agile principles are possible within the systems engineering concept, even if they are not actively supported?
- Where does incompatibility between agile principles and the systems engineering concept exist, or where does the integration of agile principles into the systems engineering concept require more extensive research?

### 2.2.3.2   Existing Agility in Our Systems Engineering Model

The capability for agility is particularly expressed in the following features of the systems engineering process model (Fig. 2.18):

- The conceptual division of the four modules, in particular, the phase and the PSC modules, facilitate adaptation of the methodology to various *sizes of projects*. *Small projects* do not need to be subdivided into several development phases with increasing concretization (from the general to the particular). A single run-through of the PSC with an immediate transition to realization (systems building) may be perfectly adequate.
- Due to the fact that also within the systems engineering concept, the responsibility of choosing the process falls to the team and can be adapted to the requirements of the project, there is an accord with the agile methods.
- Continual learning and the application of experiences are just as important in the systems engineering concept as in the agile methods. The systems engineering concept is not a process to be used stubbornly, but rather a guideline for how specialized knowledge, experience, knowledge of methodology, etc., can best be combined.

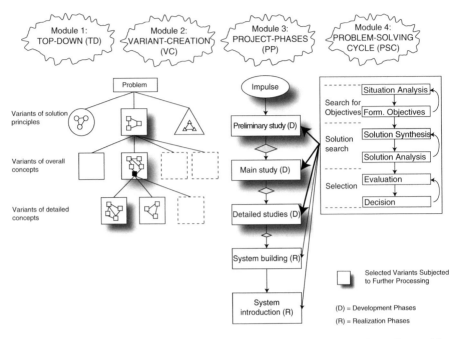

**Fig. 2.18** Relationships among the various components of the systems engineering action model (arrangement by tendency)

- The principle of variant creation on three levels (left in the illustration), combined with decisions at the end of each development phase (preliminary, main, and detailed studies, center of illustration) produces decision-making situations that allow the line of approach.
- The backward-pointing arrows in the PSC (on the right in the illustration) represent recourses to earlier steps, and can be understood as iterations.

A certain capability for agility emerges very clearly from the presentation in Fig. 2.19:

- Every detail concept worked out in detail is integrated conceptually into the overall concept at a higher level (arrows pointing upward). Unsatisfactory situations can and must be processed at the levels of the overall concept or the detailed concept.
- Even external influences that have nothing to do with the project can make an adaptation of the overall concept necessary (angled arrows in Fig. 2.19).
- The arrows pointing upward form the detailed concepts involved in the overall concept and back down (= check, modification, adaptation) support flexibility and agility, in addition to attention to external influences (represented by angled arrows).
- The broad arrows of different lengths leading beyond the detailed studies are intended to indicate a different implementation status, which, however, can also reduce flexibility (overall concept no longer adaptable, as the detailing or realization is already advanced).

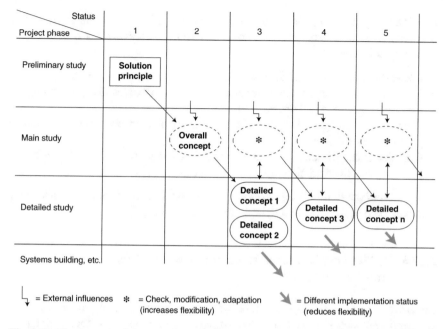

**Fig. 2.19** The dynamic of the overall conception with stepwise integration of partial results and the possibility of external influences

### 2.2.3.3    Agility That Is Easily Used in the Systems Engineering Model

The following agile principles are not really required or supported explicitly by the systems engineering model. However, they can be easily applied without contradicting the systems engineering concept[24]:

- People-oriented approaches, such as the close involvement of the user in the development team (customer on-site), regular meetings at short intervals, etc.
- The adoption of the idea of the backlog elements and Sprints (from Scrum)
- The adoption of the idea of the use cases and of the explorative prototyping for selected modules of a system that are particularly difficult to design
- The cooperation of small groups in the detailed development of concepts (based on pair programming)
- Simplicity (KISS, YAGNI, etc.)
- Strategies such as those described in Sect. 2.2.4: postponement of essential decisions, keeping options open, and choosing architectures that are relatively robust with respect to changes, can contribute agility to the *development process*
- With modular architectures and the implementation of software-type rather than hardware-type functions, the developed *systems* can be made agile, which is easily changeable or more adaptable.

---

[24] Stelzmann, E. (2011): Agile Systems Engineering.

We clearly want to emphasize, that – also when applying agile methods – it is necessary to agree upfront early on about the overall concept of the development. Explicitly, this means:

– Use early sprints to define the basic correlations and decision criteria
– Define early a system architecture and its degree of changeability depending on the kind of problem and environment
– Identify the concept defining details from the architecture and prioritize them in your product backlog
– Organize and staff your agile team so that they can work on the different levels of a problem, from architecture level to detailed parts or coding

A different possibility is the application of a combination of traditional and agile methods, such as the so-called hybrid models (Sect. 4.3.2.5).

### 2.2.3.4  Agility That Is Difficult to Reconcile with the Systems Engineering Model

According to some principles, there are greater contradictions to the systems engineering model. Further research must figure out not only how these principles can be used in the systems engineering model, but also whether it makes sense to apply them at all:

– *Development in short iterations must always deliver a customer benefit*: This principle has its origin in software development, cannot or should not be interpreted in the same manner as the development of hardware, because there, the customer benefit cannot be proven by a functioning subproduct that is ready for application. This is scarcely possible and not even expected in hardware projects (think of a vehicle and a functioning gearbox delivered to the customer). Although modern simulation methods (Hardware in the Loop, HIL, 3D printing) are very helpful, the expectations should be modified: one can take a benefit as already given if the *contributing staff members of the clients state that significant progress has been caused by the respective iteration.*
– *Neutrality toward changes:* adaptations and changes of solutions are of course easier to manage in the early phases of system development than in later ones. If, for example, a specific solution principle or systems architecture has been decided on after a preliminary study, and if in the course of the main study these turn out to be impracticable or less than optimal (clearly better variants are subsequently discovered), an adaptation of the concept may still be possible. No further detailed developments are initiated. Fundamentally changing an overall concept (a master plan) that has been decided on at the end of the main study and that will then be worked out gradually in more detail would in comparison involve much greater difficulties, especially if the design of detailed concepts is already advanced, and if the detailed design is created by various working groups that are also geographically far apart.

However, this kind of problem exists in software development as well.

## 2.2.4   Keeping Options Open as an Approach in Support of Agility

Some years ago, several members of the BWI systems engineering development team were involved in a project (planning for a production site) that was characterized by a very dynamic environment. Solutions that had just been developed had to be discarded because the starting situation and the corresponding requirements had changed drastically and unexpectedly.

The systems engineering process model, with which the people involved in the project had had good experiences in the past, was not applicable in this specific instance. There, the installed decision-making logic specifically foresees that important decisions would be made at the end of each characteristic phase for the purposes of more detailed planning (decision on a solution principle at the end of the preliminary study, on an overall concept at the end of the main study, etc.).

The simplest solution to this dilemma would have been to stop the planning until the situation had been clarified. But in many cases, people could not or did not want to do this, to avoid incurring disallowed delays, or because they feared not being able to resume the project at a later point in time (movers and shakers and key people leave the company or turn to other tasks).

In such cases, there really are just two possibilities:

- One chooses the variant that is deemed the most practicable and continue – with the declared intention of gravitating toward some other variant as soon as it becomes necessary or suitable.
- One avoids making a commitment as long as possible, but still continue with the planning by intentionally keeping options open.[25]

The latter variant is described below with the aid of Fig. 2.20:

1. A (limited) number of solution concepts are developed (S1 to S4), all of which are considered reasonable and usable, but on which no decision can (yet) be made because of the uncertain situation.
2. For every solution concept, one works up an implementation plan and investigates the solution variants for identical implementation steps.

---

[25] The approach conceived in the 1970s and then briefly described was not made public because of a lack of practical experience. The project became dormant because the "godfather" in the client company who had been involved and who had supported this project had been sent overseas on more urgent duties. Only years later did we come across a process methodology designated *set-based design*, which has certain similarities to the approach described above: "In discussing motor car design – specifically the Toyota Prius – Mary Poppendieck said: I think that the trick is to determine what is, in fact, easy to change later, and what will not be easy to change, and spend some time considering those things that are going to be very expensive to change later. And, of course, the trick is also to keep such things to a minimum – through the use of layers, services, etc." (Source: http://silkandspinach.net/2007/01/114/agile-set-based-design/, hit on 4.14.2009). See also: Morgan, J.M.; Liker, J.K. (2006): The Toyota Product Development System.

A further important parallel is the real options theory, which at the time was also unknown to us. The main features are described in Part III, Sect. 2.2.5.

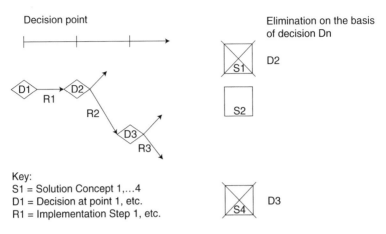

Decision point

**Fig. 2.20** Keeping options open

3. The specific implementation steps that are worked out are those similar for several solution concepts (as many as possible). In this case, a certain amount of flexibility is provided.
4. These implementation steps are brought forward in time, for they leave open the greatest latitude and the most options.
5. Thus, the first decisions do not involve concept variants, but rather implementation stages. In Fig. 2.20, this would be decision D1 over implementation step R1. The *point of no return* can therefore be delayed. It is assumed that at first, all four solution concepts, S1 through S4, remain open as options.
6. Once implementation steps have been started, which only allow a restricted number of solution concepts or require the elimination of individual solution concepts, appropriate decisions are of course necessary. These may be easier, because developments at this (later) time can hopefully be evaluated better. This would be the case in Fig. 2.20, for example, at point 2. Here it should be assumed that decision D2 precludes solution principle S1 for implementation step R2.
7. The situation is analogous for decision D3 and solution principle S4. From this moment on, solution concept S2 is used for guidance (possibly in modified form).

The real options approach described in 2.2.5 presents excellent linking options with the considerations described above. It supports this mindset and provides an approach for evaluating this type of project.

The described approach has the following *consequences:*

– The point of no return can be postponed.
– It results in an implementation sequence that may not be the best from the viewpoint of the ultimately selected solution concept. But this is accepted in the interest of possibly greater flexibility.
– The time gained is not allowed to pass idly by, but rather must focus on information gathering to facilitate the decisions that are ultimately made. It is

thus reasonable to define indicators in terms of observables that are to be followed attentively, for they may help in evaluating the usefulness of both the solution concepts and the implementation stages.
-   Therefore, the basic idea of systems engineering of creating a conceptual framework
    before working out solutions in detail is not pointless, but it is rather intensified.

However, be warned of an extensive use of this mindset: it is tempting for commissioning parties who have difficulties in making decisions and who may agree to
it eagerly because at first they are subject to fewer demands. The advantage of keeping options open is connected to the undeniable disadvantage of the increased planning effort, a potentially less than optimal process, or to the successive and unwitting
loss of possible opportunities for action on the part of the decision-makers. The
factual difficulty of the decision-making situation, rather than the insufficient resolution on the part of the decision-makers, should thus be crucial in choosing this
type of process.

## 2.2.5   Real Options Approach

The real options approach can be considered a method for evaluating decisions
(e.g., investments), in which the basic actuarial fundamentals of the option pricing
theory are applied to project evaluation.[26] But it can also be seen as a conceptual
approach that influences the process through the possibility of purchasing additional
options. We consider this added benefit to be important and have therefore placed
remarks on real options in this section. Developers and designers should be motivated to consider the additional possibilities of a (partially) reduced or expanded
implementation of concepts and address the attendant risk considerations.[27]

The difficulty that developers/planners and decision-makers face is that only
prognoses, or reasonably plausible assumptions, can be made about the future, but
of course without any certain prediction. But the future development of the
investment and the relevant environment (market, business activity, competition,
one's own business operations) must be assessed in advance. All prognoses and sup-
positions concerning future developments are fraught with uncertainty.[28]

---

[26] Description according to Wikipedia: Real Options.
      Further literature: Amram, M.; Kulatilaka, N. (1999); *Real Options*. Copeland, T.; Antikarov,
V. (2001): *Real Options*. Dixit, A.K.; Pindyck, R.S. (1994): *Investment under Uncertainty*.

[27] See the remarks in Sects. 1.3 and 2.2.4 on "Agility of Systems engineering" and "Keeping
Options Open as an Approach in Support of Agility".

[28] Presumably the Greek philosopher Thales of Miletus worked with real options as early as
600 BC. In the springtime he purchased from the local owners at a fixed price the option of using
the olive presses at harvest time. The rental price for the presses was essentially dependent on the
harvest. A good harvest meant a greater demand for olive presses by the farmers. For unknown
reasons, Thales expected a good harvest; his calculation worked out, he became rich, and hence-
forth was considered a man with visionary powers. (Source: Wikipedia).

The traditional methods for investment appraisal in preparation for investment decisions on a so-called objective basis, proceed first from secure assumptions, for example, about market and business development, the amortization period, the interest rates for capital, etc. The uncertainty factor can be included through sensitivity analyses or assumptions based on probability. After using these processes, the decision-maker has either an individual model adapted to the uncertain development, or an entire array of possible scenarios concerning the future development, and must decide on one of them. Generally, he bases his investment decision on the scenarios that strike him as most likely and most plausible. It thus becomes quite possible to incorporate uncertainty as a factor in the planning, provided that the assumptions and possible risks are known or properly estimated in advance. Changing environmental conditions during the course of the project cannot be taken into account. Thus, the decision-making rule for the investor is, "Invest today in case the investment is profitable, or never." Investment projects tend to be underestimated for reasons of caution, and thus are implemented with less probability.

An investment appraisal method that makes it possible to react to uncertain future expectations was presented by Stewart C. Myers and proved very popular among decision-makers. Myers recognized that this flexibility, opening up more room for maneuvering, can be seen as an *option*. He thus coined the term *real option*, which remains in use today. Behind this designation hides an approach to investment planning based on the option pricing theory. Although option pricing theories were previously applied to evaluation problems for financing (resources procurement), this approach broadens the application field to evaluation problems for investments, i.e., the use of resources.

### Excursus on Uncertainty and Risk by Investment

*Uncertainty* commonly designates the possibility of a deviation from an expected value or an expected situation. Positive deviation = opportunity; negative deviation = risk, danger.

The term *risk* in terms of the decision-making theory describes a situation about whose probability of occurrence the decision-maker has objective information, or that he or she can at least assess subjectively. *Objective* probabilities are achieved through empirical studies or results of equivalent decision-making situations. It may be possible to calculate the probabilities on the basis of statistical data. *Subjective* probabilities are empirical values that cannot be checked statistically or empirically, or that have so far not been validated.

*Uncertainty* in terms of the decision-making theory designates a situation in which the decision-maker really does not know how things may turn out. There are neither objective nor subjective probabilities. A rational decision between various alternative courses of action is practically impossible in such instances. And yet it would not be a good recommendation to avoid making a decision and to do nothing, because of uncertainty due to a total lack of understanding of the situations to be expected. The following outlined practices may be helpful in generating possible courses of action.

**Uncertainty and Flexibility**

As investment decisions are of course made for the future, which is uncertain, strategies must be thought of to cope with this uncertainty. Possible examples, including risk avoidance, diversification, anticipation, and the use of real options, are presented below.

(a) *Risk Avoidance*

Risk avoidance is a passive strategy in risk management. The consequence is that investments are made only in projects that offer limited risk, in known markets, in regions with politically stable conditions without great dynamics, or in fairly traditional products based on proven technologies. Another form of risk avoidance would be long-term contracts for the improvement of the capability of the prognoses with regard to future results. The risk avoidance strategy becomes more difficult, however, in an increasingly dynamic environment. At the same time, a company thus misses an opportunity for innovation (products, markets, processes, etc.), because innovation is always uncertain.

(b) *Diversification Strategy*

In the past, the diversification strategy was a widespread strategy for risk reduction. Companies sought to diversify their risk by spreading their activities over various business fields with different business activity cycles, or by spreading out over various products at different phases of the product life cycle. In spite of fluctuations in certain areas, the overall enterprise was expected to remain economically successful. This limited possible losses, but also curtailed profit opportunities. The result obtained was referred to as symmetrical risk reduction. In recent years, the diversification strategy has been recommended less frequently by investment gurus, and it no longer corresponds to the general opinion: companies should concentrate more on their *core competencies* and refrain from diversified activities. Empirical studies have also confirmed that a series of spectacular diversification strategies did not increase the company value – and that even the contrary happened. Also, highly diversified conglomerates such as ITT (telecommunications, rental cars, hotels, armaments industries) were dissolved because their overall performance lagged significantly behind less diversified firms.

(c) *Anticipation*

An active strategy for dealing with uncertainty is the attempt to *anticipate* future environmental *changes*, in other words, to recognize them early and to initiate countermeasures. This may be possible with the use of higher-performance information systems. Strategic early warning systems, decision support systems, the discovery of weak signals, etc., should increase reaction speed. With increasing dynamics, this strategy was not successful to the degree hoped for either.

(d) *Creating Maneuvering Room*

Recognizing and creating maneuvering room makes it possible to react flexibly to uncertainty and changing environmental conditions. The maneuvering room can be seen as *options*, according to Stewart C. Myers, as already mentioned, that are similar to the opportunities available to the owner of a financial option.

So, for example, the option to bring a product onto the market after a successful sale trial can be compared with a *call option*. A call makes it possible to purchase a share at an established price. In this case, the loss from an unfavorable share performance, in which the call option is not exercised, is limited by the level of the option price. On the other hand, the gain is theoretically unlimited.

This is precisely how it works with the maneuvering room mentioned earlier. The loss is limited to the level of the initial investment. In contrast, the gain from the initial investment and further investments is theoretically unlimited. Because of the different limitations on profits and losses the maneuvering room produces an asymmetrical risk profile. It works the same way with maneuvering room that allows a strategy change in the case of an unfavorable or a particularly favorable market trend, e.g.,

- *Option to terminate an investment project.* Liquidation proceeds can be achieved by the sale of tangible or intangible elements that are no longer needed – in addition to the elimination or reduction of future losses. This corresponds to a *put option*, which allows the sale of financial futures at a determined price. Here, too, the result is an asymmetrical risk profile because of the reduction of potential loss through the preservation of the potential for gain. Of course there are costs associated with creating this type of maneuvering room, e.g., from liquidation proceeds that do not cover the investment costs – just as with the purchase of financial options.
- If there is maneuvering room for *further investments* (e.g., for expanding production), it may be possible that too much time is spent observing the market trend or that the investment decision is postponed for some other reason. In this way, market shares may be lost to competitors who invested more quickly and/or get into the market more aggressively. In such cases, the costs take the form of lost earnings, which are known as *opportunity costs*.

Thus, the question for the company is not only the choice of the *right course for action*, but of course the *price for this option* too. An increase in value is possible only when the value of the maneuvering room is greater than its price. Then the question arises of when the application of maneuvering room can alter an investment decision most decisively. The answer is that the value of the maneuvering room is greatest when the following three factors come together:

- Great uncertainty over the future performance of the investment
- Flexibility of the decision-makers to react to these uncertainties
- The value of the investment without flexibility is close to the break-even point

The rationale: great uncertainty generally means that it is likely that the decision-makers receive new information in the course of the investment project. Thus, the option for maneuvering room is most valuable when the underlying situation is uncertain. This perspective is in contrast to the previous school of thought. Although in traditional investment calculations of the value of an investment diminish as risk increases, by considering courses of action with increasing risk, the value of the investment increases.

But why is the value of the option (= the price that I am willing to pay to purchase an option) highest when, according to the traditional evaluation, the investment project is near the break-even point? The answer is:

– If by the traditional evaluation methods it has already been determined that an investment will be quite profitable, it is improbable that room for maneuver offering additional flexibility will be sought and paid for at a higher price. Therefore, they have a relatively lower value.
– If it is foreseeable by traditional methods that the investment will never yield a profit, it should be suspected that even by using room for maneuver, it will likely not enter the profit zone. The value of the flexibility is low as well, for it may consist only of the passive option of termination to prevent future losses.
– Here, too, the greatest value of active maneuvering room is in the gray area of the break-even point. In this case, the presence of flexibility can tip the balance in considering an investment project to be reasonable or not. And one is more likely prepared to "buy" this course of action.

*Real Options as an Evaluation Tool*
There are several conceivable types of maneuvering room that can be integrated as an option into an investment appraisal. Only a few are described here (with partial reference to Wikipedia):

1. *Deferred option* as the possibility of delaying an investment within a specific time period. The option owner must secure the right to implement the investment at any time within a determined deadline. Losses of value can occur because of the time value of money and/or because of the loss of market shares.

   *Example:* a company wants to introduce a new product to the market. There are great costs associated with introducing the product (e.g., accompanied by a major marketing campaign). The introduction time is assessed as currently unfavorable (poor business activity, etc.). There is a fear that the product will not produce the desired sales figures and the marketing expenditure will be wasted. The decision is thus made to delay the product launch for a year and to watch how the economic situation evolves. But there is a risk that during this year a competitor who enters the market with a similar product could win a market share. If this were to occur, or if the business situation improved significantly before the year was over, there would certainly be the possibility of beginning sooner.
2. *Growth option:* the possibility of following up an initial investment with subsequent ones, e.g., expanding the existing investment or making a new one.

   *Example:* the successive development of production capacities and/or a market organization that is made dependent on market success.
3. *Termination option:* the possibility of terminating (discontinuing) an investment if a project progresses poorly. If the project is ended, losses of value occur as a result of the discontinuance.

*Example:* a product is first launched in a test market. If the anticipated sales figures cannot be achieved, there is an option of stopping the project. The initial investments are admittedly lost, but further losses are avoided.

4. *Multi-stage investment:* when one stage of a project goes well there is the option to enter the next one. Usually, the option can be exercised only at specific times (such as at the end of the previous stage). A loss of value arises from the fact that the investment is not made immediately in its entirety, and only the first stage is carried out initially.

   *Example:* a pharmaceuticals company develops a medication. In the first stage, only an active substance is to be developed that mitigates the symptoms. The R&D department is convinced that an active substance can also be developed that could completely cure the disease. As this will take longer, at first, only the agent with the relieving properties is developed and marketed. With successful sales, the company has the option to continue developing the active ingredient in a second stage. Taking this decision, the risk involves the possibility that in the meantime a competitor will develop an active ingredient for a complete cure.

5. *Expansion or reduction option:* with this option, the decision-maker has the possibility of expanding or reducing an existing project. Unlike the growth option, this does not involve a new investment, but merely the expansion or reduction of an existing project.

   *Example:* a vehicle manufacturer that develops and assembles a small series for original equipment manufacturers (OEMs) designs a new assembly line on which vehicles of various types can be assembled in any desired sequence. The number of variants can be expanded and of course reduced within certain limitations (number of variants, dimensions of the vehicles, etc.).

6. *Option of temporary closing and reopening*: if the proceeds from an investment no longer cover the variable costs to the desired extent, production should be temporarily suspended. The option involves the right to stop production and resume it later on.

   *Example:* an oil producer has the option of temporarily suspending production on a specific oil field when market prices go down. This option is exercised if the extraction costs approach, to an extent to be defined, the price that can be realized on the market. Production is suspended until the market price rises to the point where extraction is profitable.

   The possibility of deconstructing the 2008 European Championship soccer stadium in Klagenfurt (Sect. 1.3.2) can likewise be seen as an option. The selected architecture makes it possible to later reduce the capacity by about half and to use the upper part of the stands that are no longer needed elsewhere. The additional costs of realizing this architecture are the price of the option.

7. *Option for variation of input/output:* also known as *switching option*, this describes the possibility of varying the input- and/or output factors independently on the demand for products and the development of prices for production factors.

   *Example:* depending on the economic situation, a production company in the textile sector has the possibility of producing high-quality fabrics or fabrics of lesser quality (alternative input factors). If good prices can be realized for quality goods on the market, a higher-quality material is used for production. If the prices drop, raw materials of lower quality are made into more economically priced fabrics.

Of course it is possible to combine these option types, e.g., delay termination options, *compound options*, in addition to options with more than one source of uncertainty (*rainbow options*).

The option price theory is under development and is not yet widely used. Its major utility consists of the fact that it effectively supports agile systems engineering as a new design philosophy. The flexible mindset with respect to options is supported, and calculation methods are offered that make it possible to test economic feasibility on a totally new basis.

## 2.3   Outlook for the Future of Developing Products and Systems

In research and development-intensive industries such as mechanical engineering, the automotive industry, medical and electrotechnology, etc., global competition is becoming tougher.[29] High dynamics in the technology and business markets, activities arising from as yet unknown and financially strong competitors change the innovative activities of traditional companies. Innovations in shorter cycles are needed, innovation projects have to be run quicker, cheaper, and must still be of high quality. Agile frameworks such as Scrum seem to be able to release unexploited potential that is increasingly turned into practice.

A study on the question to what extent agile methods are applied in practice shows an already extensive use and – quite interestingly – 27% of topics without an IT focus. It is remarkable that most of the participants of the study do not use agile methods in the pure form, but in a hybrid and selective form.[30]

A credible scenario might be as follows. Many companies are proficient in the use of agile methods in software development and it is to be expected that hardware development is affected. If hardware can be to developed faster, more efficient, and more integrative, the following effects should be expected.[31]

*Faster development*: more rapid development enables faster reactions to market changes

*More efficient development*: the focus is not on cost reductions of their own process, but on developing better products that give more value to the customers.

*More integrative development*: the growing complexity of hardware, mainly caused by the integration with software in embedded systems, demands an

---

[29] Thanks to Armin Schulz, Stefan Wenzel, and Thilo Pfletschinger from 3DSE Munich for their inspiring workshops and specific hints.

[30] Komus, A. and Kuberg, M.: Studie "Status Quo Agile" – wie werden agile Methoden in der Praxiseingesetzt?https://www.projektmagazin.de/artikel/studie-status-quo-agile-wie-werden-agile-methoden-der-praxis-eingesetzt_1101303

[31] https://borisgloger.com/wp-content/uploads/2014/07/Whitepaper-Hardware.pdf?882268

interdisciplinary and cross-functional development approach. Developers coming from different subject areas increasingly work together, having to coordinate their development cycles, and thus jointly create new and integrated products. International competition also enforces cooperation in worldwide distributed teams.

*What effects can this agile trend have on the systems engineering methodology?*

1. The *following modules of the systems engineering concept are practically unaffected by a conflict between plan-driven and agile methods.* They have *unaltered validity and applicability:*

   - *Systems thinking* as a holistic approach to master complex interdependencies
   - *Top–down approach*, because an agile process needs a large framework – perhaps to a greater degree
   - *Variant creation* as a basic planning attitude, always asking: "what is the alternative to this idea"?
   - The *PSC* as kind of a micro-logic, asking in its simplest form: where are we now? Where do we want to go? Which ways, possibilities, and options are there? Which is best/most appropriate?

2. A certain *tension between the agile approach and the phase* model cannot be denied. The agile model wants quick iterations, which are not offered by the phase model

3. *Remedy is possible* in two different ways:

   (i) The phases are adapted to the agile needs and shortened radically in the form of sprints which go on for 2 or 4 weeks. This allows more flexibility. However, the early sprints would have to focus on the creation of a rough conceptual framework (architecture, solution principle)
   (ii) Plan-driven and agile methods are not seen as opposites, but rather as two complementary approaches, which can be merged into a hybrid method. The rather management-oriented plan-driven methods are used for rough planning (milestones, decision points, allocation of resources etc.). The elaboration of this framework is handled by using agile methods, sprints, etc.

   There is already some literature on approach (ii), where reports can be found on the practical and successful application of the hybrid approach (see Sect. 4.7).

The tasks of the *systems engineer* in a very variable and highly networked environment with different stakeholders, variable customer behaviors are:

(a) To find a coherent *product vision* and *product architecture*: for that, he needs the top–down and variant creation modules, etc.
(b) Furthermore, the *systems architectures should be open* in principle and thus be compatible for updates/upgrades even for such that have not been known when they were developed (robust architectures)

The methods and tools of *model-based systems engineering* (MBSE) can add value. To work, the agile approach needs a model-based development environment to a special degree to be able to verify and integrate products, test results, releases, etc.

Substantial *changes* are necessary in the *mindsets* and the way of thinking not only of the *development teams*, but also of their *superordinates* and/or of their *external and internal contracting authorities* or clients.

(a) The agile approach offers excellent opportunities for the development of so-called "*minimal viable products.*" This means the development of new products, with only a few features. But these products are improved and further developed step by step with the help of market and client feedback. It is important to consider that the products do not behave like prototypes that are not working very well. On the contrary, they have to work smoothly and reliably and have to deliver a customer benefit that makes themselves viable.

(b) An important cause of longlasting developments can be seen in existing contractual situations: if clients (OEMs) insist on binding contracts, based on detailed requirement specifications and fixed price agreements, the supplier has no other choice than to plan in a very detailed mode to avoid risk and contractual penalty.

(c) A more efficient approach would be to be limited to a rather vague agreement at the beginning and to specify a trust-based cooperation.[32] This would probably be experienced positively from both sides: the clients and the suppliers. However, trust cannot be ordered but has to be earned by excellent professional skills, respectful interactions, resulting in a pleasant and esteemed cooperation, and this has to happen simultaneously at different levels of hierarchies and different functional areas.

In the field of systems engineering methodology in combination with agile approaches,[33] there is certainly a need for future research, which can only be done in an interactive collaboration with practice partners.

*Finally, an Interesting and Amusing Comparison* (**Offered by P. Kruchten**)
*The agile teenager:* the agile movement is in some ways a bit like a teenager: very self-conscious, constantly checking its appearance in a mirror, accepting few criticisms, only interested in being with its peers, rejecting en bloc all wisdom from the past, just because it is from the past, adopting fads and new jargon, at times being cocky and arrogant.[34] But there is no doubt that it will mature further, become more open to the outside world, more reflective, and also therefore more effective.

---

[32] Stelzmann, E. (2011): Agile Systems Engineering.

[33] N.N. Hybrid project management: the best of both worlds. https://www.microtool.de/en/what-is-hybrid-project-management/

[34] Kruchten, P. (2011): Agile Teenager. https://www.infoq.com/articles/agile-teenage-crisis
https://philippe.kruchten.com/2011/02/13/the-elephants-in-the-agile-room/

## 2.4   Summary and Rounding Off

*Summary*
1. The systems engineering process model is a broadly applicable taxonomy/ systematic that consists of *four components* that can be combined with one another in a *modular fashion*:

   - From the general to the detailed
   - Principle of variant creation
   - Phase approach (project phases model)
   - PSC

2. The process principle "from the general to the detail" is intended to express that

   - The field of view must be first grasped and then gradually narrowed down
   - First general goals and a general solution framework should be established, whose degree of detailing and concretization is deepened in stages.

3. The principle of variant creation is intended to indicate that in principle, one should not be satisfied with the very first solution, but rather must always think in variants or alternatives. This is especially important in the early phases of a project.

4. The *phase model* (project phases) concretizes and complements the three general considerations. In combination with the principle "from the general to the detailed," it represents a type of *macro-strategy* and should stimulate a stepwise planning, decision, and realization process with predefined reflection pauses and points for decision-making and adjustments. It should not be regarded as a means of describing the chronicle of the development and realization of a solution. Instead, it should be a planning instrument that first directs attention to the results that should or must be achieved at the end of the various phases. This can give rise to a series of activities that must also carried be out. Because of the explicit challenge of involving the decision-makers in the transitions of the individual development phases, the project phase model can also be marked as a management-oriented component of the process model.

5. The *PSC* should be understood to be a *micro-strategy* that complements the macro-strategy described above.

   - It consists of the sequence *objective identification* or *formulation of objectives* (situation analysis and goal formulation), *solution search* (solution synthesis and analysis), and *selection* (evaluation and decision).
   - It should be applicable to all problems regardless of what type they are and in which phase of the project they arise – even though with a different weighting of the steps.

6. The phase model and the PSC are seen as essential components of the systems engineering process model that are conceptually separable from one another. Therefore, no new or more structured sequence of steps is extrapolated from it

(as with the SIMILAR process, the value analysis working plan, the VDI guideline 2221, etc.). The situational combination of the building blocks tailored to the specific application.

7. The user is thus free and even expressly advised to select from these building blocks the components that are best suited to his or her planning situation and to modify them as needed. The change (expansion or reduction) in the number of phases can thus be just as meaningful as an overlapping treatment or conceptual anticipation or fallback and repetition cycles. However, the basic logic of the considerations should be retained and remain recognizable.

8. The individual components of the process model can also be found elsewhere in a similar form: nearly all project-oriented process models provide a phased procedure. Those models focusing on the theory of management decisions exhibit process steps that are similar to those in the PSC. The combination of the two is original and characteristic of the systems engineering concept.

9. Of the other process models belonging to the *plan-driven models* category, engineering design methodology according to VDI 2221, the ideals concept, the prototyping concept, the versions concept, and the concept of what is known as simultaneous engineering, are presented here.

10. The *agile methods* should be regarded as alternative approaches to the *plan-driven methods*. We have addressed the represented mindset in the following way:

   – *Plan-driven* and *agile* are not to be taken as opposites between which battles of opinions should be fought out. For under certain application conditions, both of them have their justifications. These we have presented.
   – The systems engineering process concept, when not dogmatically used, as we have already stated, does indeed contain starting points for agility.
   – In addition, motivation can be taken from the various *agile methods* to further increase the agility of the systems engineering process concept as needed.
   – Agile process models will gain more importance in the future as they are driven by the technology and market dynamics
   – In our view, there are several components of the systems engineering process model that are valid for the future: (a) the top–down approach, (b) the variant creation principle, and (c) the PSC. The largest change may happen in the phase model: the phases will be replaced by sprints (length 1–4 weeks) or they will continue to exist in the form of hybrid models (see Sect. 4.7).

This development not only affects the methodology but also the way of thinking and the mindsets of the developing teams and their clients. Here, a process of learning and "getting used to" is unavoidable with intensive training in new thinking.

*Rounding Off*

We are convinced that the systems engineering process concept can serve as a viable basis for designing a problem-oriented process. It offers an extensive framework for changes or simplifications – without having to challenge its basic statements.

In the interests of a thoroughly critical reduction, a few *principles for usage* should be drafted in conclusion.

The methodology presented here:

- Is *not for its own sake*, but rather must serve the development of good solutions (the methodology should not beat problems and ideas to death)
- Does *not mean recipes* that are simply to be followed, but is rather a guideline to be used creatively and intelligently
- Is *no substitute* for talent, natural abilities, familiarity with situations, technical knowledge, involvement with the specific situation, ability to work as a team, and so forth, but rather requires these features or should guide them to a certain extent.
- Thus produces merely a *formal framework* whose useful application results only from the intellectual and character potential involved,
- Should align the required effort to the expected benefit.

In any case, systems and solutions are made or changed by people for people. This statement should be taken as a reminder for both the design of solutions and for every process intended to produce these results.

*Note:* application-oriented interpretations and more detailed presentation of individual components, steps, and interdependencies of the process model are to be found in Parts III (System Design), IV (Case Studies), and V (Systems Engineering in Practical Experience).

## 2.5  Self-Check of Knowledge and Understanding: The Process Model

1. Describe the role of the process model in the systems engineering concept.
2. What are the four components (basic principles) of the systems engineering process model?
3. What are the individual components focusing on?
4. How are these components interconnected, related to each other? Are there any logical relationships between the components that show the modularity of the concept?
5. What is the difference between project phases and life cycle phases? Does it make sense to see a difference?
6. Is it necessary to pass all project phases?
7. The term "analysis" is embodied twice in the PSC. What is meant in each case?
8. What is the difference between the steps analysis and evaluation of solutions?
9. What is the difference between actual-state-oriented and desired-state-oriented process models? How or where would you position the Hall/BWI process model?

10. What role does the client play in the PSC?
11. What is meant by the "waterfall model"? To which component of the systems engineering process model does it relate?
12. What is meant by the term "V-Model"? What are its characteristics?
13. What is the "prototyping approach"?
14. What does the term "simultaneous (concurrent) engineering" denote?
15. What are the characteristics of the so=called "Agile Process Models"? Give some important examples of that category.
16. What do you think is better: agile models or plan-driven models?
17. Can you find some agile properties in our systems engineering model in spite of its primarily plan-driven character?
18. What is the basic idea of the real options approach?
19. What kind of risk considerations would you cite as examples?

# Literature

Amram, M.; Kulatilaka, N. (1999); Real Options.
Bahill, A. T.; Gissing, B. (1998): The Systems Engineering Process
Bahill, A. T. and Madni, A. M.(2017): Tradeoff Decisions in System Design, Springer Nature
Beck, K. et al. (2001): Agile Manifesto http://agilemanifesto.org/
Beck, K. Andres, C. (2008); Extreme Programming Explained
Boehm, B. (1988): A Spiral Model of Software Development and Enhancement
Boehm, B.; et al. (2004): Balancing Agility
Büchel, A. (1969): Systems Engineering
Cockburn, A. (2004): Crystal Clear: A Human-Powered Methodology for Small Teams.
Cockburn, A. (2006): Agile Software Development. Software Development: The Cooperative Game
Copeland, T.; Antikarov, V. (2001): Real Options.
De Luca, J.; Coad, P.; Lefebvre, E. (1999): Java Modeling in Color with UML
Dixit, A.K.; Pindyck, R.S. (1994); Investment under Uncertainty.
Forsberg, K. and Mooz, H. (1991): Relationship
Haberfellner, R.; de Weck, O. (2005): Agile SYSTEMS ENGINEERING versus AGILE SYSTEMS engineering.
Hall, A.D. (1962): Methodology
Jensen, R. and Tonies, C. (1979): Software Engineering
Kruchten, Ph. (2011): Agile Teenager.
McDermid, J.A. and Rook, P. (1991): Software Development Process Models
Moeller, M; et al. (2008): Strategic Innovation
Nadler, G. (1967) Work Systems design
N.N. Hybrid project management: the best of both worlds. https://www.microtool.de/en/what-is-hybrid-project-management/
Morgan, J. M.; Liker, J. K. (2006): The Toyota Product Development System
Pahl, G.; Beitz, W.; Feldhusen, J.; Karl-Heinrich Grote, K.-H. (2007) **Engineering Design.**
Palmer, S.R.; Felsing, J.M. (2002). A Practical Guide to Feature-Driven Development.
Robins, D. (2016): Is the Hybrid Methodology the Future of Project Management? © ProjectManagement.com
Royce, W. (1970): Managing the Development of Large Software Systems.
Schwaber, K.; Beedle, M. (2001): Agile Software Development with Scrum.

Schwaber, K. (2004): Agile Project Management with Scrum

Schwaber, K. (2007): The Enterprise and Scrum.

Scrum: https://en.wikipedia.org/wiki/Scrum_(software_development)

Scrum.org: What is Scrum? A Better Way Of Building Products https://www.scrum.org/resources/
what-is-scrum

Spear, St.; et al. (1999); Decoding the DNA of the Toyota Production System https://www.lean-
competency.org/wp-content/uploads/2015/12/Spear-and-Bowen-Rules.pdf

Stelzmann, E. (2011): Agiles Systems Engineering

Sutherland J., Schwaber, K (1995). Business object design and implementation. Springer, London

VDI-2221 (1993): Methodik zum Entwickeln und Konstruieren

Westland, J. (2016): What is Hybrid Methodology? https://www.projectmanager.com/blog/
what-is-hybrid-methodology

Complete references in the "Bibliography" at the end of this book.

# Part II
# The Problem-Solving Process

# Chapter 3
# Systems Design

The problem-solving process is influenced and supported by the systems engineering principles and their two central building blocks, systems thinking and the systems engineering process model (see Fig. 3.1). *Systems thinking* supports a holistic approach to cause-and-effect relationships in delimiting, structuring, and subdividing the system. The *systems engineering process model* with its four modules supports the execution of a process with models and methods to stimulate the thought process and with planning and agreement on a method commensurate with the problem. As such, we divide the problem-solving process into two areas:

- *Systems design*, understood as engaging in substantial questions of the problem and its solution
- *Project management* as the sum of organizational and dispositive measures for planning, guiding, monitoring, and steering a project with regard to content, duration, and costs. Thereby, an important role is played not only by the ever limited available resources, such as manpower, financial, and material means, but also by the connection of the project task to the commissioning group and to the clients and users by the decision-making processes, the questions of collaboration, authority, conflicts and their resolution, etc.

In *systems design* (Fig. 3.1), we differentiate between:

- *Architectural design* in terms of developing a fundamental systems architecture (systems architecting) and
- *Concept development* as a concrete embodiment of a selected architecture

An overview of systems architecting will be looked at in Chap. 5.

For the topic of systems design, we once again pick up the building blocks of systems engineering philosophy (systems thinking and the systems engineering process model)

---

The original version of this chapter was revised. The correction to this chapter is available at https://doi.org/10.1007/978-3-030-13431-0_17

© Springer Nature Switzerland AG 2019, Corrected Publication 2021    101
R. Haberfellner et al., *Systems Engineering*,
https://doi.org/10.1007/978-3-030-13431-0_3

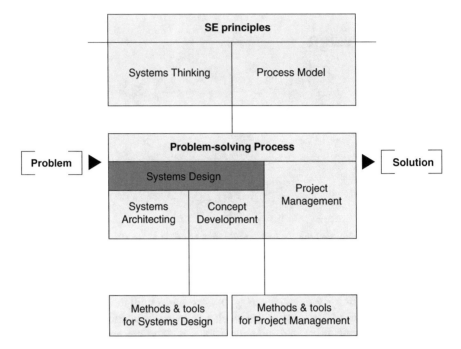

**Fig. 3.1** The problem-solving process as an overall term for systems design and project management

that were described theoretically in Part I, Chaps. 1 and 2. Now, we concentrate on aspects of their implementation, and on in-depth and situational considerations.

We have chosen this two-tiered treatment primarily for didactic reasons. On the one hand, we did not want to unnecessarily inflate the treatment of the model. We believe that an earlier treatment of restrictions and directions for implementation would have more likely hindered an understanding of the model. On the other hand, the deliberations that follow presuppose a knowledge of the overall model; thus, background knowledge of the content of the individual components is required.

## 3.1  On the Application of Systems Thinking

The following sections describe several cases in which systems thinking is applied to the problem-solving process. These thoughts are also taken up again and enlarged upon in other chapters.

### 3.1.1  *Discourse on the Formation of Elements and Relationships*

Figure 3.2 shows one possibility for the structure-oriented observation of the system "corporation." On the one hand, elements (components) of a system can be formed in very different ways (component categories); on the other hand, a wide diversity in types

| Element Categories \ Relationship categories | Flows | Flows | Flows | Flows | Physical connections | Physical connections | Physical Connections | Physical connections | Organizational connections | Organizational connections | Causal relationships | Social relationships |
|---|---|---|---|---|---|---|---|---|---|---|---|---|
| | Material | Energy | Information | Values | Paths, roads | Rails, tracks | Power lines | Communication lines | Work steps | Management relations | | |
| **Physical, spatial units, for example** | | | | | | | | | | | | |
| • Sites | X | X | X | | X | X | X | X | | | | |
| • Buildings | X | X | X | | X | X | X | X | | | | |
| • Rooms | X | X | X | | X | X | X | X | | | | |
| • Work areas, work places | X | X | X | | X | (X) | X | X | | | | |
| • Machines | X | X | X | | (X) | (X) | X | X | | | | |
| • Assemblies, modules | X | X | X | | | | | | | | | |
| **Organizational units, for example** | | | | | | | | | | | | |
| • Areas of responsibility | X | (X) | X | X | | | | | X | X | | |
| • Cost centers | X | (X) | X | X | | | | | X | X | X | |
| **Functions/task, for example: planning, decision making, aligning, inspecting…** | | | X | X | | | | | X | X | X | |
| **Process elements, for example: material-, information inputs, testing, processing…** | X | X | X | X | X | X | X | X | X | | X | |
| **Problem components, for example: leadership, qualification, motivation, quality …** | | | | | X | | | X | X | X | X | X |
| **Groups of people, for example: qualification levels, residents, foreigners. women / men of lower, middle, upper class** | | | X | | | | | | | X | | X |

**Fig. 3.2** Examples of element and relationship categories. Note: Several of the frequently used combinations are marked. The categories may be further differentiated as required

of relationships is possible (relationship categories). The combination of a component category with a relationship category yields a starting point for an aspect of the system.

Thus, one could examine the organizational units of a corporation with respect to its material, information, value, and energy flows. However, one could also examine information flows among functions, positions, tasks or locations.

The actual method of the approach chosen depends on the type of problem to be examined and on considerations regarding appropriateness. There are structures that afford a more in-depth look into the manner of functioning or the way in which problems relate to each other, and others that afford less insight. For example, if there are managerial problems, one examines organizational elements with regard to their relationships with, for example, information flows, work sequences, assignment of tasks, competence, responsibility, etc. But it would also be meaningful to analyze management levels and the existing social and human relationships, behavior, making use of their competences/power, etc.

There is no recipe for the right choice. However, it is known that "one-dimensional" observations yield insufficient insights and that any additionally recognized structure can promote an insight into the relationships and an understanding of a problem.

## 3.1.2  Problem Area and Solution System

In systems engineering, systems thinking is applied to the problem area *and* to the solution.

Systems may be designated as problem areas where problem relationships are suspected, and as such should be examined. This is what a situation analysis concentrates on primarily, but in terms of holistic thinking, further areas are usually involved.

*Example*: if the quality of a product leads to complaints, the *corporation*, and its quality-related connections would have to be examined as a problem area. The focus would then be on the engineering drawings, on strength calculations, tests, production, the materials and, external parts employed, assembly, the specific application of the product by users, customers, etc.

The *solution* may lie in design changes to the product, other materials, better instruction and motivation of the production operators, changes in responsibility and/or procedures/processes, remuneration systems, more consistent quality management, etc. Potentially, only simple or selected changes may be necessary, but these would not become apparent without a systematic analysis of the problem area.

The problem area and solution area relate to each other, as illustrated in Fig. 3.3. Above all, the problem area is, at the beginning, at the forefront of the analysis, whereas later, focus switches to the solution area and its components, its structure, the different design variants, the assessment of its effectiveness, the advantages and disadvantages of the individual solution proposals, etc.

The place between the "problem area" and the "solution system" may be considered to be the "opportunity space."

The distinction between the result or *product* (in terms of the system that is to be designed) and the *organization* (as a designing system) is an interesting one. With the first – corresponding to the systems engineering process model – one has to differentiate between the objectives (requirements) and the solution concepts for the product or object to enable a comparison or an assessment of alternative solutions with regard to the objectives. In this respect, there is an essential difference, regarding the method, depending on whether the cause of a qualitatively substandard product is to be sought in "incorrectly" defined objectives, or if the product was not designed according to (correctly) defined objectives – for whatever reasons (inexpertness, unresolved accountabilities, etc.). This leads to a differentiation between a process system and an activity system that in a broader sense can also be described with a process (project)-based organization or a corporate organization.

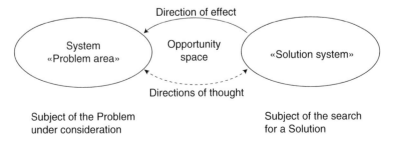

**Fig. 3.3** Relationship between the problem area system and the solution system

### 3.1.3  Application of Systems Thinking to the Structuring and Analysis of the Problem Area

An examination of the broad problem area usually occurs at the beginning of the problem-solving process (see Part II, Sect. 2.1.4.2 Situation Analysis). In many cases, the starting point is the following two questions:

1. How is the problem area working today, what is happening there?
2. What are the difficulties or the unused chances?

Question 1 can be illustrated by, for example, a flow chart or a process diagram. Process steps would thereby be represented as elements; logical consequences of the steps, information flow, material flow, etc., would be represented as relationships.

Question 2 can be examined and represented in a system-oriented fashion according to the rules of "networked thinking." Here, beginning with the difficulties observed, one should consider cause-and-effect relations, and for this purpose develop nets that contain all significant causes, consequences, side-effects, etc. If existing difficulties are thus analyzed, the danger of merely curing the symptoms is reduced. In addition, this can result in starting points for possible solutions and insights regarding the potential effects of interventions. This can be illustrated by different methods, such as bubble-charts, input-process-output (IPO) charts, in structured matrices (DSM), etc. In doing so, it is important to remember that every model illustration is only a simplified depiction of reality and needs to be supplemented or interpreted with additional information. Figure 3.4 is a small example of this approach.

### 3.1.4  Demarcation (Boundary) of the Problem Area

To draw a boundary of the problem area is a significant step in every project. In the example of a quality problem, one could limit the examination to one's own operation and try to find quality influencing factors in respect of machines or intermediate products. As such, one would accept customer complaints and demands as facts, without first checking. But one could also examine the situation of the customers or suppliers before resorting to expensive measures within one's own operation. More effective results may be achieved through simple indicators or measures outside of a hastily defined problem area.

If this type of deliberation were carried out, one would extend the system boundary of the problem area. This is not without danger because some people pursue a policy of looking for the causes of problems that principally lay outside of their own field of responsibility. The questions of attitude and expenditure are driven in connection with the problem boundary. (The customer may be delighted to have one of our specialists with whom to discuss the problem).

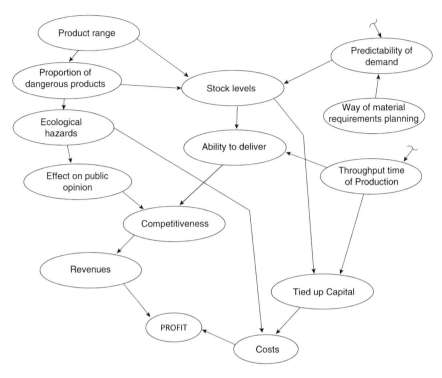

**Fig. 3.4** Networked thinking – excerpt from the example "warehousing"

Within what boundaries should the problem area now be considered? Finding the right system boundary is made more difficult by two opposing tendencies: the more comprehensively the system to be examined is defined, the greater the effort required for the analysis and the search for a solution. The narrower the limits are drawn, the more constrained will be the discovery of further possibilities of a solution or intervention. However, a more comprehensive look at the problem field does not necessarily mean that the solutions themselves have to be of a more comprehensive nature. The knowledge of having many connections and their associated insights only partially under control can actually inspire solutions, albeit in small steps.

A reasonable boundary has to be found by using intuition and experience, coupled with a critical scrutiny and a weighing up of the situation.

### 3.1.5  Systems Thinking and Solution System

Systems thinking is also important in the working steps that are oriented toward a solution. When applying system-hierarchical thinking, it makes sense, at the beginning of the search for a solution, to look at the solution and the solution building blocks as *black boxes* and to first define the functions, along with the related inputs and outputs. Subsequently, one models the solution system according to the aspect of the mode of operation and functional relationships at the lower levels, the components/modules/connections (interfaces) between the components, etc., and the allocation of functions to the elements/components (= architecting). Thus, the idea of a system hierarchy is not only applicable to modeling a problem area, but also to the search for a solution, because the solution must of course be successively detailed.

*Five General Design Principles*
In connection with systems creation, a *consideration of general design principles* is recommended. Five of these are introduced below. Further design characteristics of a good systems architecture, such as ease of integration, scalability, and decentralization, may be found in Sect. 5.3:

1. *Ideality*: this is to be understood as the ratio of the sum of all productive functions and the sum of all undesirable functions and effects. It is based on the conviction that an ideal system should consist only of productive functions. This originally derived from the fundamental evolutionary model in the → TRIZ,[1] that all systems develop in the direction of an ideal state. For systems design, this means that the complexity of a system should be kept to a minimum. An example would be BMW's iDrive concept, by which many different comfort functions in the car can be easily operated with a single push–turn knob instead of having to design and operate individual buttons for every function. Another example is the use of touch displays where the view area is also the handling area, using the same interface.

   An ideal system also uses pre-existing resources/elements from higher level systems and the system surroundings.

   Example: head-up displays in automobiles where the front windshield is used to place information within the driver's field of vision.
2. *Independence*: this principle states that the effect of variable design parameters should be reduced as far as possible. This principle is derived from axiomatic design,[2] according to which each function of a system or each functional requirement should be fulfilled by an independent design parameter. Here, *design parameter* can mean a physical principle, a technical parameter, or even a component. The idea is that when it is required to change a parameter or a component, other components and their ability to function should not, as far as possible, be influenced. For example, in certain major software systems, the individual system functions are deliberately implemented in different, distinctly separate software modules.

---

[1] TRIZ = Theory of Inventive Problem Solving. The classic: Altshuller, G. (1984): *Creativity*.
  Further: Clausing, D.; Fey, V. (2004): *Effective Innovation*.
  Schulz, A.; Clausing, D.; Fricke, E. and Negele, H. (2000): *Development and Integration*.
[2] Suh, N. (1990): *The Principles of Design*.

3. *Modular construction*: this principle signifies that subsystems (system building blocks) should be formed and defined such that they cover functions that are as clearly defined as possible and that are re-useable. Here, one must be careful to draw the boundaries of systems and subsystems in such a way that the interfaces/relationships to the outside are as simple as possible and kept to a minimum (loose coupling) and those directed inward are as strong as possible (strong cohesion). As such, tuning and coordination are simplified. This principle figures in both software and hardware systems. The advantages consist of:

   – A reduction of complexity because the system becomes more manageable
   – A greater chance to use modules repeatedly in different solutions
   – The opportunity to subsequently replace modules with better, that is, more efficient ones
   – Simpler maintenance
   – The possibility for more economical production and storage, etc.

4. *Piecemeal engineering*: this principle (according to K. Popper) states that, particularly with large and complex systems, one should beware of performing major changes in large steps, whereby the effects of those changes are not clear. Here, it is both acceptable and reasonable to broadly define the problem area and to develop a comprehensive solution concept. Its realization, however, should be carried out in smaller steps that can be more easily reversed, should they prove to be inexpedient or even wrong.

5. *Minimal prejudice*: this principle states that, when in doubt, one should give preference to the solution that presents the most scope for further development and is thus the least prejudicial.

Here, it is important that the solution has a certain robustness in the face of changing conditions and/or contains the appropriate flexibility to accommodate necessary changes/adaptations.

We have limited ourselves to five general design criteria whose application should be fundamentally considered in every search for a solution. Further design criteria are introduced, particularly in Sects. 6.3.2.3 and 6.3.6. References to further literature can also be found there (for example, Altshuller, Baldwin, and Clark, Pahl and Beitz).

Finally, we should also refer, in terms of the holistic approach to systems thinking, to the higher level, *simultaneous consideration of solution and problem area*. Solutions should be thought through to a conclusion during the early design stage, and they should be repeatedly implanted into the problem area during the course of development. This helps to review and assess:

– Whether the interfaces, i.e., the connections between the solution system and the problem area, have been clearly designed
– Whether the solution system, when embedded in the problem area, achieves the changes that were originally intended

- Whether undesirable side effects could arise in the problem area that had not been considered up to this point, but that could still be avoided during the design stage

### 3.1.6 System-Oriented Thinking and Teamwork

Working in a team is of great significance, particularly for problems that cannot be routinely analyzed or processed. It is not only the scope of the work that makes teamwork necessary, it is also the knowledge that familiarity with the situation, expertise, and creativity are required to both determine the problem and find a solution. A team provides better opportunities for a more comprehensive view.

When those persons who have to deal with the real problem can work out the structure of the problem solution within the team – in terms of the systems engineering model introduced in this book – they will always be able to return to this as a basis later on. Experience has shown that in this manner, a better common understanding of the problem and enhanced team spirit develop. This in turn is a prerequisite for good solutions.

In this context, we should point out the usefulness of graphic system diagrams (bubble charts, flip charts or pin boards). This makes it possible to develop descriptive "maps" of the problem area or the solution systems, thus allowing these to be better discussed, analyzed, expanded, or reduced.

### 3.1.7 Systems Thinking and Project Management

The more extensive an undertaking and the closer it comes to completion, the more important and complicated the institutional allocation of tasks and responsibilities become for its treatment. Subcontracts have to be assigned; tasks have to be outlined. An unclear formulation of tasks leads to overlaps and gaps – if this matter is not sorted out autonomously by the team. Work packages are therefore often defined with the aid of a → work breakdown structure. The work breakdown structure relies in turn on the system approach; thus, it seems reasonable to appoint subproject managers for the design and realization of components (= subsystems). As such, it also makes sense to appoint integrators responsible for the interaction of components, who supervise the cross-functional system aspects.

Examples: in a large company organization project, there are fields of responsibility for functional subsystems such as sales, purchasing, manufacturing, and also for system aspects such as logistics/material flow and information systems. These can be shown clearly in, for example, a matrix organization (see Sect. 4.3.2.3).

Or: the developers of different product parts, components or function groups meet periodically with assembly experts in so-called installation meetings, to evaluate and discuss results – currently only available in digital, virtual form – with regard to their ease of integration.

## 3.2  On Implementing the Systems Engineering Process Model

In the following, we consider certain aspects of the four modules of our process model that we consider important with respect to implementation.

### 3.2.1  Implementing the Process Principle "From the General to the Detail"

The basic idea of the process principle *from the general to the detail* has already been explained in Sect. 2.1.1. It consists of applying this principle to both the problem area and the search for a solution. Before beginning with a detailed analysis of the problem area (the starting situation), one should first structure it by initially posing the questions: what are the important elements of the system for us? What are their mutual relationships? How do they relate to their environment?

Also, in the design of solutions, one should start from general reflections or solution concepts, which are further detailed and substantiated during the process. In this way, complexity can be more easily mastered. *From the general to the detail* can also be described by other phrases that characterize the same basic idea with varying nuances: "top–down," "from the rough to the smooth," "from abstract to concrete," "from the incomplete to the complete," "from the framework to the form," and many more.

This basic principle stimulates a line of thought "from the outside to the inside." Its usefulness in the new design of solutions or in redesigns on a larger scale cannot be called into question. Modifications and additions to this principle, in terms of the prototyping approach (Sect. 2.2.1.6), an agile process model (Sect. 2.2.2), or the versions concept (Sect. 2.2.1.7), have already been mentioned.

In special situations, such as when it concerns the improvement (*melioration)* of existing solutions that should not be inherently changed or called into question, the inverse could make sense. The line of thought would then be "from the inside to the outside," "bottom–up", or "from detail to an (improved) whole." In such circumstances a knowledge of details or the improvement and optimizing of portions is of special significance. As a consequence, the whole should be improved.

We return again to certain aspects of this consideration in Sects. 6.3 (Search for Solutions) and 6.5 (Special Cases and Situational Interpretation).

### 3.2.2  Implementing the Principle of "Variant Creation"

The principle of creating variants has already been indicated as an important component of the systems engineering method in Part I, Sect. 2.1.2. This principle, however, is often contravened when it comes to practical application. This can be outlined using a few typical cases.

*Case 1  No variants are created*, one chooses the first thing that springs to mind – and one is happy to have found a solution. This is often justified through lack of time, but one should consider that this approach contains significant risks:

- The chance of an (even) better solution is dispensed with.
- One runs the risk of introducing fundamental alternatives to the discussion at a later point, by which time much detailed work would have already been done and when it may be time to begin the realization process. Thus, time pressure is further intensified, and the consequences can be stubbornly clinging to a single solution, uninformed discussions, and finally even postponement of deadlines.

*Case 2*  The planning group agrees that *there is no alternative* to the solution presently under discussion. This, however, is seldom the case, for there is rarely only one solution to a problem. Although the solutions may be appropriate or achievable to different degrees, they should at least come under consideration for a time. Such an assumption could also be associated with a "loss of face": when, for example, a person not belonging to the planning team (client, decision maker, concerned party, other experts, etc.) brings a wholly reasonable alternative to the discussion that had not previously been considered.

*Case 3  Apparent variants* are created that do not differ in principle but do differ in detail. Or when the favored variant is compared with alternatives that have not been thought through, or are even intentionally inferior. Although the methodology may ostensibly have been applied, it does not actually contribute anything substantial. Here, the same basic risks apply that were mentioned in Cases 1 and 2. Moreover, such a course of action also touches on questions of the credibility, professionalism, and ethics of the planning team.

The cases outlined here are designed to call to mind that, although methodology presents a formal framework, it is the personal potential, contributed by the participants, that yields the benefit. When a colleague presents you with "the best solution," always inquire what the alternatives were or are – and always be prepared for the same question if you bring up suggestions for a decision. The principle of building variants should cause neither tense situations nor needless delays; rather, it should be an expression of open-mindedness and a lack of bias.

### 3.2.3   Application of the "Phase Model"

The following partial aspects of the phase model described in Part I, Sect. 2.1.3 are treated in depth:

- The problem of *concept decisions* following important phases and the inexorably related cost/benefit considerations
- The question of the amounts of *expenditure* to be allotted to the individual phases
- The question of *integrating partial solutions* within the framework of an overall concept

- The *dynamics of an overall concept* in terms of a timing change
- The frequent necessity of an *overlapping procedure*
- The role that could be played by *immediate measures*

### 3.2.3.1   Concept Decisions

Formal concept decisions, which determine the direction for the next steps in development or realization, must always be made at the end of a developmental phase (preliminary, main, and detailed studies; Fig. 3.5. It has already been pointed out that it is important for the client to be aware of the significance of these decisions (Part I, Sect. 2.1.2 Variant Creation).

Timely consideration should be given to working toward these decision-making points, not only by the project team, but also by the client and the decision-making bodies appointed by him or her (project committee, steering committee, etc.;

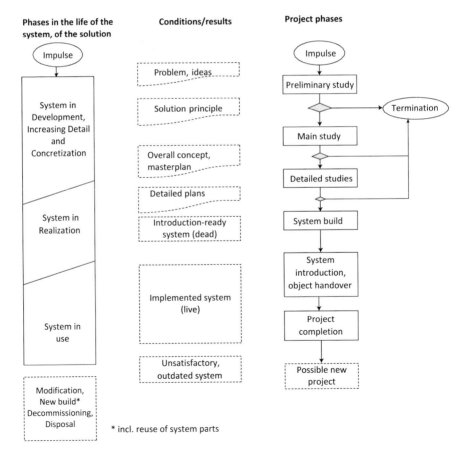

**Fig. 3.5**  Phase concept – basic version

see Chap. 4 Project Management). The right track is created for further work. Additionally, *learning opportunities* become available for both parties:

- The *client* is required to acknowledge project progress at this point in time and to confront different solution options. In cases in which the client consists of more than one person, they become aware of their differing values and have an opportunity to clarify these and to make them known. Also, the client becomes acquainted with impending changes.
- The *project group* is encouraged, at least during the transition from one phase to the next, to become aware of how the solution appears from the client's perspective. Details recede into the background for a time, whereas the overall context has to be moved to the forefront. This can benefit the usefulness and acceptance of the solutions. As such, concept decisions offer the group a great chance to highlight its activities from the point of view of project marketing.[3]

Cost/benefit considerations play a major role in concept decisions alongside considerations of function and fitness for purpose.

With regard to *cost*, two components must be distinguished:

- The generally nonrecurring costs of development and realization (including expenditure for project management)
- Recurring operating expenses comprising, besides allocation for investment costs (via depreciation), personnel costs, materials and energy expenses, etc.

The *benefit* side likewise consists of two components:

- The *operating benefits* that can be expected when a solution, or even just individual parts of it, are ready for use
- The *planning benefits* that can occasionally arise in terms of an increase in know-how, even if the development is stopped and its realization dispensed with

In many instances, cost can be expressed quantitatively and in monetary terms. In the case of benefit, nonquantifiable qualitative aspects, whose weighting is often a matter of judgment, play a role alongside the quantitative aspects.

Benefits are usually much harder to quantify than costs, especially in a conservative culture. Therefore, trust in benefit is usually much less than trust in cost. This should be taken into account when deciding on the concept. The opposite may also apply: a future solution may be associated with the expectations of benefits that have not been scrutinized.

Each project-oriented decision is characterized by the fact that the expected *costs* must be offset by a proportionate *benefit*. A *time delay* with regard to the availability of information must be accepted as a given in the assessment of cost and benefit: *development costs* have to be ascertained or budgeted for at the *beginning* of a development phase (preliminary, main, and detailed studies), whereas the expected *benefits* can only be estimated when the planning results are available in the form of

---

[3] Project marketing = knowing what is important for the customer (according to Peter Drucker); see Sect. 4.2.4.

detailed concept designs – thus, not until *toward the end* of a phase. (Note: the actual benefit can be determined once the project has been realized or it cannot be reliably determined at all).

It is therefore easy to justify the *fitness for purpose of the phase model*, with its incremental planning and decision-making process. Owing to the express requirement to consider future costs and benefits at the end of important phases, the chances increase that projects that promise little success can be terminated in a timely fashion, i.e., while the expenditure on development incurred to date is still within tolerable limits. A different approach, and one that would not be recommended, would be to structure the development process without regard to timing, and to make a comparison of cost/benefit considerations not in phases, but only at the end of the development process.

As the development of solutions progresses, it is further significant to note that in general, the increase in relevant knowledge is not linear over the course of a project; rather, it flattens out. This leads to the question of the *marginal benefit of developmental costs*, as the recommendation to create variants gradually at each system level should not lead to an endless development process. At each level, one must consider whether any additional benefit that might be achieved through further development has not already been overcompensated for by the planning expenditure it entails.

### 3.2.3.2    The Course of Expenditure During the Different Phases of a Project

Although it is not possible to list generally valid operating figures concerning the expenditure trend during individual project phases, we nonetheless consider the trend depicted in Fig. 3.6 as characteristic in its basic structure. Expenditure for the completion of a preliminary study or a main study is, in proportion to the expenditure to follow, still relatively modest: 2–5% for a preliminary study, 10 to a maximum of 20% for a main study. The concrete values naturally depend on factors such as the type of project, the extent of existing knowledge, the required investments, etc.

The trend of expenditure in the realization phases is significantly influenced by the extent of investments to be made: if, as in the case of software projects, the primary concern is with the employment of labor and the demands placed on IT resources for programming, the expenditure apportioned to the system build will be comparatively low. For a project in the building industry or in machinery or plant engineering it can be very high. This leads to a trend in the curves as shown in Fig. 3.6.

Without investigating these differences in more detail, two generally valid conclusions can be drawn:

– Preliminary and main studies reduce the risk with little expenditure. They should therefore be neither poorly budgeted nor executed in haste.
– With projects that involve large and expensive investments in the systems build phase, the proportion of the total that is required for planning and development

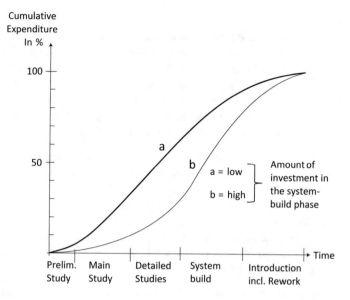

**Fig. 3.6** Expenditure trends in project phases

is comparatively low. For that precise reason, it makes little sense to be overly frugal in the planning stage.

It should be mentioned in passing that reworking, repairing, or servicing solutions in cases of careless or incompetent planning involves a rapid increase in expenditure.

### 3.2.3.3   Integration of Partial Solutions

The following considerations assume that the detailed concepts developed during the course of detailed studies must fit the overall concept. To check this, they must already be embedded mentally and repeatedly within the framework of the overall concept at the development stage. This can be described as a *successive integration process*. After all, the demarcation of subsystems, task areas, and system aspects, which was carried out in the main study, had as its chief purpose the creation of manageable and operable units. Therefore, it always had to be clear that the artificial separation of subsystems can only be maintained for a limited time. Special attention had to be given here to the relationships and interfaces with the surrounding system, whereby an effect-oriented approach was notably at the forefront.

In detailed studies, i.e., in the gradual transition to a structure-oriented approach, in terms of an increasingly concrete arrangement of details, the structure and functioning method of detailed solutions becomes clearer and more concrete, along with the type and scope of their relationships with other detailed solutions and with the environment. These results represent *additional knowledge*, which has to be integrated into the overall concept. Through the integration process, it should be ensured that different detailed concepts do not begin to lead a life of their own, but that they are placed, according to plan, within the framework of an overall concept. This does not rule out that some adjustments to the overall concept may prove necessary.

Based on this understanding, the general process model "from the general to the detail" is expanded insofar as now it must be called *from the general (or the whole) to the detail and back again.* In the course of individual phases of development (preliminary, main, and detailed studies) it may emerge that:

– Existing problems cannot be overcome with a solution based on the original objectives, because essential influencing factors, which were initially not known or simply ignored, have changed in the course of the project.
– Interfaces, dependencies, and relationships that were not known or thought of earlier are discovered during the detailed treatment.
– The original objectives are basically considered justifiable, but they cannot be successfully realized in view of economic, psychological, technical, political, social, ecological, etc., factors, and must therefore be changed.

Such new insights can ensue in the course of the preliminary study, the main study, the detailed study or, in an extreme case, only during the system build, and they lead to corrections or modifications corresponding to the curve representing the knowledge of a system (Fig. 3.7).

Here, it should be noted that, on the one hand, knowledge regarding a system is not simply zero at the beginning of a project (the user knows the current situation, the system designer has expertise and a knowledge of methods at his disposal), and on the other, that complete knowledge about an emergent system can never be reached. This is because a system, as a rule, changes, as do the demands put upon it, in a "live," i.e., a changing environment. In risk assessments of systems, one speaks

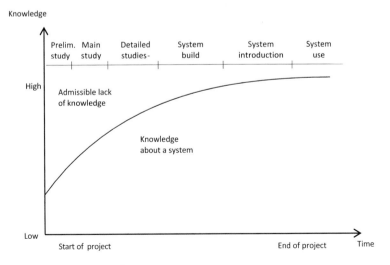

**Fig. 3.7**  Course of knowledge about a system

of a "permissible level" of a lack of knowledge. This never falls to zero, although with particularly safety-relevant systems, the permissible level of lack of knowledge must be relatively low. This is achieved through adequate safety additions (which in essence are in fact "uncertainty additions"), particular analysis efforts (including various simulations), redundancy of system parts, insurance agreements, etc. Thus, one can attempt to reduce the lack of knowledge or, as the case may be, make possible negative effects controllable. However, one cannot entirely eliminate uncertainty when it comes to complex systems.

### 3.2.3.4   Tendency Toward Decreasing Innovation During the Course of a Project

The degree of innovation of a solution should successively decrease during the course of the phases, and known solution elements or practiced procedures should be used to an increasing extent. This is because the character of the innovation generally involves two types of components:

- A component consisting of existing and proven solution elements with known characteristics that are combined into a new type of whole (complete concept).
- An innovative component that results from the fact that individual solution elements have to first be newly developed (for example, in additionally required detailed studies).

The greater the second component, the less one can reliably estimate the result that is to be supplied by the working group assigned to the solution search – and the more the difficulties in coordination between the working groups are magnified. Greater attention must therefore be paid to the coordination tools that help to ensure the necessary agreements with respect to objectivity and timing.

It may therefore often make more sense, to the extent that this can be justified from the point of view of the desired overall solution, to rely as much as possible on known and proven solution elements and, when in doubt, to aim for less ambitious solutions. Even so, there are no grounds for thinking it is possible to eliminate all coordination problems entirely, because in practice additional adaptations or expenditure for the development of intermediate links are almost always required.

Or stated differently: if a particular innovation is regarded as an important competitive tool in part of a system, one is neither able to nor wishes to dispense with this innovation. On the whole, however, one should be careful and use predominantly proven solutions for other functions, so that the total complexity remains controllable.

### 3.2.3.5   Dynamics of the Overall Conception

The following reflection is directly related to the one above. The process of developing complex solutions often does not occur in such a way that the overall concept is "frozen," and only then are successively detailed concepts developed.

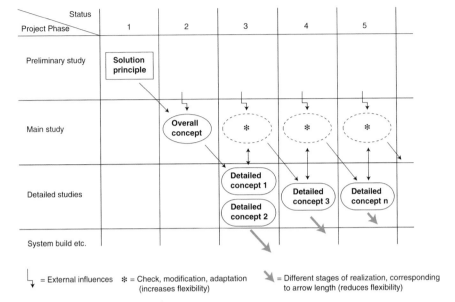

**Fig. 3.8** Dynamics of the overall conception (this model has already been illustrated in Fig. 2.20 in connection with the topic "Agility in Systems Engineering")

As a rule, the sequence proceeds in a similar way to the format illustrated in Fig. 3.8. Detailed concepts are defined on the basis of the overall concept selected at the end of period 2; thereafter, or in parallel with this, possible effects on the original overall concept are reviewed and then detailed concepts 3 and 4 are developed, etc. Thus, adjustments of the overall concept may prove to be necessary or desirable[4] for instance, for the following reasons:

– *Internal influences*, such as a better insight into problem relationships and solution possibilities during the course of developmental activities (arrows from bottom to top).
– *External influences*, i.e., unforeseen developments in the project surroundings, such as new and more efficient technological possibilities, unexpected activities of the competition, changes in the financial and profitability situation; changes in target, procurement, and employment markets; unexpected difficulties with currently employed technologies, new legal regulations, changes in leadership personnel, etc., illustrated as external lightning bolts.

It is evident that overall concepts arrived at earlier may require an overhaul. The likelihood that this will happen grows with the extent and duration of a project. Here, the following problems may arise:

---

[4]This integration step corresponds – at least partly – to the ascending line in the Vee model (Sect. 2.2.1.2).

- Overall concepts are basically at risk of aging. They are based on the level of knowledge current at the time when they were chosen.
- The more prolonged the planning, the greater the risk that planning results will be voided by the passage of time (better detailed insights, diverse external changes).
- As the project advances, opportunities for the overall concept to exert influence become fewer, because for reasons of efficiency, one has to strive to safeguard the development results achieved (detailed concepts) or the realization steps initiated in the interim; that is, to change them as little as possible (see the next Sect. 3.2.3.6).

Basically, this trilemma cannot be solved, but some general *behavioral suggestions* may be given:

- *Beware of super-integrated solutions* that are too large and take several years to carry out, especially when they are located within a dynamic environment and are subject to uncertain conditions. In case of doubt, it is preferable to lean toward smaller solutions that facilitate a quicker benefit.
- *Implement flexibility in overall concepts, pursue "agile systems"* (Sect. 1.3): Keep options open regarding later adaptations, expansions, dismantling, etc. Plan for modular building blocks that can later be replaced with better or more efficient ones. Be open to or plan for opportunities for expansion or reduction.
- Consciously plan for flexibility or possibilities of multiple use, even though this may require somewhat increased investment.
- Consciously make partial introductions to provide interim benefits.
- Dispense with optimizing unproductive details.
- Postpone decisions based on uncertain premises for as long as possible – as long as this is compatible with the logical process (agile systems engineering, for example, in accordance with Sect. 2.2.4)

This can ultimately lead one to accept the ideas of the versions concept (Sect. 2.2.1.7), and thus dispense with further improvements of the overall concept or, as the case may be, to defer them until later versions. The project manager will arrive at this point in practically every project. We are convinced that the decision on a certain variant of the overall concept does not necessarily mean that the overall concept is frozen. To benefit from individual and/or collective learning effects, this "concept freeze" should, as far as possible, take place successively and at a later stage within the project.

We believe that these considerations have very much to do with the "agile methods" (see Sect. 2.2.3). Even when using these methods, there is a need for an overall concept as an orientation framework for development. *Early sprints* should be aimed at developing architectures that make overall contexts clear and allow the steps for the development of the respective detailed concepts to be prioritized – for example, in the form of *product backlogs*. Also, the configuration and organization of the teams should be allowed to work at various levels of concretization: at the level of the overall concept and at the level where the details are located that determine the overall concept.

### 3.2.3.6  Temporal Overlapping Procedure

The sequence described in the explanation of the phase concept in Sect. 2.1.3; detailed studies – system build – system implementation, does not have to refer to the overall concept, but can in fact refer to single, partial solutions that are worked out and realized in overlapping stages. This is illustrated on the left-hand side of Fig. 3.5 by means of the slanted lines among system development, realization, and use.

It is therefore quite conceivable that certain partial solutions can be implemented or are already in operation, while others are still in the detailed study phase. However, this overlapping should not go so far that a system build is started before the conclusion of the main study, which in turn would supply the overall concept that determines the functionality and realization principle of the system.

Several causal complexities, notably in organization or IT projects, demonstrate the expediency of advancing the development of detailed concepts in a staggered manner and not across a broad front:

– The usually limited capacity of the project group, but also of the later user, to absorb innovations.
– Limited budgets that do not allow great advancements for development and realization.
– Methodical and logical considerations that result in first defining the particularly important solution components, with which the rest have to be aligned. The conscious limiting of the degrees of freedom in the creation of the remaining solution components can significantly reduce the complexity of the developmental process.
– The possibility of obtaining useful interim effects by first putting single solution building blocks (partial solutions) into operation while others are still being worked out in detail.

Of course, this type of approach has repercussions for the considerations posed earlier regarding the dynamics of the overall concept (Fig. 3.8): the more frequently detailed concepts are found at the realization or utilization stage, the fewer the opportunities for any subsequent influence. Although this may be an advantage from the standpoint of labor economics and with regard to the progress of the project, it can also be perceived as a disadvantage in the sense of forfeiting a better solution.

Another consideration pertains to situations in which previous thought has to be applied to be able to assess the sustainability of a particular concept or solution principle. This could be the case, for example, when test drilling is required in a construction project to be able to assess the load-bearing capability of the subsoil. If there is a risk that a certain solution principle might fail when this question is posed, it makes sense to conduct such examinations in a preliminary study and not to defer them until the detailed study phases as details requiring clarification. In this case, the phase model must be modified (Fig. 3.9). The fields shown under the project phases as black appendages and the slanted transitions under the life phases on the left reflect this permissible overlapping process.

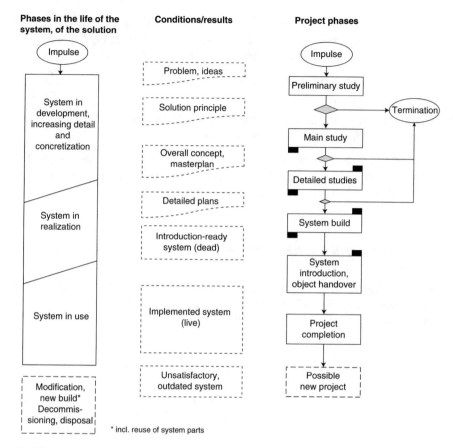

**Fig. 3.9** Phase model – modified version (see Fig. 3.5)

### 3.2.3.7 Immediate Measures

Immediate measures play an important role in many organizational and/or technical projects. Even at an early stage, often during the preliminary study, they represent scope for action and as such opportunities, but also risks.

*Opportunities* arise particularly when immediate measures can alleviate or even mitigate a problematic situation. This can be an advantage both psychologically and objectively: psychologically, because immediate measures create goodwill and confidence in the efficiency of the project team; and in objective terms, because it may provide the time required for an orderly, i.e., deliberate and systematic handling of the situation.

Possible *risks* are, for example, that the eventual solution may be unfavorably prejudiced by the immediate measures, in that a course of action would be taken prematurely or by obstructing alternatives. Moreover, the work of implementing immediate measures can hinder or considerably delay the work of the project group.

However, one can also imagine situations in which it makes sense to be satisfied with "quick" results and to postpone or even dispense with basic improvements.

Therefore, the following *recommendations* should be observed:

- Primarily realize those immediate measures that:

  Have little expenditure
  Bring as visible and tangible an effect as possible, "quick wins"
  Prejudice later solutions as little as possible

- Optionally, leave the execution of immediate measures to other people, but remain in contact with them (participate in the interim success, become better acquainted with the operating mechanisms and the reaction of the system, etc.)

### 3.2.4  Implementation Aspects of the "Problem-Solving Cycle"

The following sections primarily deal with the implementation aspects of the problem-solving cycle (PSC) as a whole and the interaction of the individual partial steps. An in-depth look at the individual steps is given in Chap. 6.

#### 3.2.4.1  Focal Points of the Single Partial Steps of the PSC

Figure 3.10 shows the focal points of the individual partial steps of the PSC, arranged according to different types of reasoning. *Conditions* are represented by the numbers 1 through 6, and *mental or working steps* by letters a through f. Their meanings:

1. Current condition with current means or technical procedures (= instrumental, "how" or "with what" level)
2. Current condition, raised to the functional, task-oriented level ("what?") and illustrated functionally. Enables an assessment of condition 1, for example, with regard to the practicality of current working means for current tasks
3. Purpose or justification level for the actual condition. What is the purpose of the current tasks or functions (why)?
4. Purpose or justification level for the target condition.
5. Requirements of the target condition with regard to function. Central point of formulating objectives (forgo or add tasks). Source: actual condition and additional demands.
6. Possible fulfillment (how, with what?) of the desired functions in the target condition.

A *situation analysis* primarily covers the analysis of the functions of the ACTUAL condition (*what?*) and, on this basis, the assessment of current aids or instruments

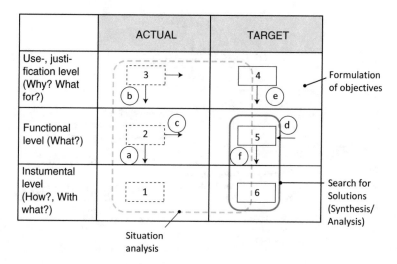

**Fig. 3.10** Different reasoning levels in problem-solving. (**a**) Assess current means on the basis of current functional demands. (**b**) Justify current functions. This process is not always very productive, directs attention too much toward the past. (**c**) Use current functions as a source of future functional requirements, but also consider a possible elimination of functions. (**d**) Additional functional demands that have nothing to do with the actual condition. (**e**) Justification of future functions. (**f**) Fulfil new functions by new means

(*how, with what?*) – looking from **2** to **1**. In the background and above everything is the use or justification level (*why?*) – looking from **3** to **2**. For the purposes of our problem definition (the difference between ACTUAL and TARGET), the situation analysis does not satisfy itself solely with the ACTUAL condition, but should at least touch upon all three levels of the TARGET condition – steps **3** to **4**.

The *formulation of objectives* must concentrate on the functional level (primarily the TARGET condition) and of course also on the use or justification level, particularly with reference to the TARGET condition – parts of **2** and also of **4** and **5**.

The *search for solutions* (synthesis/analysis) finally comprises the search for instrumental solutions (*how, with what?*) for defined functions – steps **5** to **6**.

The focal points of these reflections on situational analysis, the formulation of objectives, and the search for solutions are illustrated by the bordered fields in Fig. 3.10.

### 3.2.4.2   Information Flow Between the Partial Steps of the PSC

The information to be transmitted between each of the partial steps of the PSC can be seen in Fig. 3.11. The connecting arrows express what information should be prepared as a basis for each of the following steps.

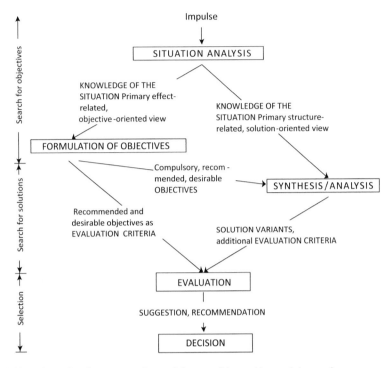

**Fig. 3.11** Information flow among the partial steps of the problem-solving cycle

### 3.2.4.3 Anticipation or Regress (Repetitive Cycles, Iterations)

The depiction of the problem-solving cycle up to this point may give the misleading impression that it concerns a linear process that would have to be executed exactly according to the stated sequence of steps and would finally lead to an optimal result.

This impression would be neither intended nor realistic, and therefore a few additions and limitations are appropriate. On the one hand, anticipation is necessary and desirable; on the other, a true problem solution in the form of a linear process is not attainable; *repetitive cycles*, also known as *iterations*, are often necessary and reasonable.

The problem-solving cycle should therefore neither exactly describe the actual process, nor should it constitute mandatory instructions. It should rather be an aid to orientation and represent a reasonable compromise between an idealized linear sequence (manageable) and a realistic, universal behavioral pattern (complex).

*Anticipation*
The individual steps of the problem-solving cycle are to be regarded as a series of activities that have different purposes and contents. This is concerned with different individual aspects that, although they may temporarily occupy the foreground, still allow one to aim at the next step and the one after that.

A targeted and adequate situation analysis is only possible, for example, if the planner is aware that he or she must use it to create a basis for the formulation of objectives. Also, he or she must come up with reference points for intervention or opportunities, for solutions, which by rights belong to the synthesis/analysis step of the process.

This pragmatic view should not, however, tempt one to completely skip individual process steps or to grossly neglect them, or indeed to arbitrarily change the sequence.

### Regress and Repeat Cycles

Often, one has to return to earlier steps and modify their results. This does not have to be viewed negatively as design processes often take place iteratively and if iterations lead to improvements, they are naturally, in principle, to be welcomed and accepted.

The most important repetitive cycles in the PSC are discussed below (Fig. 3.12), whereby a distinction is made between rough and fine cycles:

- Rough cycles – the step back passes over the individual sections of the search for objectives, the search for solutions, and the selection.
- Fine cycles – the cycle transpires within the sections of the search for objectives, the search for solutions, or the selection.

### Rough Cycles

The following rough cycles are conceivable as examples (the numbers cited refer to those in Fig. 3.12):

1. From the search for solutions back to the search for objectives
   One or more of the mandatory objectives originally decided upon (for example, performance, costs, deadlines) may prove to be so restrictive that no useful solutions can be found. The requirements have to be revoked wherever this is easiest. Thus, an additional and complementary analysis of the situation may prove necessary.
2. From the selection back to the search for solutions
   The evaluation may reveal that solution variants are still insufficiently developed or that they cannot (yet) be assessed with regard to certain criteria. Or: at the decision stage, the client comes up with new requests for the design of variants, which leads, for example, to the creation of new variants by combining elements from available variants.

3. From the selection back to the search for objectives
   If at the decision stage, the client is not satisfied with any of the solutions offered or introduces new considerations with regard to system boundary, target definitions, etc., this can lead to a renewed mental run-through of the PSC in which earlier solution suggestions can then be incorporated.

### Fine Cycles

4. Within the search for objectives, i.e., between the formulation of objectives and situation analysis.

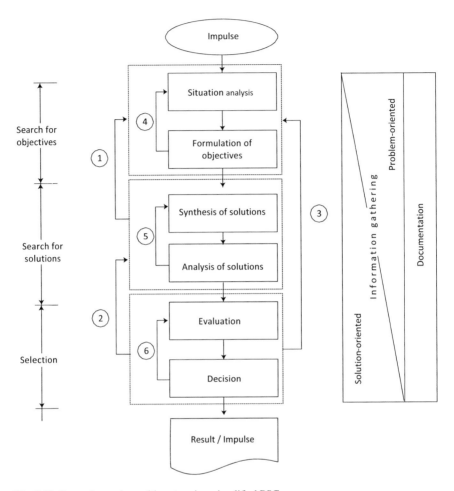

**Fig. 3.12** Reversion and repetitive steps in a simplified PSC

Newly emerging objectives may make additional situation analyses necessary. The understanding of the problem must be rendered in more detail or improved (a closer look at the actual condition and/or so-called paradigms/good examples).

5. Within the solution search.

An iterative solution search takes place in the alternation of synthesis (constructive step) and analysis (critically checking step) of the concept variants.

6. Within the selection.

During the decision stage there may be repercussions for the evaluation phase, for example, modifications to weighting and criteria, assimilation of new criteria, etc.

The repetitive cycles and mental regress presented here are limited to a certain problem area within a certain project phase and thus to a defined problem-solving cycle.

However, there are also cycles of a higher order that point to higher system levels (understood as an adaptation of higher-order concepts) and thus, among other things, to earlier project phases. This has already been referred to in Sects. 3.2.3.3 (Integration) and 3.2.3.5 (Dynamics of the Overall Concept).

### 3.2.4.4 Expanded Problem-Solving Cycle

The problem-solving cycle is extended below such that separation of tasks between client and the project group can be made and depicted graphically. Thus, a relationship is created between the process model and the institutional project management (project organization). In addition, the problem-solving cycle is augmented by components from the functional project management – in particular, the project planning. This is reflected in an additional process element, so-called "process planning," which comes into effect at two points in the problem-solving cycle (Fig. 3.13).

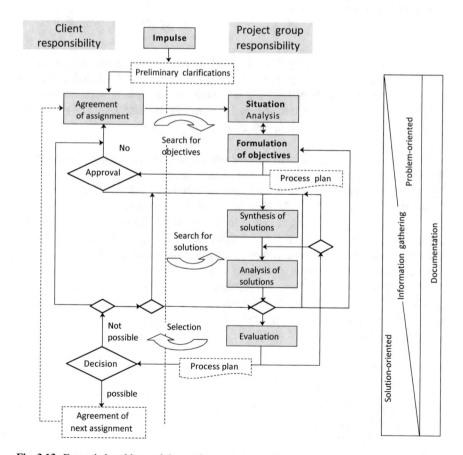

**Fig. 3.13** Expanded problem-solving cycle

Only those elements of the PSC that are new with respect to Fig. 3.12 are described below.

*Client*
The client is responsible for the higher-order system level. In respect of very important projects in industry, the management board is the client and the decision-making body for the overall concept, although other types of committees may be responsible for detailed concepts (for instance, the steering committee responsible for the overall concept could be the client and the decision-making body for the development of detailed concepts).

It is the task of the *client* in each case:

– To participate decisively in the demarcation of the problem area and the design area, and in the formulation of objectives
– To provide the necessary allocation of resources to the project group
– To decide upon the solution variant that is to be selected or to determine an appropriate decision-making body
– To create momentum, as required, i.e., to support the project and its proponents.

Difficulties with regard to a rapid execution of the project often arise when the client and the decision-making body are not one and the same, or if the decision-making body is not constituted soon after the beginning of the project, because then the danger exists that at the time of the decision, the decision-making body assumes different notions regarding the appropriate system boundary from other objectives and values.

Such situations are particularly prominent in projects within the public sector. In such cases it is especially important to conduct a careful analysis of the influencing factors as part of the situation analysis and to consider whether the composition of the decision-making body depends on the type of solutions proposed. If this is the case, one should try to bear in mind the potential decision-making body during the synthesis/analysis. This lessens the risk that the reality or the demands of the client and decision-makers are missed in the planning. Nonetheless, a cyclical process sequence, that is to say, a repetition – possibly several times – of single operational steps or their sequences, is sometimes unavoidable. If these steps signify progress, in the sense of ultimately leading to better solutions and not merely resulting in extra costs or needless delays, then this may be accepted as part of a learning process.

*Project Group*
The project group should:

– Find appropriate system boundaries, system objectives, and process objectives in consultation with the client
– On this basis, develop functional draft solutions
– Analyze these designs to the extent that an evaluation and a rationally justifiable selection are possible

*Preliminary Clarifications*

Before an assignment (project assignment) can be agreed upon, it is usually necessary to conduct some preliminary clarifications. These serve to elucidate the starting situation, the intentions, and the extent of the problem. They are necessary for estimating the costs of executing the task and for preparing a project assignment. This applies above all to the earlier phases (preliminary study). In later phases, the assignment can usually be derived on the basis of existing information. In that case, preliminary clarifications can be dispensed with or kept brief.

*Agreement*

This concerns an agreement – preferably written – between the client on the one side and the project group on the other.

The less clear the starting point and the intentions, the more advisable it is to agree upon a project mandate solely for the next phase. Thus, the risk for the client can be reduced.

The further the project has progressed, the more concrete and detailed the project assignment can and must be (for example, an assignment for developing a solution building block within the framework of the overall concept).

*Approval*

The objectives and criteria underlying the development of the system should be approved by the client, possibly in connection with a rough process plan. This should achieve a mutual matching of objectives and values between the project group and the decision-making body.

However, the objectives and evaluation criteria, declared as valid in this step, should not be regarded as being fixed for all time. The project group and client go through a learning process; they gain a better insight into problem relationships and solution possibilities, which can make it appear necessary to supplement, modify, concretize, and, if need be, even disregard objectives and criteria during the course of the system development. Under no circumstances should changes be made unilaterally, that is, without informing the other participants. Well-functioning communication is therefore an indispensable prerequisite.

The continuous line between commission, situation analysis, formulation of objectives, and approval (Fig. 3.13) represents the process of the *search for objectives* (or the substantiation of objectives in later phases), in which the project group and the client are equally involved.

*Selection*

The choice of system, that is, the decision regarding the solution variant to be chosen and potentially further worked on, cannot and should not be the sole responsibility of the project group. If the client leaves the decision and selection to the project group, one has to assume that he or she has succeeded in transmitting his or her perceived objectives, limiting conditions, and subjective standards.

However, with complex systems this prerequisite generally does not happen, and it would be asking too much of the client to expect him or her, in the stage of formulating the objectives, to be capable of formulating the requirements of a solution that is unknown to him, in such a way that another could make the choice for him.

In reality, and especially in the development of complex systems, it therefore seems more practical if the client:

– Defines and formulates requirements of the solution in conjunction with the project group
– Adjusts his objectives and values during the course of development (learning process) and communicates them to the project group
– Selects from a number of concrete solution variants the one that comes closest to his ideas and expectations; therefore, in the decision phase, arguments and perceived values may be taken into account that were not (explicitly) considered during the course of system development – either because they were too difficult to formulate or because they were simply forgotten

The decision-making body often makes its decision such that it is dependent upon measures and process steps that must be triggered in respect of this. In such cases, evaluation results have to be supplemented with specific process plans.

Note: the *decision* determines and legitimatizes the subsequently worked on variant.

The following aspects are important with regard to the collaboration of the client and the project group:

– As far as the development of complex systems is concerned, neither the client nor the project group has precise ideas of what the result will look like and how to approach it in detail; both are objectives of a gradual working process.
– The client can greatly facilitate development through a balanced collaboration. This should be done when important courses have to be set that do not have to be on a purely objective level. Sensible interim decisions can considerably limit the range of variants and accordingly reduce expenditure on development.
– The client needs to be able to make these interim decisions and the selection that follows the evaluation process without being "bulldozed" by the specialists. Also, the client has to continually interact with the emerging solution. This presupposes a mutual exchange of information that should not be limited to a declaration of objectives and to the concluding transmission of planning results.

*Reversion and Repetitive Cycles*
The reversion and repetitive cycles shown in Fig. 3.14a–g correspond to those in Fig. 3.12, but additional considerations are made possible in light of a more differentiated representation:

– Cycle (a): objectives emerge from the interplay between a situation analysis and the formulation of objectives.
– Cycle (b) expresses the common search for objectives between the project group and the client.
– Cycle (c): iteration analysis → synthesis. The process of the solution search, understood as an improvement of solutions as a consequence of the critical analysis.

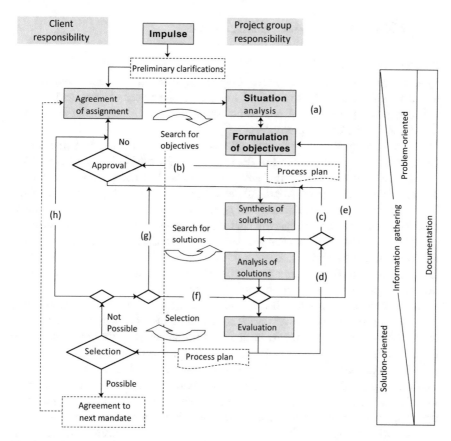

**Fig. 3.14** Reversion and repetitive steps in an expanded PSC

- Cycle (d): ideas for improvement arise from a comparative evaluation.
- Cycle (e): agreed targets cannot be met. A return to the formulation of the objectives should be made – with the required approval of the client for a possible modification of the objectives.
- Cycle (f): disagreement between the project group and the client regarding the favorite(s). A new evaluation with an altered or newly weighted catalog of criteria possibly needs to be made. The focus is not on the solutions per se (these remain unchanged), but on differences in their assessment.
- Cycle (g): new solution (possibly a combination of desirable properties of various solutions). The process begins anew with the synthesis.
- Cycle (h): none of the solutions satisfy the client. A return to the search for objectives (as far as a relationship of trust still exists) is made. Possible causes include: the search for objectives not taken seriously enough; misunderstandings between the client and the project group; the client retrospectively changed his expectations, the project group knows nothing about it, as there is insufficient communication.

## 3.3   Model-Based Systems Engineering

**Baseline Situation**

Systems nowadays become more complex because of customer and market demands; simultaneously, the time to market becomes even shorter, and steadily increasing cost pressure and increasing demands on quality are additional challenges for many product development projects.

There is a long-standing trend to support development processes by using model-based approaches. This was one of the reasons for the establishment of the *Object Management Group (OMG)* in 1989, which some 100 companies have joined, among them IBM, Apple, Microsoft, etc. The best-known developments of the OMG are *Common Object Request Broker Architecture (CORBA)*, designed to facilitate the communication of systems that are deployed on diverse platforms (which enables communication between software written in different languages and running on different computers). Another important development in the MBSE context is the Unified Modeling Language (UML), which allows modeling and documentation of object-oriented systems in a standardized syntax.[5]

This brings us closer to MBSE, which leads away from a *documents-based development* to a *model-based* one. The disadvantage of the traditional practice (documents-based development) is that it is accompanied by a large number of describing files (using Word or Excel), which can hardly be kept up to date and joined or integrated with other files. On the other hand, data and models can be maintained consistently when using a model-based development approach. Using models, the system or product to be developed can be better understood, analyzed, detailed, documented, communicated, and transmitted for further processing.

The MBSE approach has been forced and actively pushed by INCOSE (International Council on Systems Engineering) since 2007. The primary goal was increased productivity by minimizing manual interventions for the transmission of concepts, especially when working in different locations.

**MBSE Definition**

In 2007, the MBSE was defined in the International Council on Systems Engineering (INCOSE) SE Vision 2020 as[6] "the formalized application of modeling to support system requirements, design, analysis, verification, and validation activities beginning in the conceptual design phase and continuing throughout development and later life cycle phases. MBSE is part of a long-term trend toward model-centric approaches adopted by other engineering disciplines, including mechanical, electrical and software. In particular, MBSE is expected to replace the document-centric approach that has been practiced by systems engineers in the past and to influence the future practice of systems engineering by being fully integrated into the definition of systems engineering processes."

---

[5] http://www.omgwiki.org/MBSE/doku.php

[6] http://www.omgwiki.org/MBSE/doku.php

Furthermore, "Applying MBSE is expected to provide significant benefits over the document-centric approach by enhancing productivity and quality, reducing risk, and providing improved communications among the system development team."

An MBSE system model comprises all information created in the above-mentioned development steps. It consists of elements that represent requirements, design concepts, test cases, and interdependencies between the elements.

Thus, MBSE may be considered a summarizing name for an interdisciplinary approach that combines best practices of traditional systems engineering with powerful virtual modeling techniques.[7]

Using our systems engineering concept (Fig. 3.1), we place MBSE into the box Methods and Tools for Systems Design (left foot of the Systems Engineering manakin): it supports systems design (consisting of systems architecting and concept design) and should be inspired by the systems engineering principles, placed at the head of the figure and consisting of systems thinking and the systems engineering process model.

With its methods and tools, MBSE is an indispensable prerequisite for the use of agile concepts (see Sect. 2.4), because working on an "agile" project particularly requires a model-based development environment to be able to integrate and verify products, test results, releases, etc., quickly.

### SysML

As mentioned above, the systems modeling language, SysML, was developed from UML as a standardized graphical modeling language (Weilkiens, Tim (2008): Systems Engineering with SysML/UML; Delligatti, Lenny (2013): SysML Distilled). Nowadays, SysML is the modeling language that is most widely used for modeling complex systems in systems engineering. The set of diagrams as defined in SysML consists of a subset of diagrams derived from UML 2 and complemented by SysML-specific diagrams. There are different types of diagrams: the group "structure diagrams" represent the static aspects of a system, whereas the group of "behavioral diagrams" represent dynamic components.

Besides SysML AP233 is another important component of MBSE.[8] It is designed as a neutral information model for the exchange of data between systems engineering, systems architecture description and related tools. SysML uses this AP233 data exchange protocol for data exchange between tools such as product data management, computer-aided development, computer-aided systems engineering, and computer-aided software engineering.

SysML supports the requirements for modeling: structure modeling, behavior modeling, and parametrics modeling.[9]

---

[7] Diskussionsforum   MBSE:   XING   https://www.xing.com/communities/posts/was-ist-mbse-1000989991#33792101

[8] http://homepages.nildram.co.uk/~esukpc20/exff2005_05/ap233/index.html

[9] Zingel C. (2018): AVL List, MBSE/SysML-Workshop, Institute of Automotive Engineering, TUG, V1.0.

**Miscellaneous**

Purpose of MBSE and expected specific advantages[10]:

- Combination of "top–down" (management aspects) with "bottom–up" (engineering aspects)
- Giving all stakeholders access to information on products and their development, and keeping data consistent within complex applications
- Facilitating change management, risk, and impact analyses
- Allow for early verification, validation or tests
- Allow for frontloading approaches with usage contexts, use cases, etc.
- Allow for bidirectional traceability of requirements in the process chain

More advantages, connected with the move from document-based systems engineering to model-based MBSE are:

- Function-based development instead of component-based
- Quality and process support
- Facilitation of collaboration and information exchange
- Increased speed and efficiency
- Support in mastering complexity
- Facilitation of the reuse of know-how

## 3.4   Self-Check of Knowledge and Understanding: Systems Design

1. What is meant by the term systems design?
2. By which categories can we make elements and relations within a system?
3. What do we mean by saying that systems thinking may be applied to a problem and to the solutions?
4. How does the curve of the development expenses ideally run over time? What are the conclusions to be made?
5. How does the curve of knowledge of a solution ideally run? What about the curve of admissible ignorance?
6. Do you think it is possible that the knowledge at the beginning of a project may be zero?
7. Do you think it is possible that the admissible ignorance tends toward zero over the course of the project?
8. Is it possible to change/modify the overall concept after the decision at the end of the main study?
9. Are immediate measures in contrast to a systematic systems engineering approach?
10. Explain the different thinking levels of problem-solving with the help of Fig. 3.10.

---

[10] Ibid.

11. What kind of information is acquired in the single steps of the PSC and is passed over to the next steps?
12. Does the flow of information in the PSC run linear or are there feedback and repetition cycles? If yes, which?
13. What is meant by the term MBSE?
14. Which benefits are expected from MBSE? What are the expectations?
15. What is meant by the term SysML? Are there any relations to MBSE?

# Literature

Altshuller, G. (1984): Creativity as an exact Science.
Clausing, D.; Fey, V. (2004): Effective innovation.
Delligatti, Lenny (2013): SysML Distilled
Fricke, E.; Schulz, A. (2005): Design for Changeability.
Haberfellner, R.; de Weck, O. (2005): AGILE SYSTEMS Engineering versus Agile SYSTEMS ENGINEERING.
Schulz, A.; Clausing, D.; Fricke, E. and Negele, H. (2000): Development and Integration of winning technologies.
Suh, N. (1990): The principles of design
Weilkiens, Tim (2008): Systems Engineering with SysML/UML
Zingel Chr. (2018): AVL List, MBSE/SysML-Workshop

# Chapter 4
# Project Management

## 4.1 Terminology and Overview

The place of project management within the framework of the systems engineering concept can be seen in Fig. 4.1. Project management is part of the problem-solving process, which we divide into two parts:

- *Systems design* is understood here as dealing with the substantive issues of the problem and its solution
- *Project management* as the sum of organizational and directed measures for planning, leading, monitoring, and steering a project in respect of content, duration, and costs. A number of factors play an important role here, such as limited manpower; financial and material resources; the connection between project content with, on the one hand, the client, and on the other, the customers and users; the decision-making process; teamwork; authority; conflicts and their resolutions, etc.

It is important to note that the division made here is a conceptual one and is by no means a division of personnel in the sense of assignments to different persons. In general, there is a single individual, who, in the role of project manager, has to deal with questions of systems design and who cannot distance himself from questions of project management (costs, schedules).

Note: in large and multilayered projects in which systems design/systems engineering aspects are of greater relevance, this may lead to a special person representing this matter (see Fig. 4.4).

---

The original version of this chapter was revised. The correction to this chapter is available at https://doi.org/10.1007/978-3-030-13431-0_17

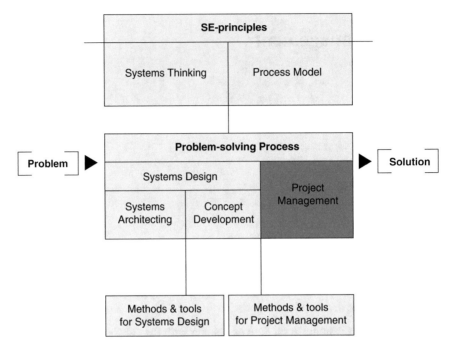

**Fig. 4.1** Project management within the framework of the problem-solving process

### 4.1.1  What Is a Project?

A *project* can be understood, in most cases, to be an extensive, complex undertaking that requires a series of tasks for its implementation and, in turn, special organizational arrangements for preparing and executing these tasks.

A project has the following characteristics:

- It has objectives (results) that are defined or to be defined.
- It has a time limit; thus, a discernible beginning and end.
- It has a certain uniqueness and distinction for the organization concerned (corporation, administration, authority); it is therefore not a continuous operation carried out in a uniform or identical manner (not a routine matter).
- It has a scope that requires division into different subtasks that are connected and interdependent.
- Generally, there are several people or departments within the organization, and often also outside of it, that participate in its implementation.
- The subtasks to be carried out may compete for resources (personnel, finances, materials, etc.), not only amongst themselves, but also with other tasks outside of the project.
- It is frequently associated with a degree of uncertainty or risk in respect of achieving the project objectives, adhering to cost or time limits, etc.

Given these restrictions, tasks that fall under the categories of small, short-term or minor tasks are excluded from the term *project*.

*Typical examples of projects* that match the above characteristics are:

- Building projects: plants, office buildings, homes, streets, bridges, dams
- Development of new products: vehicles, drive trains, measuring instruments, tooling machines inclusive of software
- New and customized designs of machines and facilities: pumps, turbines, trucks, engines
- Organizational projects: organizational structure, process structure, shift models, quality assurance
- IT projects: development and introduction of new application systems, production planning and control, computer-aided design, outsourcing, insourcing
- Restructuring of production areas: introduction of flexible manufacturing systems, computer-integrated manufacturing concepts
- Entry into new markets

### 4.1.2 What Is Project Management?

Management can be understood as the process of *decision-making and decision implementation*. This process can be further divided into subfunctions: planning (thinking ahead), deciding (choosing between different actions), arranging and controlling, organizing (clarifying structures, areas of responsibility and processes), and staffing (the right person for the right job).

Organizations (enterprises, service providers, administrations) are usually structured in such a way that the day-to-day running functions smoothly, and as such, routine processes can be conducted as efficiently as possible. Within the framework of systems engineering, however, the focal point is not the routine operation of systems, but rather their redesign or new design (innovation processes), for which the most expedient organizational form is the project.

Experience shows that extensive and complex undertakings (= projects) can be carried out successfully only when suitable organizational provisions have been made. As a rule, this means an organizational separation between the treatment of the innovation process and that of routine processes. This does not, however, exclude the possibility that personnel may be occupied with project tasks in addition to routine tasks.

As such, project management can be seen as an umbrella term for all planning, monitoring, coordinating, and steering measures that, going beyond problem-solving in its actual sense (systems design), are necessary for a redesign or a new design of systems. Thus, the focus is not on the system (the solution) itself but on the process (the partial steps) necessary for achieving a solution and on the required resources (manpower, finance, material, time, etc.), their deployment, and effective coordination. The conceptual distinction between systems creation and project management can be seen in Fig. 4.2.

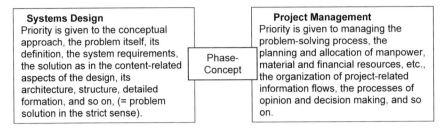

**Fig. 4.2** Distinction between systems design and project management

**Fig. 4.3** The iron triangle
in project management

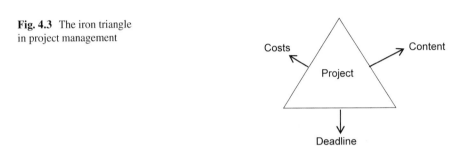

### 4.1.3   The "Iron Triangle" of Project Management

One of the fundamental concepts in project management is the so-called iron triangle (also called devil's triangle; Fig. 4.3). It illustrates that *requirements* (the stipulated achievement of content-related objectives in terms of functionality, quality, etc.), *costs*, and *timing* are to be simultaneously fulfilled, while existing in a state of tension with regard to one another. Achieving more in terms of content usually costs more and requires more time. In most cases, budget cuts also entail a reduction in achieving the stipulated objectives with regard to content. Note: quality or risk is sometimes also mentioned as an additional factor.

One should be careful about accepting responsibility for projects that are extremely challenging with regard to all three demands, given that each by itself lies at the limit of an excessive demand: that is to say, ambitious objectives with respect to content, very limited budgets, and an extremely tight schedule.

### 4.1.4   The Tasks of Project Management

In Chap. 12, under the title "Typical Weak Areas in Projects," one can find a catalog of stumbling blocks and traps; this is recommended advance reading.

The central managerial tasks of the project management team, consisting of project manager and steering committee (see Sect. 4.3.1), are:

- Defining the problem and tasks
- Stipulating/clarifying objectives, the logic of the process, the course of the project (including milestones, decision points)
- Procuring, implementing, allocating, and coordinating, in a goal-oriented fashion, personnel, financial, and material resources
- Leading the project group internally, networking its activities outward/upward (information, coordination)
- Monitoring and steering the course of the project with regard to content, schedules, and costs

Thus, one can divide project management formally into:

- A *functional dimension* with the tasks of starting, maintaining (keep running) and concluding projects
- An *institutional dimension* comprising the creation of formal organizational units and their establishment within the parent organization (project manager, project team, steering committee, etc.)
- An *instrumental dimension* as the sum of all methods and tools for planning, monitoring, and steering projects in respect of schedules, capacity, and costs
- A *personnel dimension*, which entails the desired personal qualities required in good project managers and team members, aspects of communication, collaboration, behavior, assertiveness, and conflict resolution

## 4.2   The Functional Dimension of Project Management

### 4.2.1   Start-Up Work

Start-up work is required not only at the beginning of a project but also, in modified form, during its lifetime (start of new phases, beginning of the realization or introduction).

*Examples of activities*: agreeing upon the project assignment (including objectives, budgets, schedules), appointing a project manager (the *driving force*) who, to a significant extent, is to carry (or help to carry) the burden of the start-up work and organizational planning in configuring the personnel of the project group and the decision-making bodies (steering committee, etc.), organizational incorporation into the parent organization, choosing the organizational model, and preparing the WBS[1], (= work breakdown structure, i.e., structuring the desired outcome), planning its resulting tasks[1], roughly structuring the project course (phases, processes, activities, schedules, costs, personnel allocation)[1], organizing information and documentation, making resources available (personnel, financial, workspace, etc.), organizing project kick-off (start meeting), etc.

---

[1] = Roughly at first; in greater detail later on.

## 4.2.2  Operational Work (Keep Running)

After a project has started and been put into motion, attention must be given to the following *activities*: assigning tasks, actions, and responsibilities in an ad hoc fashion and planning them in more detail (especially those activities marked[1] above), managing, moving the project forward, supervising, and steering the project (monitoring schedules, contents, costs, planning, and introducing corrective measures), generating progress reports and conveying them to the steering committee (knowledge of status also benefits the team!), coordinating and managing internally, coordinating and reporting outward or upward, conveying ideas and results, resolving conflicts, preparing and effecting decisions, making independent decisions, etc.

## 4.2.3  Project Conclusion (Finalizing, Completing Projects)

After the work on content, concepts, and implementation is complete, attention must be given to the following *activities*: user instruction, training of users, operating and service staff, documentation of the solution (important for service and maintenance), handover of the project, approval of the system, possible improvements, settlement of accounts, review, lessons learned (project objectives achieved, costs, deadlines, debriefing), end of project celebration.

## 4.2.4  Project Marketing

Project marketing means leading a project in such a way that it remains attractive and important for the key stakeholders, the client, later users, and project collaborators – and, of course, it means not forgetting to communicate this: "If you do something good, advertise it." Paraphrased, according to Peter Drucker, one could say that *"marketing is knowing what is useful for the customer."*

A project manager has *three types of customers*: the *client* as decision-maker (financer), the operators/*users*, and the *project team*. The interests of these "customers" are different.

The *client as an external customer* wants the agreed services to be delivered by the agreed delivery date, for the agreed price, and with the expected quality. Internal customers (e.g., development of an new e-vehicle, or an IT solution), for example, expect the project team(s) to work efficiently; the scope of the project to be aptly decided and contained; clearly differentiated proposals and ideas to be presented; work to be done in such a way that the solutions are accepted; costs and deadlines to be adhered to; and, should difficulties arise, a timely report to be submitted.

The interests of the *operators/users* may be expressed thus: the new solution should have visible advantages that are not too distant in the future. If there are disadvantages, they should not be so serious as to outweigh the advantages.

The operators want they themselves and their cooperation to be taken seriously; they should be able to contribute their ideas to the solution. However, it should also be pointed out to them, in a professional manner, that their wishes and demands have limits or consequences; they should be able to have confidence in the quality of the solution; to participate in the success of the project, etc.

The interests of the *project group* usually involve the project making discernible progress (things are moving ahead, we are on the road to success), and that the work and contributions of the team members are being recognized.

A project manager should reflect from time to time on how well he is serving these and similar interests of his clients.

## 4.3   Institutional Project Management

Institutional project management comprises above all the *project team* (consisting of project manager and team members), the selection of an *organizational model* (which serves to regulate the project manager's authority in accessing the project team members), and other participating *committees or bodies*, such as the project or steering committee.

### 4.3.1   Participating Committees/Bodies

The committees and bodies participating in a project can be broken down, as shown in Fig. 4.4, into client, project committee, project manager, and project members.

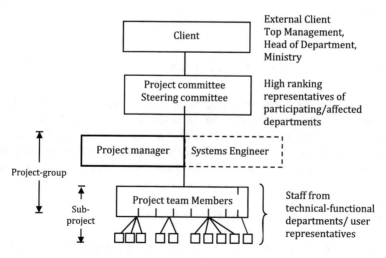

**Fig. 4.4** Committees and decision makers in a project

The *client* has the task of placing the formal project assignment or, as the case may be, to approve a proposed formulation of the assignment, in addition to the formulation/approval/agreement with regard to design and procedural objectives.

The *project committee* (also known as the steering committee) is a highly qualified and superior-ranking body appointed by the client. It has the following tasks: to act as the contact point for concept decisions; to monitor the course of the project regarding its contents, deadlines, and costs (it is thus the recipient of progress reports); and finally to support the project outward and upward, giving it momentum or avoiding setbacks.

The *project manager* is the lead person (driving force) appointed by the client, with whom the project assignment is agreed, and who has to ensure that the project's objectives are achieved within the agreed costs and time limits.

Each project needs a project manager, ideally with a systems engineering mentality and some systems engineering knowledge.

In large and multilayered projects in which systems design/systems engineering aspects are of high relevance, this may lead to a *special person* (systems engineer, see Figs. 4.2 and 4.4) playing this role.

*Project team members* are seconded to the project with the task of compiling and/or realizing concepts/solutions, with respect to task description within a project (including data and information acquisition).

*Line management* (e.g., technical/functional departments) may be represented within a project at the *working level* through the project team members they provide. They may also, however, be represented at the *opinion-making level* in that they may be members of the steering committee and as such take part in decision-making. If the line management has a particular interest in a solution, this can be to the benefit of the project (efficient work, impulse to act). But it can also be a handicap, for example, when team members working within the project are supposed to represent the particular interests of the parent department – which may be to the detriment of others.

## 4.3.2  Forms of Project Organization

The way in which the project is conducted, as in either adjacent to or within the line organization, is dependent on agreements between the respective project management and the line organization. There are a number of models available that give the project manager authority and the project team a certain degree of freedom to configure the project work, notably with regard to allocating and coordinating resources (Fig. 4.5).

### 4.3.2.1  Classic/Task Force Project Organization

In a classic project organization, project members are separated from their regular departments and consolidated in a new organizational unit under the leadership of the project manager, who temporarily becomes their line manager. The project

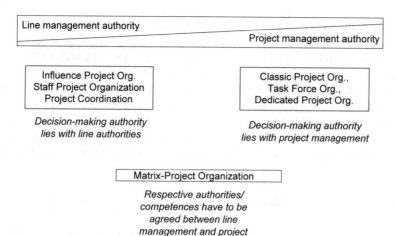

**Fig. 4.5**  Forms of project organization and allocation of authority

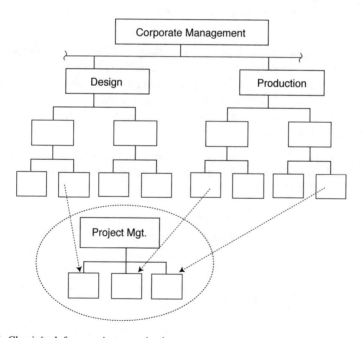

**Fig. 4.6**  Classic/task force project organization

manager then takes on relatively far-reaching powers regarding authority and direction (Fig. 4.6). Team members must be in the project full-time; otherwise, removing the project team members from their home department would make no sense.

*Advantages*: powerful, efficient organization.

*Disadvantages*: possible difficulties with recruitment and re-incorporation. Where do the team members come from at the beginning of the project? Where do they go when it comes to an end?

As this organizational form requires employees to be employed full-time, organizational hybrids are necessary for employees who are only required on a part-time basis. A full-time core group would thus work according to the principle of a classic project organization, and other staff, who are not required full-time, would be assigned according to the principle of the matrix or influence project organization.

This form of organization is only suitable for large and long-term projects in which it is worthwhile changing the organizational assignment of individuals. It is also occasionally chosen, with success, for projects that have become critical and when effective organization is required, and as such is also known as a *task force*.

### 4.3.2.2   Influence Project Organization (Staff Project Organization)

In an influence project organization, the project members remain in their functional departments. Additionally, the position of a project coordinator is created, who, with regard to staff, has no direct authority within the project. Rather, he is assigned to coordinate the project work and can only *influence* the work of the project members (Fig. 4.7).

*Advantages:* a high degree of flexibility regarding staff deployment, and no organizational changes required. At the same time, staff can easily be assigned to other projects or tasks within the line organization (but this only makes sense when none of the tasks is full-time; otherwise, there is a risk of excessive workload).
*Disadvantages*: it is less effective because the project manager has no authority and cannot give direction.

This organizational form is better suited to smaller projects, but also for internal projects (such as organizational changes) in which the project team should not be viewed as an external body.

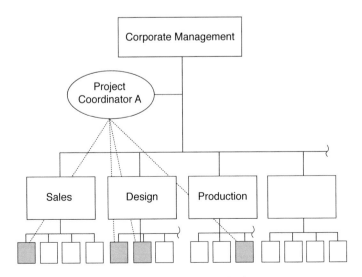

**Fig. 4.7**  Influence project organization (staff project organization)

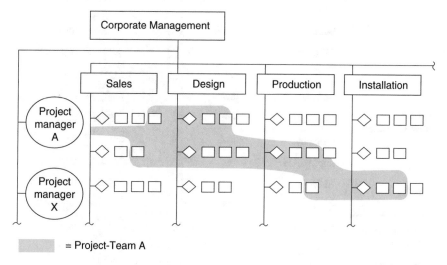

= Project-Team A

**Fig. 4.8** Matrix project organization

### 4.3.2.3   Matrix Project Organization

A distinguishing feature of the matrix organization is the dual or multiple assignment of project staff: in project matters, the team member is assigned to the project
manager; for other activities, he remains within the line management organization
or is assigned to other projects (Fig. 4.8). The matrix organization attempts to marry
the advantages of the influence project organization (no organizational changes, no
reassignment of staff from the line organization) with those of classic project organization (effective owing to the authority assigned to the project manager).

The project manager's authority over personnel and other resources is not unlimited within the matrix project organization, and this has to be shared with other
project managers or with the line management organization, which can lead to conflicts in authority.

*Advantages:* effective, efficient, flexible, with the added possibility that staff not
   required full-time may take on multiple assignments.
*Disadvantages*: risk of conflicts in authority between line and project management
   due to staff being assigned to multiple tasks.

### 4.3.2.4   Suitability Ranges of Organizational Forms

The tendency of valid suitability ranges is shown in Fig. 4.9 for each of the organizational forms. From this, it is apparent that the suitability range of the *influence
project organization* is primarily dictated by project characteristics located in the
area on the left, namely small projects of short duration, etc. The suitability range of
the *classic project organization* (task force) lies in the area opposite; large projects
of long duration with intense time pressure, etc.

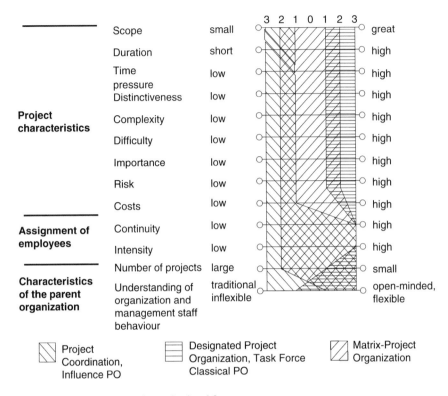

**Fig. 4.9**  Suitability ranges of organizational forms

The *matrix project organization* shows the greatest suitability range: all characteristics can be quite pronounced from the far left to the far right. The bottleneck is found with the last characteristic: the high degree of openness and flexibility with regard to understanding the organization and management that has to be present to manage multiple assignments of staff.

### 4.3.2.5  Hybrid Forms of Project Organization

A variety of hybrid forms are conceivable, and are also found in practice, between the organizational forms already specified, such as:

– A *classic project organization only for a core team.* This is automatically the case when a project manager is already the line manager of several members of staff whom he takes with him into the project team.

– A *change of the organizational form is possible over the course of the phases,* for example, when a classic or matrix project organization is in effect in the *development phases* and is then handed over to a standard organization for the *realization,* which is then accessible for project management only in the form of an influence project organization. An example would be a large facilities project: an effective development in the form of a classic project organization would be handed over to the production department, which also has other contracts and also processes other orders. Project management monitors interim dates and milestones, but no longer has direct access to those carrying out the operations. The converse can also make sense: *development* of a new organizational solution by means of a matrix or influence project organization (broad support from the line management is desired). The *realization* (e.g., programming) is handed over to a classic project organization, which is located externally and possibly even abroad (India).

- A *change* in *personnel* (project manager or project team member) is possible if this has been planned in advance. A change in personnel may also be necessary in cases of unsuitability or unbearable tension or friction. However, it is always irksome when project team members are unexpectedly removed as a result of the decisions of others outside of the project; loss of know-how and the incorporation of replacement members places an additional burden on the core team and can put strain on their manpower.

The tendency of suitability ranges is shown in Fig. 4.9 for of each of the organizational forms. From this, it is apparent that the influence project organization is rather more suitable for small projects and that the classic project organization is better suited to large projects that also require the full-time use of a core team over a longer period. The matrix project organization would inherently be a very versatile one, were it not for the difficulty of dual assignments. This is a matter that can only really be negotiated by mature organizations and individuals who are open to discussion. (Note: people who look for arguments or seek conflict are best served within the matrix project organization.)

## 4.4   Instrumental Project Management

Important instruments and techniques, notably in connection with planning and monitoring projects, include:

- Work breakdown structure (WBS) for structuring tasks
- Linear responsibility charts for dividing and assigning tasks to individuals
- Network plans (CPM, PERT, DSM) and bar charts (Gantt charts) for planning and monitoring project deadlines, timing calculations, determination of the critical path, etc. These can also be carried out with IT support
- Resource planning, for example, on the basis of operational figures from completed projects
- Time–costs–progress diagrams
- Progress reports

## 4.5  Personnel Aspect of Project Management

For this aspect, attention is directed toward the project manager him- or herself and at the members of the project team.

### 4.5.1  Project Manager Requirements

The selection committee and project group may each have their own expectations of the project manager. The *selection committee* looks for individuals who:

– Are interested in the project, demonstrate their commitment, and thus, in a sense, already offer their services
– Are uncomplicated, systematic, task-oriented, broad-minded, open to discussion
– Are resilient – as opposed to oversensitive
– Are assertive and convincing, even without (sufficient) formal competence
– Have a technical overview, without trying to be the best specialist in each subject area
– Can give sufficient time to the project
– Are considered to be good, fair, and persuasive negotiators, both internally and externally
– Can be pivotal in communication with users
– Enjoy the confidence of the selection committee that they will have the project under control

Desired qualities from the point of view of the *team members* are:

– Technical competence in essential project issues
– An eye for essentials
– The expectation of a clear policy and clear agreements externally (project assignment) and internally (project meetings)
– Personal conviction of the usefulness of the project and its prospects for success
– High level of personal engagement, role model
– Good personnel management and moderating skills for communicating progress and results and conveying a collective mood, a "we" feeling and a "can-do" atmosphere
– Ability to market the project externally
– Fair and share success

### 4.5.2  Successful Teamwork

The success of *teamwork* is influenced by:

– The group dynamic process of team building (team members are not "forced" into projects they do not consider promising. Projects are "sold" by constructive argument)

- High level of commitment and interest of the team members in the task
- Good technical qualifications (availability of resources – even if these are not plentiful)
- Good leadership and moderating skills
- Ability to resolve conflicts
- Discernible progress

*Productive teams* display the following characteristics:

- Mutual recognition as partners
- No rigid roles – roles may be rapidly changed according to the situation
- Silence does not necessarily mean agreement: viewpoints are inquired by asking "what do you think? I didn't hear your opinion," "speak your mind," "give us your opinion," etc.
- Listening is just as important as speaking
- Differences of opinion are not simply a nuisance, they may also be sources of information
- Unproductive, hair-splitting arguments are rare
- Criticize ideas: yes – blame others: no
- Decisions are not necessarily made by a majority vote but by reasoned argument; team members who do not agree with an idea are invited to come to terms with it (I see that you are not a particularly ardent supporter of this concept, but could you at least live with it?)
- No secrecy, no forming of cliques within the team

## 4.6 Characteristics of Successful Project Management

The results of an empirical research study are summarized in Part V, Chap. 14, under the title "Characteristics of successful project management". It cites, on the one hand, the criteria for determining success and, on the other hand, the features that are characteristic of successful projects.

## 4.7 Agile Project Management?

As already mentioned in Sect. 2.3, the global competitive pressure transforms the logic of innovation processes in virtually every industry. Innovation in shorter cycles is the current keyword. Agile management frameworks such as Scrum seem to be suitable. Many companies are applying agile methods in software development and try to transfer this method to the development of hardware. Hardware also has to be developed faster, more efficiently, and more integratively than before.

*Project Management Is Affected as a Whole*

The stronger the trend toward agile system architectures[2] and agile processes[3] the more project management as a whole is affected. If at the beginning of a project only a rough task description exists and the detailed specifications of a product will emerge step by step during the course of the project, the consequences are obvious: rigid and detailed specifications concerning performance, price, quality, and delivery date cannot be agreed by the parties, because all parameters may be altered by the iterative approach. The collaboration between clients and suppliers has to be reflected by modified mentalities of the parties concerned. Strict "claim management" is impossible and has to give way to an intensive, trustworthy cooperation.[4]

This also requires changes in the composition of persons who are active in the development process: of course, the *design team* but also the *decision makers*, who have to agree to concept alterations or modifications. The decision makers have to engage in development to a greater degree or give decision-making authority to the team. It is not self-evident that design teams have the strength and the will to make concept decisions – even if they were factually reasonable, they may fear criticism from their superiors. Note: the hierarchy offers a protective function, as individuals with a higher position are less vulnerable when making decisions.

*Management, Decision-Making Bodies, Steering Groups Are Affected*

In traditional project management, the scope of supply and performance, deadlines and cost are planned and managed with regard to these parameters.

Without doubt, this shows some advantages for management: the performance to be expected, deadlines, and project course are known and may be bindingly agreed. Resources may be allocated. Planning, control, and management are intuitively understandable for all management levels.

Previously, in Sect. 2.1.3, we called the phase model the management-oriented component of the systems engineering process model because it allows joint agreements at management levels, joint decision-making between developers and clients. In an agile approach, this is only given for short-term prospects and is almost impossible for the longer term because of the agile cycles and iterations.

This means that traditional management is not suitable for projects with unclear requirements. If there is no way out from this dilemma, great efforts are the logical consequence. Even changes that seem to be evident and justified may cause ponderous processes of shaping opinions and decision-making at different levels of the hierarchy: postponements of milestones and deadlines, alterations of budgets and resources allocation, etc. Or they are not made at all, despite being reasonable.

---

[2] See Sect. 1.3.

[3] See Sects. 2.2.2, 2.2.3, and 2.4.

[4] There is even a Deutsches Institut für Normung (DIN) Standard for it. DIN 69905 defines claim management as "Monitoring and appraisal of deviations or changes and their economic consequences to determine and enforce financial demands."

*Hybrid Project Management as a Way Out?*
Recent developments that try to link the advantages of both methods could offer a way out. They are called "hybrid methods".[5] Although *agile project management* gives the project team the optimal framework for the creation of the scope of performance and service, *traditional project management* represents the demands of top management. By the combination of an *agile approach* at the *operative levels* (team manager, work package manager, team members, etc.) and a *traditional approach* at the *decision-making level* (steering committee, program management, contracting authorities, etc.), hybrid project management tries to combine the advantages of the respective approaches. The project manager has to ensure a smooth connection between the two levels.

By using a traditional phase model, a conceptual framework is to be established that has to be filled out in an agile and creative mode.

The idea of hybrid project management may also be interpreted as the subprojects being processed in parallel in different modes – traditional or agile.

For a more detailed discussion of this topic, see the relevant literature in the footnote below.

## 4.8  Self-Check of Knowledge and Understanding: Project Management

1. Explain the term "project". What are the typical characteristics of a project?
2. Give typical examples of projects.
3. What kinds of activities would you not call a project?
4. Explain the term "project management."
5. What is the difference between "systems design" and "project management"? Does it mean different jobs for different persons?
6. What kind of message should be transported by the "iron triangle"?
7. Which tasks are summarized under the term "project management"?
8. What do we mean by institutional organizational aspects of project management?

---

[5] Inspired by the following publications:

- Rodov, A.; Teixidó, J. (2016). Blending agile and waterfall
- Conforto, E.C.; Salum, F.; et al. (2014): Can Agile Project Management Be Adopted by Industries Other than Software Development?
- Agile Project Management: Best Practices and Methodologies
- Hoda, R.; Noble, J.; Marshall, S. (2008): Agile Project Management.
- Karlesky, M.; Van der Noord, M. (2008): Agile Project Management (or: Burning Your Gantt Charts).
- Howell, D.; Windahl, C.; Seidel, R. (2010): A Project Contingency Framework Based on Uncertainty and Its Consequences.
- Bashir, M.S.; Qureshi, M.R.J. (2012): Hybrid Software Development Approach For Small to Medium Scale Projects

9. Which ideal typical forms of project organization do you know?
10. What are the characteristics, the advantages, and disadvantages of these forms of project organization?
11. Which type of project organization is suitable for which type of project?
12. Which methods, techniques, tools for planning, and control of projects do you know?
13. What are typical demands of a project manager? From the management perspective, from the perspective of the team members, etc.?
14. What are typical attributes of a high performing team?
15. In which direction may project management evolve in response to the emergence of agile methods?

# Literature

Altexsoft: Agile Project Management
Berkun, Scott (2008): Making Things Happen.
Brown, S. F. (2004): Toyota 's Global Body Shop
DeMarco, T.; Lister, T.R. (2003): Waltzing With Bears
Edivandro C. Conforto, Fabian Salum, and others (2014): Can Agile Project Management Be Adopted by Industries Other than Software Development?
Fossa, C. E., e. a. (1998): An Overview
Fricke, E.; Gebhard, B.; Negele, H.; Igenbergs E. (2000): Coping with Changes
Goldratt, E. M. (1997): Critical Chain
Hoda, R.; Noble, J.; Marshall, S. (2008): Agile Project Management
Howell David; Windahl Charlotta; Seidel Rainer (2010): A project contingency framework
IPMA (ed. 1997): Managing Risks
Karlesky, M.; Van der Noord, M, (2008): Agile Project Management.
Keplinger, W. (1991): Merkmale erfolgreichen Projektmanagements
Kerzner, Harold R. (2013): Project Management. A Systems Approach to Planning, Scheduling, and Controlling
Kloppenborg, T.; Petrick, J. (2002): Managing Project Quality
Kruchten, Ph. (2011): Agile Teenager.
Lyon, D. D. (2000): Practical Configuration Management
Morris, Peter W. G.; Pinto, Jeffrey K.; a.o. (2012): The Oxford Handbook of Project Management
PMI (2013): A Guide to the Project Management Body of Knowledge (Pmbok Guide).
Robins, D. (2016): Is the Hybrid Methodology the Future of Project Management?
Rodov, A. and Teixidó, J. (2016). Blending agile and waterfall
M. Salman Bashir; M. Rizwan Jameel Qureshi (2012): Hybrid Software Development Approach For Small To Medium Scale Projects
Schuyler, J. R. (2001): Risk and Decision Analysis in Projects
Westland, J. (2016): What is Hybrid Methodology?
Wysocki, R. (2014): Effective Project Management: Traditional, agile, extreme. John Wiley

For details of the bibliographic references, see "Bibliography" at the end of this book.

# Part III
# Systems Design as Systems Architecting and Concept Development

# Chapter 5
# Systems Architecting

As in our earlier treatment, we divide systems design into systems architecting and concept development (Fig. 5.1).

The terms architecture and systems architecting are used in many disciplines. Thus, one speaks of vehicle, IT, hardware or software architectures. Hence, the content of the terms often remains unclear, either because it has not been defined at all, or because it is a synonym of the term structure. (Note: for structure as an arrangement of elements and their relationships to one another, see Sect. 1.1.2.3). However in the latter case, one has to ask oneself why architecture is needed as an additional term.

The term *architecture* denotes something that goes beyond the term *structure*; namely, the *allocation of functions to the elements of a structure*. Thus, the architecture of a system can be seen as a kind of *solution principle*; it is distinct from other solution principles, has important advantages and disadvantages, and focuses on the process of systems creation. For example, the architecture of a product or a product family is considered to be good when the products are expandable, adaptable, and robust.

Here, the reflections of Ed Crawley (professor at MIT in Cambridge, MA, USA) and his group are very useful (see in detail in Crawley et al. 2016). We adopt his approaches to the extent that they are required to understand the systems engineering concept.

Based on the work of Ed Crawley[1] and other authors, for example, Ulrich (1995) and Rechtin (1991), we define *architecture* as:

- The allocation of functions to elements
- The arrangement of these elements in a structure
- The definition of the interfaces between these elements and with the system environment
- Creation of a defined value

---

[1] Crawley, E.; Cameron, B.; Selva, D. (2016): Systems Architecture.

---

The original version of this chapter was revised. The correction to this chapter is available at https://doi.org/10.1007/978-3-030-13431-0_17

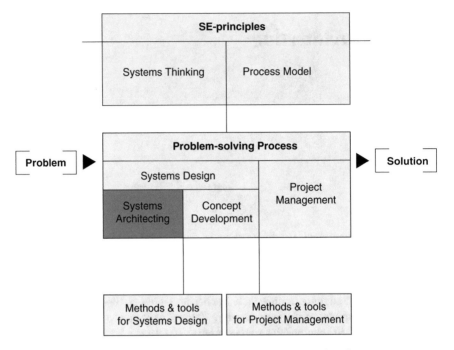

**Fig. 5.1** Architectural design within the framework of the systems engineering concept

## 5.1   Examples of Architecture Variants of Systems

The following historical example of the *Apollo Moon Landing Mission* from the 1960s reveals quite clearly what the term *architecture* means; it shows that the correct architectural decision was intrinsic to its success. At the time, there were three architectural variants:

(a) *Direct ascent*: direct transfer from the Earth to the Moon with an extremely powerful, as yet to-be-developed rocket (Nova), direct landing, and return to Earth.

(b) *Earth orbit rendezvous*: two modules were to be sent into the earth's orbit and assembled there, to then fly directly to the Moon. The existing rocket technology and the Saturn V under development could have accomplished the goal. This variant would have required the landing of an enormous space ship on the moon.

(c) *Lunar orbit rendezvous*: with this variant, the spaceship was to be sent into the Moon's orbit, the command module was to remain in the Moon's orbit, whereas the two-stage lunar landing module was to land on the Moon. The lower stage of the lunar landing module was to remain on the Moon and only the upper stage was to lift off from the Moon to dock with the command module in orbit and from there begin its return to Earth.

The third variant initially had very little support at NASA. However, it was selected because it showed crucial advantages: no new, more powerful rocket had to be built, whose development would have been time-consuming and would have presented technical problems, and a much lower weight was required to land on the moon, meaning that a lower weight had to be transported off the Moon again. As such, the total fuel consumption of the mission could also be reduced.

This decision made it possible to concentrate on further technological development activities: not a new launcher, but rather a new, important, and critical architectural element that allowed the execution of the docking maneuver in lunar orbit.

Another example is automotive drive train architecture. A front-wheel drive car has a different architecture from a rear-wheel drive car. Even though both drives, having four wheels, front and rear axles, transmission, etc., have the same basic *structure*, their architecture is different.

With front-wheel drive cars the function of propulsion is allocated to the front wheels, whereas with a rear-wheel drive it is allocated to the rear wheels. This results in different driving performance. Thus, rear-wheel drive is used particularly for sporty cars, as the decoupling of the functions of steering (front wheels) and propulsion (rear wheels) leads to, among other things, more precise steering and better driving dynamics.

The combination of front engine plus rear-wheel drive makes for an almost ideal 50:50 weight distribution between the front and the rear axle, as vehicle parts that are not dependent on their location within the vehicle, (e.g., the battery) can be optimally positioned. In contrast, the architecture of a front-wheel drive car with front engine leads to a front-heavy weight distribution and thus to unfavorable driving dynamics and, when the motor is installed longitudinally, a longer front overhang.

Systems architecting is the first step in the process of systems creation, in which the basic architecture of a system/product is established in terms of a given solution principle. A more detailed concept design, along with its individual subsystems and components, can then be built on this. The architecture of a system, however, not only determines the essential components of a system, but also their structural and functional relationships. It is only an understanding of these relationships that enables effective and efficient management of complex development projects.

The selection of an architecture pre-defines important properties of a system/product, such as the extent of functional performance, changeability, possible derivatives, component standardization, the ideal developmental organization, and much more.

In the case studies presented in Chaps. 8 and 9 (home construction and airport planning) we emphasize those definitions that can be designated as architectural decisions.

## 5.2    Relationship of Function and Form to Architecture

*Function* and *form* are fundamentally different from one other:

- *Function* refers to what a system does and thus to the benefit or the value[2] that a solution or a solution approach should bring.
- *Form* is in a sense the actual carrier of the function.
- The elements and their mutual arrangement (structure) represent the *form*.

---

[2] Value is viewed here as a very basic cost–benefit ratio.

**Table 5.1** Distinction between form and function

| Function | Form (elements and structure) |
|---|---|
| What a system does/could do | What a system is/could be |
| Creates behavior | Is aggregated and decomposed |
| Is a source of benefit/value | Is a source of costs |
| Requires form | Enables function |

Source: E. Crawley, MIT Course Material

Often, when concentrating on the form, which is seen as what is concrete and tangible (elements and structure), one may neglect its actual connection to value creation, that is, function. The functional view of a system is therefore the central starting point of systems architecting; it determines the system's purpose and value.

– *A function cannot be implemented without form.*
– *A form without function creates no value.*

*Function* – usually in an abstract formulation – is what the system does or should do to fulfill objectives, and it must be conceived and worked out by the architect (Crawley). It should initially be conceived independently from a possible later design, and as such be neutral with respect to a solution. It is then the *form* that determines the way in which the function(s) is/are realized. Form is a product or system property that is actively designed by the architect to enable function. Form is what is implemented (built, written, programmed, produced etc.) (Table 5.1).

An architecture can and must be observed from different perspectives (architectural views). Thus, one can speak of different views of an architecture, but also of its functions and form.

In *automotive engineering*, for example, one differentiates among a geometric, a functional, and an informational view of the architecture:

Thus, the *geometric view* of the electrical system of an automobile (data bus system, cables, steering devices, etc.), shows mostly the spatial allocation of the constituent components: how are cables routed, that is, laid out? An *informational* view, for example, the depiction of data flows, would express a different perspective and other attributes. But in each case, it concerns the same system. Therefore, it is important not to confuse the view of an architecture with the architecture itself.

Software development distinguishes, for example, among four views (logic, process, physical, and developmental view) of an architecture (Kruchten 1995).

The same reflection forms the basis of the idea of system aspects (looking at something through a different pair of eyes), which was treated in Sect. 1.1.2.9.

## 5.3   Task and Meaning of Systems Architecting

Within the framework of the problem-solving process, systems architecting is the first step toward translating the defined requirements, needs, and objectives, as derived from a problem, into a rough sketch of a solution. In Sect. 2.1.5, Fig. 2.7, we called this a decision for a particular *solution principle*.

Architecting can be repeated at every level of the product hierarchy, because with subsystems, too, their architecture must first be designed or determined before they can be developed in detail.

Example: in addition to the previously mentioned overall architecture of an automobile, architectural concepts can of course also be found at each subsequent level of decomposition: the architecture of an engine, a drive train, an all-wheel drive, etc.

Another example would be the avionics[3] of an airplane. This too has its own architecture, but is still a part of the overall architecture of the aircraft. Systems architecting can thus be run iteratively – across the decomposition of a product or the detailed concepts, across different developmental phases and problem-solving cycles (system – subsystem – components).

The basic principle of how architectural design could be systematically derived on the basis of customer requirements harks back to the architect C. Alexander, among others. He recognized that the systems to be developed became increasingly complex and that a responsible designer could no longer approach a problem merely intuitively (Alexander 1964). From this, he deduced the necessity of a methodical process that would enable architectures to be designed to satisfy customer needs in a sustainable fashion. The goal should be to establish the "goodness of fit" between the system to be developed and its environment (context).

The objective of systems architecting is therefore to develop an architecture that fulfills a previously defined value or purpose/objective (Crawley 2009; Rechtin and Maier 1997). Thereby, the useful function that a system carries out or should carry out in the context of its environment can be described as its purpose. Whether or not the purpose was assigned to the system deliberately has no bearing on this concept. An architecture can later prove to be very good or very bad, even if no conscious architectural decision was made.

*Example*: city planning in Vienna. The partial removal of fortifications had the aim of creating grand boulevards for the monarchy, such as the Ringstrasse.[4] Thus, in terms of systems architecting, one element (Ringstrasse) is allocated a certain function (to become a grand boulevard). This function is no longer required today; the architecture is nonetheless useful because the broad Ringstrasse nowadays serves as an important traffic carrier.

A further thought: systems architecting, in terms of the systematic process "from the general to the detail," is the process in which the solution space is constrained for the first time by an architectural decision; this process can and should enable, by way of feedback, a comparison between solution concepts and objectives and demands. Such a formal development of an architecture is especially worth striving for with complex, novel, or highly interconnected systems. Thus, the architecture is seldom based on a fixed, consistent set of objectives but rather on the fundamental benefit/value it is meant to create (for example, performance, scalability, stability,

---

[3] Avionics: the sum of electric and electronic devices on board an aircraft, including flight instruments (Wikipedia).

[4] Today we know, that the width of the boulevard (more than 50 meters) was also favored by army advisors because it impedes the erection of barricades. After the 1848-revolution in Vienna this risk was seen mainly in conjunction with domestic riots.

versatility, and much more). In systems architecting, the customer or contracting authority and the systems designer/architect should discuss, substantiate, and come to a binding agreement on these goals to generate the maximum value.

The creation of an architecture is subject to the basic principles, that is, the fundamental approaches of the systems engineering problem-solving process that have already been described in Sect. 3.2. It can be seen as the result of certain phases in the system development models introduced here. Thus, systems architecting is initially independent of the process model selected for the development of a system.

The question of whether the system should be further developed later on, whether parts/subsystems of the system should be re-used, or whether the architecture should be the basis for a whole product family, or even over several generations, assumes an important role in systems architecting. Should a defect already be present in the development of an architecture, one that limits flexibility or quality or results in a cost-intensive structure, then this not only negatively influences the expenditure and time required for the development of a single product, but it can also have a lasting effect on the competitiveness of a whole product family.

Moreover, an architecture also significantly influences later opportunities for re-using components (commonality), for allocating development tasks, for limiting sub- or pre-assemblies, for the sequence, and thus also the standardization of integration processes, in addition to the ability to update products. In particular, the architecture predetermines a large portion of the costs.

## 5.4  Characteristics of Good Architectures

When is an architecture a good architecture? This can really only be answered in the context of the defined objectives. A good architecture is one that meets the required objectives. This entails meeting the needs of customers and users. Beyond that, one can broadly say that for a good architecture – the basis of later products – it is important that:

- It enables competitive products
- It takes into account the fulfillment of strategic business objectives
- It ensures compliance with present and future laws and regulations
- It can be operated and serviced efficiently and is sustainable
- It is scalable and adaptable with little effort
- Further products can be developed from it within given schedules and resources
- It is "elegant"

The property stated last may sound somewhat strange; however, the experience of many product and systems architects – including the authors – has shown that an architecture that at first glance appears inconsistent and, precisely as said, not "elegant," will also lead to later errors, problems or unnecessarily complex products.[5]

---

[5] Wernher v. Braun: "The eye is a good architect."

The above list also shows that architectures must be aligned to the future. This may be because more products or derivatives of a product family will be derived from this architecture, because it has to comply with future legal requirements, or simply because it requires high initial investments. Different authors have described this aspect using the term *changeability* (e.g., Fricke and Schulz 2005; Ross et al. 2008). It requires – in addition to the five general design principles already named in Sect. 3.1.5 – three additional important principles that should be take into consideration:

*Integrability*

Integrability means that allowance should be made for aspects of compatibility or inter-operability of components and subsystems and, with this, the use of open, standardized, or common interfaces. This enables, with little effort, the exchange of components or technologies of an architecture during the course of its life cycle; it also enables the distributed development of individual subsystems of an architecture – that is, by several contractors. This characteristic is especially important in the context of a "System of Systems" (Sect. 1.1.2.6), where one's own product/system has a common operational connection with systems of other manufacturers. One example is the universal serial bus, which standardizes the integration with external peripherals and the related data-exchange; another is Bluetooth.

*Scalability*

The scalability of an architecture means that the scope of a function and/or the performance of a system/product can be expanded or possibly even reduced. This can be achieved by connecting several identical elements of an architecture with each other, or supplementing them, to scale performance properties or functions. A single element of an architecture can also be scaled by expanding or reducing its characteristic properties.

A good example is *Ariane 4*, the European launcher whose payload capacity was expandable with additional booster rockets according to customer demand – in contrast to Ariane 5, for which this is not possible and whose maximum payload capacity is always available, even when not required, resulting in unnecessary higher costs.

Another example would be *batteries*, e.g., those for hybrid or electric vehicles, whose performance can be expanded by adding cells and thus be scaled according to the desired driving range of such a vehicle.

*Decentralization*

It is characteristic of a decentralized architecture that certain functions, including control or information processing, are not assigned to a central element; rather, they are distributed over several elements. As such, necessary decisions can be taken during the operation of a system at the point of the best information or the greatest knowledge. For example, for time and safety critical functions, decisions concerning systems behavior must be made without long delays via data buses or gateways, but rather directly at the actuator level (e.g., actuator in the control loop that opens or closes a valve).

Another example is current fly-by-wire flight control systems, in which rudders or elevators are not controlled by the mechanical transfer of the pilot's lever movements (by steel cables, push rods, or hydraulic systems), but by actuators locally mounted on the respective rudder and activated by centrally controlled electrical signals.

The advantages that a decentralized architecture may offer are speed, security against outages or malfunctioning, easy changeability, etc.

The opposite of decentralization is centralization, which could also be useful, for example, when it saves resources.

## 5.5   Architecture and Innovation

Particularly with regard to economics, it is important to ask how long an architecture can be in use and when the change to another one becomes necessary. Major technological innovations usually entail a change in the systems architecture. Henderson and Clark (1990) differentiate among "radical," "incremental," "modular," and "architectural innovation."

*Architectural innovation* can be seen as the reconfiguration of an existing system. This pertains to both function and form; the new product is developed on the basis of a new architecture and the change does not necessarily need to be triggered by technological developments.

*Incremental innovation* refers to the further development of a product with the same architecture, but the focus is on single, better performing subsystems, for example, in connection with the use of new technologies.

*Modular innovation* means that only the technical concepts of a module change, whereas the actual architecture or the relationships among the elements remains the same. Simply put, one technical module is exchanged for another. This was the case, for example, in the change from analog to digital telephones.

*Radical innovation* is a far-reaching architectural innovation, most often prompted by advances in the underlying technologies.

*Architectural innovation* can be a crucial core competence of a company[6]: when new products, because of a new architecture, require a different allocation of functions to components and a different arrangement of components, or when new architectures on the market enable completely new product functions and there is a risk that competitors will quickly and successfully implement them. In that case, the existing knowledge of a company may rapidly become worthless.

---

[6]Therefore, it is not surprising that companies that develop complex products, allocating the architectural responsibility to a separate organizational unit.

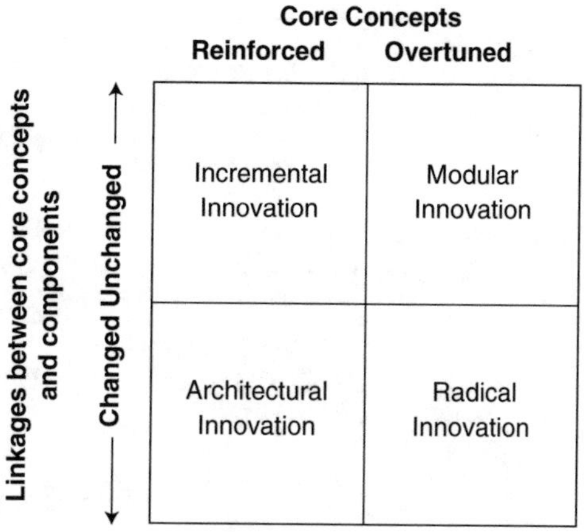

**Fig. 5.2** Architectural
innovation in context with
other innovations
(according to Henderson
and Clark (1990):
Architectural Innovation)

|  | **Core Concepts** | |
|  | **Reinforced** | **Overtuned** |
| *Unchanged* | Incremental Innovation | Modular Innovation |
| *Changed* | Architectural Innovation | Radical Innovation |

*Linkages between core concepts and components*

The basic messages of the Henderson/Clark model (Fig. 5.2) are:

– The distinction between *incremental* and *radical* innovation says something about the extent of the innovative step.
– Those innovations between *architectural* and *modular* identify the levels at which this innovation takes place: the level of the overall architecture or the component level.

It is not easy to decide the right time to introduce a product onto the market on the basis of a *new architecture*. It requires a systematic evaluation of opportunities and risks, based on good access to customer requirements.

Maintaining the existing architecture makes it possible to continue to make use of investments that have already been made. A new architecture can develop new functions and potential customers.

Changing too rapidly to a new architecture can be just as detrimental as waiting too long. The market may not yet be ready for innovations; and may not (yet) appreciate them. The products based on the old architecture could still be salable for a long time. On the other hand, a protracted development based on existing architectures can lead to the loss of the competitive edge. Thus, innovation capability and competitiveness depend to a great extent on the skills of those involved in the architectural development and on the right architectural decisions.

Figure 5.3 shows the functionality of technologies, over time and effort, in an S-curve. At the beginning of a technological development cycle, one is usually glad to be able to master to a certain extent the basic functionality and it is not expected to develop quickly. As things progress – also through increased application and experience – significant functionality improvements can often be achieved. Toward the end of the S-curve, increased functionality is not really possible, regardless of the level of expenditure. This is the latest point in time when a decision with regard to technology has to be made.

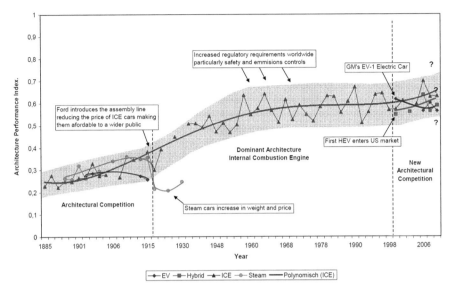

**Fig. 5.3** Development of drive train architectures in automotive engineering. Architectural S-curve based on Gorbea et al. (2008)

The architecture of the automotive drive train also traverses such an S-curve, which necessitates the changeover to new architectures, such as electrical propulsion. By around 1920, the internal combustion engine format of the drive train architecture had prevailed over the steam engine and electrical drive architecture.

## 5.6   The Role of Systems Architects

If systems architecting has the above-stated significance for successful systems design, it is logical to think about the particular role of those persons who create or make decisions regarding architectures. Here, the function of the architect is not necessarily tied to a single, specific person who is seconded as an addition to the project group. It is rather that this function is tied to an intellectual approach, to a role that involves making deliberate reflections regarding architecture and maintaining a view on them – a role carried out either by an experienced member of the project team or by a small team.[7] However, some companies assign specific architectural tasks to designated persons.

---

[7] "The greatest architectures are from a single mind" (Frederick Brooks).

This statement does not pertain only to architectures; it is also valid for all larger strategy decisions within a company. Truly good and sustainable strategies have most often been developed and decided upon by a few individuals – and not in democratic gatherings where they are more prone to be endlessly discussed or burdened with unnecessary compromises.

It is the task of systems architects to understand the objectives of a system in terms of functions that bring value for customers, to define these functions, to evaluate the benefits of a system, and to determine the boundaries of a systems approach. Their task requires them to work closely with customers and clients, market researchers, etc., or – depending on the nature of the project – they may be required to discern, discuss, understand, and stipulate the objectives and demands of a corporate strategy.

During this phase, there is usually a great deal of uncertainty with regard to information. Therefore, architects must be in a position to interpret corporate and marketing strategies, regulatory constraints and competition analyses, and to enter into a dialogue with customers or their representatives. They will neither demand nor stipulate a fixed set of requirements – as is generally the case with many of those engaged at the start of a project. Rather, they will try to develop a common understanding with their partners before moving on to the development of a concept.[8]

Thus, along with objectives and functions, architects also develop concepts that translate functions into a form, that is, they assign certain functionalities to the elements, define the interfaces among the elements and their arrangement, etc.

Therefore, it is also important to understand that architects, on the basis of the required technical expertise, not only have to be generalists looking at the bigger picture, but also specialists in reducing complexity, resolving uncertainties, and focusing the creativity of the developers and project members (Crawley).

## 5.7 Self-Check of Knowledge and Understanding: Systems Architecture

1. How would you define the term "structure" of a system?
2. How would you define the term "architecture"?
3. Give your own examples of architectural variants
4. Differentiate between the terms "form" and "function".
5. At which level of a product hierarchy should an architecture be developed?
6. Give a few examples of principles of good design characteristics for good architectures.
7. Sketch the Henderson/Clark model for architectural innovations in the context of other innovations.
8. What is the objective of architectural design?

---

[8] Note: imagine that you want to build a house. You would immediately dismiss an architect who demands that you give him a fixed set of objectives and requirements at the start, which you would then no longer be allowed to change. It makes much more sense to develop some detailed objectives with an experienced architect and – on this basis – agree upon a suitable architecture (foundation, style of house, rooms, heating system, insulation, etc.). You can then hire a general contractor to implement it.

# Literature

Alexander, C. (1964): Notes on the Synthesis on Form

Baldwin, C., Clark, K. (2000): Design Rules, Volume I

Beam, W. (1990): Systems Engineering, Architecture and Design

Crawley, E. (2009): Systems Architecting.

Crawley, E.; Cameron, B.; Selva, D. (2016): System Architecture: Strategy and Product Development for Complex Systems

Dori, D. (2002): Object-Process Methodology

Fricke, E.; Schulz, A. (2005): Design for Changeability

Gorbea, C.; Fricke, E.; Lindemann, U. (2008): The Design of Future Cars in a New Age of Architectural Competition

Haberfellner, R.; de Weck, O. (2005): AGILE SYSTEMS Engineering versus Agile SYSTEMS ENGINEERING

Henderson, R.; Clark, K. (1990): Architectural Innovation.

Kruchten, Ph. (1995): Architectural Blueprints

Rechtin, E. (1991): Systems Architecting

Rechtin, E. and Maier, M. (1997): The Art of Systems Architecting

Ross, A.; Rhodes D.; Hastings, D. (2008): Defining changeability.

Steward, D. V. (1981): The Design Structure System

Ulrich, K. (1995): The Role of Product Architecture in the Manufacturing Firm

Weilkiens, T. (2008): Systems Engineering with SysML/UML

For details of the bibliographic references, see "Bibliography" at the end of this book.

# Chapter 6
# Concept Development

Concept design (Fig. 6.1) involves developing a selected architectural design in a more concrete and detailed fashion. This may make it necessary to make architectural decisions at the subsystems or system element level. *Example*: overall vehicle architecture, drive train architecture, safety systems architecture, etc.

Both the development of the architecture and the concept can and should use the logic of the problem-solving cycle (PSC) :

- *Search for objectives*, comprising situation analysis and formulation of objectives: where do we stand? Where do we want to get to? Why?
- *Search for a solution*, including the partial steps of synthesis (building creatively) and analysis (critical, checking); what possibilities are there for getting there (variants)?
- *Selection*, including the steps of evaluation and decision; which option is the best, most appropriate, and should be pursued further (i.e., planned in more detail or executed)?

The steps of the PSC (situation analysis, formulation of objectives, synthesis/analysis, and evaluation/selection) are discussed in depth in the following sections.

## 6.1   Situation Analysis

Situation analysis is the first step in the PSC that has to be preceded by an impulse – initiated for whatever reason (Fig. 6.2).

The impulse can be more or less concrete. At the beginning of a preliminary study, it can be a fragmentary, perhaps even contradictory, description of a situation that is perceived as either *desirable* or *problematic* (i.e., inappropriate, not perfect,

The original version of this chapter was revised. The correction to this chapter is available at
https://doi.org/10.1007/978-3-030-13431-0_17

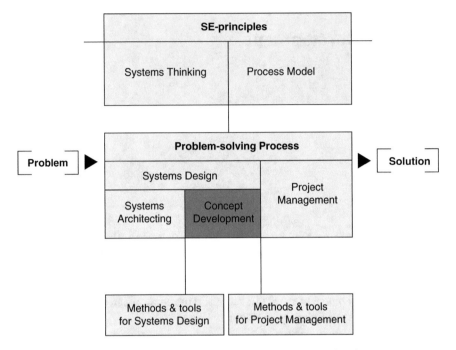

**Fig. 6.1** Concept development within the framework of the systems engineering concept

uncompetitive, risky). At the beginning of a detailed study, it may be the clear task to develop a solution module that is defined in terms of its function, performance characteristics, and interfaces.

The insights and results developed in the situation analyses should:

–   Enable recognition of the embedding of the system within its environment
–   Heighten understanding of the problem for all parties concerned and help to clarify requirements
–   Help in the definition and formulation of objectives
–   Prepare for the development of solutions

### 6.1.1  Purpose and Terminology

The *purpose* of situation analysis is:

–   To make the situation with which one is confronted "easier to grasp," i.e.,

    –   To understand problems and their nature and to investigate causes and their relationships (problem)
    –   To understand new ideas or solution approaches for as yet unidentified applications and their logic (opportunity)

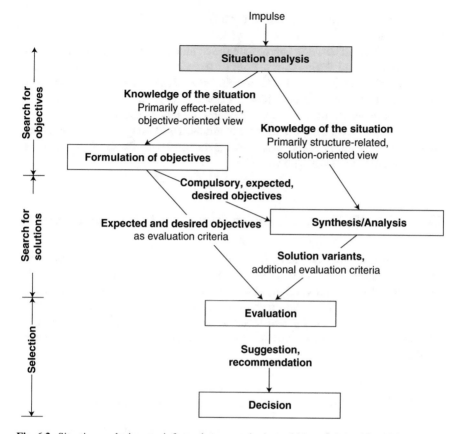

**Fig. 6.2** Situation analysis as an information source in the problem-solving cycle (PSC)

- To recognize the aims and tasks and their initial situation, i.e., to clarify the requirements and line of thought for the nature and extent of the desired or required developments and changes

  - To structure and define the problem or area under examination
  - To set out the intervention and design areas for the solution search
  - To create an information base for the subsequent steps of defining objectives and searching for solutions (Fig. 6.2).

The term *situation analysis* is related to a *situation assessment*, as found in military vocabulary, and to *diagnosis* from medical terminology.

A situation assessment consists of the intellectual dismantling (analysis) of circumstances (location, situation), the compiling and ordering of relations, and the determining of causes. The aim is to gain the necessary information for one's own plan of action, i.e., information about desirable alternative conditions and possible ways of achieving them.

In medicine, one tries to find, on the basis of existing symptoms, combinations of symptoms, so-called syndromes, on which to base a diagnosis. This diagnosis then determines the basis for possible therapies. The situation analysis can take on such a character.

As mentioned earlier, it is appropriate to separate the area to be examined from the intervention and design areas. Both of these areas are to be elaborated in a situation analysis, whereby one can use the approaches, model deliberations, and illustrative techniques of systems thinking (Part I, Chap. 1).

The area under examination will inevitably be larger than these, to be able to assess the requirements, opportunities, and limitations of integrating the subsequent solution. The definition of the design area is the result of a "skillful" restriction to what is suitable, necessary, and workable.

## 6.1.2   Guidelines and Principles for the Analysis of Situations

The difference between the ACTUAL condition and the notion of a TARGET condition was identified as a problem earlier, however vague that difference may be. Without any previous knowledge, idea, or vision of a TARGET condition, there is seldom cause for dissatisfaction with existing conditions. There is no awareness of a problem or an opportunity, and therefore no need for action. The definition of problem and solution areas is tied up with this thought. It will be taken up again here and dealt with in depth.

### 6.1.2.1   Factors That Influence the Understanding of a Problem

The problem area encompasses the actual condition that is embedded in its environment, the solution area, and the – often still unknown – target condition. The solution area also has an environment; it consists of the technical, economic, organizational, social, and ecological context in which the solution is to be sought and realized, but mostly out of the more developed environment that now encompasses the actual condition.

Subjective impressions, appraisals, and perceptions play just as much a role in the assessment of the actual condition as in that of the perceived target.

Barriers may arise in the development of a solution, that is, in the transformation of the actual condition into the target condition or, as the case may be, in the removal of differences between them. Although there may be emotional or personal reasons for these barriers, in addition to a fear of the necessary engagement (time, work, money, risk), they should primarily be understood as *information deficits*:

– Too little insight into the problem area and its constraints
– Too little familiarity with the solution area and its constraints (vague or threat of the unknown)

The focal points of activities in a situation analysis have to be geared toward this.

#### 6.1.2.2   Operational View of a Problem

The desire to understand systems with a view to developing approaches requires an investigation of:

- The impulses for desiring to act, for wanting to engage with an existing situation, and in the expectation of a new or improved solution
- The operation of the relevant system and its key influencing factors
- The relevant parts of the system environment
- The strengths and weaknesses of a system
- The causes of these strengths and weaknesses; what are the opportunities and risks for the system in the future?

Recognizing strengths and weaknesses, opportunities and risks presupposes a *previous knowledge* of average, normal, and attainable conditions. *Further possible sources for ideas* regarding targeted or desirable states are: guidelines (visions), different types of role model, objectives from higher levels, requirements from previous examinations, comparisons with related circumstances or systems, conclusions by analogy, expectations based on theory, intuitive expectations, etc.

### 6.1.3   Different Approaches in a Situation Analysis

In illustrating systems thinking, it has been pointed out that one-dimensional approaches and representations cannot sufficiently cover complex situations. This statement is also true here. In the performing of situation analyses, there are *four different types of approach* that should be emphasized. The object to be observed is structured and examined under the following aspects:

- Systems-oriented
- Cause-oriented
- Solution-oriented or
- Time- or future-oriented

#### 6.1.3.1   Systems-Oriented Approach

Specifically, this relates to the following systems-oriented intellectual and operational steps in connection with the analysis of the problem area:

- Establish the system and its environment and separate them from one other and also from nonrelevant areas (the tendency should first be to set the boundaries rather broadly)
- Develop structural models for systems, parts of the system, and relevant parts of the environment
- Take a structure-oriented approach (formation structures)

These steps are supplemented by intellectual and operational steps that can be described as function-oriented:

- Black-box approach
- Process-oriented approach (process structures)
- Determination of the characteristics of elements
- Identify current and, if applicable, earlier influencing factors
- Develop functional models: rising from the instrumental (concrete = HOW) to a functional (abstract = WHAT) level of observation
- Identify greater relationships and interactions between the system and its surroundings

The considerations originating from systems thinking help to structure the initial situation. The following are some examples:

1. *Effect analyses* are based on a rough view of the system and should allow for a quick introduction to the problem situation. Here, the focus is especially on the effect of the system as a whole and the types of inputs and outputs, but not on the internal structure (see Part I, Sect. 1.2). This is basically a *black-box approach*.
2. *Structure analyses* prepare, with the help of terms such as elements, relationships, superior ranked system, subsystems, environment, etc., the internal build-up of the sequences and processes, display them in their relationships, narrow them down, and thereby render them transparent and manageable.

   Here, a conceptual distinction can be made between a formation structure and a process structure:

   In the *formation structure*, the focus is notably on the (static) construction of a system in terms of its functional or spatial composition, and with this representation and analysis, one can gain an initial overview of the system. The following are helpful:

   - Organization charts that reveal the structural characteristics and the hierarchical composition of a company/authority, etc.
   - Layouts or site plans that reflect the type, size, and formation of different functional areas and departments
   - Configuration illustrations of machines or facilities
   - Exploded views and parts lists that show the structure of a product
   - Thematic maps, etc.

In contrast, the *process structure* provides insights into the (dynamic) processes of a system. The functional mechanisms are more easily recognized and problems and their causes can be better localized. Examples are:

- Process maps
- Material flow diagrams (stating amounts and frequencies)
- Work flow diagrams, process descriptions (Fig. 6.3)
- Macrostructures of processes
- Microstructures, etc.

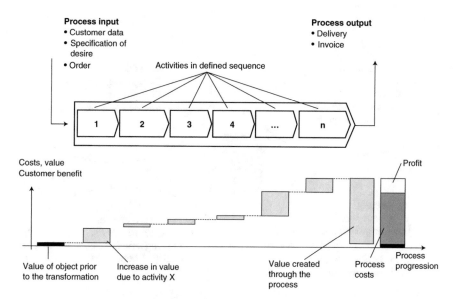

**Fig. 6.3** Activities of a process for creating value and customer benefit. (According to Schantin 2004: Macro modelling)

3. *Analyses of influencing factors* (environment-oriented approach) help to determine and compile not only the sources, type, and extent of external influences on
systems or plans (the ideas and requirements of involved parties or those affected
with regard to the effects, behavior, and performance of systems), but also the
reciprocal influence of systems and surrounding elements and, where applicable,
the effects on third parties.

Influencing factors, depending on the type and condition of a system, can be
of different kinds: natural, legal, political, macro (overall) economic, financial,
personnel-related, social, technical, ecological, emotional, etc. This is illustrated
in Fig. 6.4 with an example from the medical field.

Each system is embedded in surroundings with such influencing factors.
However, being embedded does not mean that it simply exists in a state without
relationships, but rather that it functions because of and/or in spite of these
factors and the resulting relationships.

Influencing factors should be examined to ascertain if they are of a passive; active;
supporting and strengthening, or hindering, diminishing, quantitative or qualitative
type? Are they direct or indirect, individual or joint (combined, synergistic), linear or
interlinked (influenced by third parties, neutralizing)?

Influencing factors are seldom apparent; their presence and effect are often only
revealed through symptoms. Additionally, it should be noted that they normally do
not remain stable; rather, they continue to develop either autonomously or because
of the influence of outside forces.

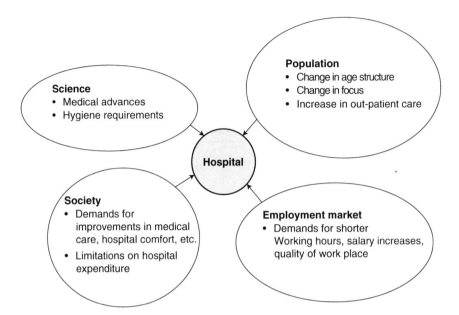

**Fig. 6.4** Analysis of influencing factors. Example: hospital

The environment is also a source of knowledge for designing new systems or for redesigns. Its influences provide ideas for changing a system, point to opportunities and limitations for new structures, and reveal resources for realizing solutions; and they also supply facts for evaluations and decisions.

The influencing factors of a system and its environment that are considered in a situation analysis must be relevant for defining and solving problems.

By highlighting various *aspects of the system*, different views can be taken of the same system (see Sect. 1.1.2.9, Fig. 1.6).

In a hospital, for example, these might be:

– Material flows in connection with medical care (medicines, equipment, medical reports, etc.)
– Physical care (food, bed linen, etc.)
– Movement of human traffic (patients, staff, visitors)
– Allocation of treatment facilities, equipment, personnel resources
– Information systems with regard to results of examinations, therapies, patient administration

The result of the situation analysis should provide the project group and the client with a uniform view of the object under consideration (the system to be created), its essential components, its boundaries, its influencing factors, etc. Thus, a systems-oriented view serves to prepare or support a cause-oriented view.

### 6.1.3.2   Cause-Oriented Approach

The purpose of the cause-oriented (diagnostic) approach is to:

– Identify and describe symptoms of an unsatisfactory situation, a looming threat, or opportunity
– Collect and organize these symptoms and check them for completeness, duplication, or inconsistency
– Assign them to the appropriate elements of a statement of the facts, and possibly to detect and connect elements that are still hidden
– Reveal backgrounds

Ultimately, this should establish possible causes, causal chains, and interconnections. Those causes that are of greatest interest are those that reveal opportunities for action.

It is rather seldom that causes and effects relate to each other in an exclusively linear and single-step (direct) fashion (Schematic 1 in Fig. 6.5). It is possible even in comparatively simple situations for single effects to have several causes or for single causes to have multiple effects (Schematic 2). Often, direct causes are themselves in turn only effects of causes whose alteration appears necessary for improving the situation. Multiple interconnected relationships then become apparent (Schematic 3), and feedback is an essential effect mechanism of complex systems (Schematic 4). One form of a systematic intellectual approach for assigning causes is the fishbone (Ishikawa) diagram. In a quality investigation, the major components

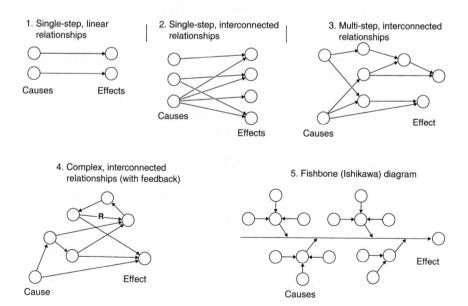

**Fig. 6.5** Approaches to cause–effect relationships

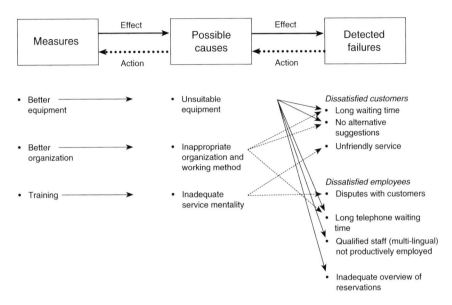

**Fig. 6.6**  Consideration of failures–causes–measures (example: seat reservation system of railway company)

(first-order causes), for example, of a manufacturing process, are applied in a fishbone structure to the process axis (manpower, material, machine, measuring device, etc.), and the influencing factors (second-order causes), such as type, condition, maintenance, pre-operation, etc., are illustrated for each of these components (Schematic 5).

A simple approach to improving existing solutions, using the following basic logic, can be quite helpful (Fig. 6.6):

1. We note; what failures are frequently mentioned or have been documented elsewhere, what failures can we see ourselves, etc. – and write them down. Thus, we create a catalog of failures that are made transparent and can be analyzed and discussed objectively.
2. We reflect on possible causes, discuss them, and establish connections between causes and failures.
3. The result is a catalog of measures that we can consider possible, sensible, etc.

It is important that we clarify and differentiate between the direction of our *actions* and the direction of the *effects*, which are currently working in opposite directions (Fig. 6.6). It makes no sense to speak of measures before one is clear about failures and their causes.

The illustration itself is inconclusive. It merely serves the purposes of visualization; it has to be interpreted and supported by examples or numbers. But in spite of its simplicity, it can serve amazingly well by introducing disciplined thought and reasoning to a confused discussion.

System dynamics (see Part I, Sec. 1.4) is an efficient method of illustrating and modeling complex cause–effect relationships, as it also enables the modeling of dynamic relationships, i.e., those that change over time.

### 6.1.3.3   Solution-Oriented Approach

A solution-oriented (therapeutic) approach should focus on opportunities for inter- vention and design and on their boundaries. It is often required for gaining an under- standing of the problem in the first place and realistic ideas regarding the objectives (the problem as a difference between the ACTUAL condition and ideas regarding a TARGET condition). Solution-oriented approaches should not, however, get out of hand in the situation analysis, but should always take into consideration that the actual search for a solution should not take place until the later stages of a synthesis/ analysis.

In many cases, this is just about preparing a grid for target-oriented collecting and selecting of information during the development. A solution-oriented approach should also differentiate between a functional view (what?) and an instrumental view (how, with what?) – see Sect. 3.2.4, Fig. 3.10.

- *Function analyses*: a situation analysis encompasses its *functional* level (focus: WHAT?) in addition to the use or justification level (focus: WHY?) alongside the instrumental view of the ACTUAL condition (focus: HOW? WITH WHAT?).
- A *solution-oriented* approach to the situation analysis also contains an initial "thinking ahead" regarding the TARGET condition on the three levels (HOW/ WITH WHAT? WHAT? WHY?), as shown in Fig. 3.10.

### 6.1.3.4   Time or Future-Oriented Approach

As the formulation of solutions is not generally concerned with a reconstruction of the past, but with a conscious shaping of the future, future developments can be assessed.

The system-oriented examination, as with the cause-oriented examination, is to be superimposed, beginning with the current condition, on a development-oriented examination of the problem and its surroundings and solution area. It should look at the following questions:

- How will the situation develop in the short, medium and long term if there is no intervention (development of the problem area)?
- What developments are to be expected in the surroundings? What does this mean for the solution and intervention areas?
- What does this mean for the urgency for a solution?
- What effects do possible interventions and solutions in the area under examination have? In what direction and in what way do they have an effect?

With these considerations, it is intended to reduce the uncertainty regarding future developments in the problem and solution areas in addition to their surroundings, or at least to make it easier to understand. Relevant statements may also be obtained in the form of prognoses based on past or current trends and facts, because many changes do not occur erratically or chaotically, but rather gradually and sometimes even regularly.

Predictions, of course, do not yield deterministic results. Generally, the longer the period of the prognosis, the lower the accuracy. Nonetheless, long-term prognoses can be meaningful as they focus attention on possible future developments and scenarios.

Thinking about future developments should serve to create:

– A better understanding of the problem and its urgency
– A basis for establishing objectives
– A "correct" rating of the later solution

### 6.1.4   Boundaries of the Problem Area, Solution Area, and Area of Intervention

As mentioned earlier, it is necessary to distinguish between the boundary of the problem area and those of the solution area, along with the area of intervention.

The boundaries of the *problem area* should be drawn such that both the areas where problems occur (where they are visible) and the primary causative factors are in focus. This could lead to an unexpectedly large and also unwanted expansion of the *investigation area* (Fig. 6.7).

The *problem area* (problem area, area under investigation) is therefore that area within which problem relations are suspected and scrutinized in accordance with the appropriate level of observation.

At the beginning of a project, the boundaries of problematic systems are often not clear and have to be approached cautiously. They need not coincide with physical, organizational, or similar boundaries of an object. Even so, in many cases it can prove advantageous to adopt such empirical boundaries as initial working hypotheses to be able to identify the relevant area in the first place. A closer examination of relationships with the surroundings may then, however, lead to a later change to the boundaries of the area under investigation (a narrowing or an expansion).

The *area of intervention* is that part of the problem area within which opportunities for solving problems are recognized, established, and deemed plausible. Here, a competent narrowing down is important.

The area of intervention is, for example, limited by means of the following considerations:

– Where does one intervene because one is entitled, has been asked, or otherwise has influence?
– Where are there potential technical and organizational solutions?

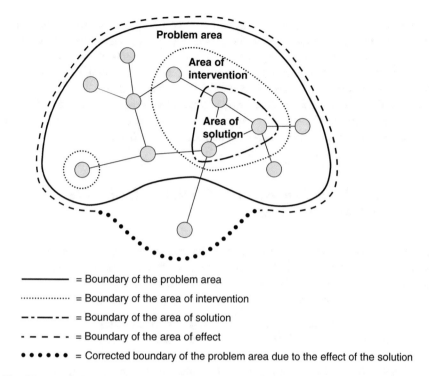

= Boundary of the problem area

················· = Boundary of the area of intervention

— · — · — = Boundary of the area of solution

- – – – - = Boundary of the area of effect

• • • • • = Corrected boundary of the problem area due to the effect of the solution

**Fig. 6.7**  Boundaries of the problem area, intervention area, solution area, and area of activity

- How urgent is it to realize a solution?
- Where and how can one expect a good cost/benefit ratio?

The nature and extent of the problems identified on the one hand and the available framework on the other (engagement skills and objective, personnel, financial resources, time) will therefore be essential defining criteria.

Different strategies for effecting changes may result.

- Situational improvement: minor corrections, elimination of particularly disturbing failures
- Redesign: major changes
- New design: replace the old solution with a new one

If in doubt, the principle of subsidiarity should be applied; interventions are undertaken at the lowest possible level at which they can promise sufficient effects without significant drawbacks.

The *solution area* is generally more closely defined, as it arises in connection with the conception and definition of the effective solution. In the area of intervention, more practical solutions are to some extent still open. The solution that is to be developed later is effectively located in the solution area.

The *area of activity* comprises that part of the problem area in which effects are expected once the solution has been implemented. This area generally extends beyond the solution area. It may even be necessary to rethink the boundary of the problem area, owing to potentially negative effects that, for example, had not been foreseen at the outset. However, one should already try to define the problem area during a situation analysis in such a way that it includes the potential area of action to avoid uncontrolled interventions and to detect any unwanted side effects as early as possible (see the remarks on concept analysis in Sect. 6.3.4).

## 6.1.5  *Identifying Boundary Conditions and Limitations of Design Freedom*

In addition to organizational, spatial, and other system boundaries, the area of intervention is also influenced by existing boundary conditions. *Boundary conditions* are understood to be determining factors, mostly originating from the surroundings, which cannot be or can no longer be influenced by planning, such as:

– Governmental directives, regulations, contractual agreements
– Previous substantive decisions, but also those concerning time frames and financial resources
– Planning results for relevant parts of the system and its surroundings
– Immutable facts regarding the ACTUAL condition would continue to affect the TARGET condition
– Physical phenomena that must be observed, etc.

Thus, boundary conditions may arise from the surroundings, from higher level systems, from upstream project phases, or from the ongoing study, and are often externally imposed on the planners.

Certain matters of fact can be recognized at the outset as unalterable boundary conditions, whereas others are the result of decisions. This second type of boundary condition is subject to the dictates of fitness for purpose (as with the establishment of organizational, spatial, and other system boundaries). On the one hand, they must be justifiable on the basis of logical argument; on the other hand, their effects on the course of systems creation should not run counter to the purpose of the system. It may also be necessary to call the prescribed specifications of the client into question if they interfere with finding a truly satisfactory solution (see Sect. 3.2.4.3 Anticipation or Regress (Repetitive Cycles, Iterations)).

Some boundary conditions influence the demarcation of the problem area; others come into effect as limitations to the scope for a solution during the solution search.

In addition, there are limitations or existing conditions of a less imperative nature, which have to be taken into account later, when objectives are formulated and solutions sought, for example:

– Institutional regulations to be observed:

  Organizational guidelines
  Guidelines for executing projects
  Procedures for loan approval

– Available personnel resources
– Company standards that have to be complied with, etc.

In addition, there are self-imposed restrictions (either conscious or unconscious) that influence the assumed or available design freedom. These include issues such as the extent of the desired changes and the design principles to be used, in addition to technical and methodical know-how and the ethical attitude of the planners, designers, system architects, and concept designers.

## 6.1.6   Openness Toward Objectives, Neutrality Toward Solutions and Transparency

The guiding principle in performing a situation analysis is to render, as objectively as possible, an interpretation of a situation that is supported by facts, is open in regard to objectives and neutral in regard to solutions; and then to present this in a transparent manner. Later decisions, particularly those of a fundamental nature (at the end of the preliminary study or possibly the main study), require this solid base.

Situation analyses cannot meet these requirements if there is a fixed view of the problem, or if the client and/or the planners are biased. By taking on a predetermined and uncritical view of the problem, the contractor runs the risk of supporting a change of course toward a purely instrumental corrective action.

Suggested solutions that are developed on such a basis, possibly using specified means, may even be poor, as they build upon an inadequate or unsuitable definition of the problem.

The results of situation analyses should be transparent and easily comprehended with regard to the sources of information. Conclusions drawn from them should be clear and credible, as they represent the basis for the formulation of objectives and the search for solutions.

## 6.1.7   Techniques for Situation Analysis

The techniques to be applied can be divided into techniques of:

– Acquiring information regarding present, past, or expected conditions
– Processing information
– Presenting information

In particular, techniques for presenting information aid the assessment of past, present, and future situations by making circumstances, hypotheses, and approaches to solutions transparent.

There are a variety of techniques that may be applied to different steps within the PSC. They are briefly outlined here, particularly with regard to their relevance in connection with the situation analysis. For further descriptions, we refer the reader to Part VI, Methods, Techniques and Tools (Chap. 15), and the literature cited therein.

### 6.1.7.1  Techniques for Acquiring Information

These include in particular[1] ➔ interviews, in which people are questioned who one feels are capable of supplying information regarding present conditions, desires, and future developments, or even solutions. Depending on the kind of questioning, a distinction is made among standardized, nonstandardized, and semi-standardized interviews.

Similar functions are fulfilled by a ➔ questionnaire, which has the advantage of being less expensive in the case of extensive surveys. On the other hand, a questionnaire usually yields less differentiated information and impressions.

So-called ➔ observation techniques are applicable only when conditions can be detected and evaluated by visual inspection. A ➔ work sampling study is an observational procedure based on statistical sample surveys, which requires little effort.

A roundtable, for example, in the form of ➔ brainstorming can provide the comments and ideas of several people simultaneously. In this instance, reciprocal information and proposals are especially valuable. The ➔ card technique achieves a similar result.

➔ Polling panels or ➔ Delphi surveys make it possible to obtain information regarding expected future conditions or developments by interviewing different experts. The ➔ scenario method pursues a similar goal.

➔ Prediction methods, such as ➔ correlation or ➔ regression analyses, can support an estimation of possible future developments, insofar as the appropriate data are available.

The ➔ analogy method enables the transfer of insights and findings from unrelated areas.

Internet research, supported by efficient search engines (Google and others), is becoming increasingly important. Here, however, the volume and credibility of the information and data can increasingly become a problem.

➔ Checklists can support or regulate the acquisition of information.

Information can frequently be derived from existing documents and notes, i.e., secondary surveys.

---

[1] An arrow in the text ➔ refers to a corresponding key word in the encyclopedia, Chap. 16. The particular method/technique/tool is described and explained there.

As the acquisition of information can be very expensive and sometimes not particularly productive, it is advisable to develop an ➜ information acquisition plan that is adapted to each problem.

### 6.1.7.2  Methods and Tools for Processing and Representing Information

Here, one must of course first name all the methods and techniques mentioned and described in connection with systems thinking, such as:

- *Black box illustration* for reducing complexity by limiting observation, for the time being, to effects and inputs and outputs
- System illustrations (elements, relationships, boundaries, environment, etc.), for example, in the form of *bubble charts*
- *System-hierarchical* tree diagrams in terms of supra- and subsystems
- Implementation of the idea of different *system aspects* (observation through different lenses)

In addition, there are a multitude of methods and techniques borrowed from different specialized fields that are capable of rendering deeper insights:

So-called ➜ ABC analyses enable one to target focal points, as in the Pareto principle (or 80:20 rule; for example, 80% of the difficulties are caused by 20% of the cases).

➜ Process diagrams, ➜ process analyses, etc., create insights into concrete processes and possibly related flaws or bottlenecks.

Matrix illustrations, for example, in the form of cause or influence matrices or allocation matrices, make it possible to structure a set of facts clearly and establish them in mutual relationships.

Charts, matrices, ➜ and histograms enable a visualization of the facts and circumstances by subdividing and arranging them.

In addition, every kind of plan, drawing, diagram, or chart is obviously a form of illustration and thus facilitates analysis.

Therefore, spreadsheet programs simplify the manipulation of charts (correction, addition, expansion, reduction, etc.) in a convenient manner.

➜ Key indicators of various types compactly describe existing or desirable conditions and enable ➜ benchmarks or plausibility reflections.

➜ Polarity profiles can be used to characterize situations in respect of fulfilling or developing important properties.

So-called ➜ security-management, ➜ risk analyses can not only be implemented by assessing solutions, but also applied to existing conditions.

The methods of ➜ correlation- and ➜ regression calculations not only serve the procurement of information, but they also illustrate the information by characterizing its actual or suspected development.

The IT field offers modeling techniques, such as ➜ UML or SysML, which can also be used to show actual situations.

### 6.1.7.3  Types of Information Procurement/Acquisition

*Concerning the Degree of Detail*
The phase model recommends a process from the general to the particular. Of course, this is also pertinent to information procurement. In practice, however, it is not always easy to find the right degree of detail.

The extent and degree of detail of the information to be procured and the requisite expenditure are in fact correlating factors. Therefore, one should take as a guideline the motto: as little as *necessary*, i.e., as little detail as necessary instead of as much as *possible*, i.e., as much detail as possible.

This can, however, produce contradictions. If one is satisfied by working with relatively general and global information in the preliminary study (which makes sense), one might consequently have to ascertain, in more depth and detail, the same content during the later phases. This may cause unnecessary overheads and test the patience of those persons who (have to) serve as information sources. Therefore, it is quite reasonable in particular cases to gather the required information about a circumstance only once, if possible, and then in detail. It is easier for experienced individuals to recognize and assess this correctly.

*Primary or Secondary Sources?*
Information or records from *primary sources* are those that are generated in the context of surveys (oral, written), observations or specific inquiries with regard to a concrete investigation or study. They usually make it possible to receive more precise answers to the questions posed. The drawback is that they require more expenditure.

Hence, it is advisable, before beginning a larger inquiry, to ascertain whether or not the desired answers could be derived, at least partially, from existing records. These may have been produced for any number of reasons and without any connection to the present purpose; they are referred to as *secondary sources*.

Therefore, one should ask oneself not only what information is needed to what degree of detail, but also how best to get this information with the least amount of expenditure and yet with sufficient reliability. Before the start of information gathering, it is generally worthwhile to hold an exploratory talk with those persons who are familiar with the problem – and possibly also the solution field – and who can help with creating an ➔ information procurement plan. They will be all the more willing to cooperate if one is able, as one should be, to justify why and for what the information is needed.

Information sources can be rendered accessible by means of an extensive survey (for example, over a certain period of time) or in the form of random sampling. By and large, extensive surveys are considered, for reasons of expenditure, only when the records have been stored in an IT system. Otherwise, samples have to suffice, whereby a random selection must be guaranteed.

Given the options of today's standard software for evaluating data (for example, spreadsheet programs), it is recommended to compile primary volumes as far as possible (for example, in the industrial sector, annual consumption, price, inventory per item) and only then, in a second step, to calculate the derived values (for example, revenue per item, inventory range, etc.). This enables data to be evaluated in a more nuanced manner, whereas the primary volumes reveal regularities that can no longer be reconstructed from the derived values. In the most general sense, multi-dimensional evaluations and illustrations yield deeper insights.

## 6.1.8  Procedural Steps in a Situation Analysis

### 6.1.8.1  Use of Working Hypotheses

Experienced planners enter into situation analyses with a kind of premonition. It would be reasonable for them to articulate this premonition in the form of working hypotheses – thereby rendering it verifiable. Often elements and relationships, cause–effect relations, environmental influences, etc., can initially be shown only qualitatively (for example, graphically, by arrows) and, at best, justified as plausible, but not proven in a concrete sense.

If one observes two principles with this procedure, it will be more effective than it first appears:

- Drawing up graphic illustrations forces one to reveal one's own opinion about problem fields, their elements and relations, and so make the opinion a subject for discussion. Arguments can then be made for or against it, and one is stimulated to search for the facts.
- One's basic attitude has to be nondogmatic. The point is not to prove hypotheses, but to view them as opportunities, to substantiate, and thereby to test them. If other structures (factors and relations) prove to be more plausible because of new information, the original hypotheses must of course be discarded or corrected.

### 6.1.8.2  Action Steps

We order the previous reflections according to a simplified sequence of steps. Certain steps may of course be skipped in a specific case when they prove to be unnecessary, or the situation is clear anyway, or at least looks that way at the outset.

*1. Analyzing the starting position and the scope of the task*
First, it has to be worked out how the originator of the project or the commissioning party perceives the situation and which expectations they envision for the solution. It may also have to be taken into account that the targeted problems are merely symptoms in a more comprehensive problem field. This could lead to an expansion of the area to analyze and, as a consequence, of the scope of the task as well. In particular, it concerns the following tasks, by way of example:

- Crystallizing important aspects of the initial situation (difficulties, inadequacies, chances, etc..)
- Looking at the impulse for engaging in the problem
- Discovering models or ideal concepts

- Clarifying unclear terms
- Clarifying free spaces for development
- Identifying processes and contact persons for procuring information

*2. Roughly structuring and delimiting the baseline scenario*
Information collected during initial inquiries makes it possible to roughly structure the baseline scenario and its surroundings. This can result in a problem map that reveals, in the sense of a working hypothesis, the problem field, its system-building elements and relationships, and the relevant surroundings and their boundaries. Thus, the area of investigation is defined. Discussing the problem map and the problem boundary with the commissioning party will prove expedient.

*3. Taking analysis deeper*
The task here is to collect and structure concrete facts and data, using the different techniques for procuring, preparing, and representing information.

This not only creates an information base for a quantitative evaluation of the situation, it also lays the foundation for a correction or refinement of the previous working hypotheses. The previously mentioned methods of approach (system-, cause-, solution-, time-, and development-oriented) create a deeper understanding of the problem and more refined, meaningful structures.

*4. Delimiting the area of intervention*
The area of intervention or design must be defined. Which parts of the problem field should and may be changed? With what intention? What are the expected effects? What are the existing approaches to a solution? This step, too, should be agreed upon with the commissioning party.

*5. Consolidating the results*
Essential core assertions, such as determinations of the actual condition, articulation of established problems, difficulties, opportunities, expected developments and the rationale for them, tendencies in the solution field, etc., should all be consolidated, with the help of graphic illustrations, charts, etc.

This step generates a solid basis, both credible and confidence-inspiring, for the formulation of objectives and the search for solutions.

*6. Iterative procedure*
Of course, these steps cannot be traversed in a strictly linear sequence. Instead, as demanded in each case, they must pass through the appropriate feedback and repetitive cycles.

*7. Documentation*
Results of a situation analysis must be documented so that other planners, commissioning parties, participants, or persons concerned can reproduce and continue to use the hypotheses, conclusions, and calculations made. These results will facilitate the work of the planner when he has to reopen the facts and circumstances. Documentation is the basis for monitoring the development of essential factors that describe the problem or solution field. Moreover, it is easier to manage personnel changes in the processing teams when work results have been retained.

## 6.1.9    The Varying Significance of Situation Analysis in the Phase Sequence

Because various project phases have different tasks, there is also a change in the scope of the observed area and in the degree of detailing where existing or obtainable information potential has to be fully developed and processed. Therefore, in a situation analysis, the main purpose, scope, and degree of detailing vary widely during the different phases of a project.

In a *preliminary study*, the boundaries of the problem and solution field and of the intervention area are, for the most part, more open to choice; in a *main study*, the selected solution principle already imposes restrictions; and in *detailed studies*, the boundaries are already defined relatively precisely (the overall concept and master plan are already established).

This successive narrowing is, conversely, linked to an expansion: as a rule, there is an increase in knowledge about the problem, about the solution possibilities, and about opportunities for intervention (Fig. 6.8).

The increase is especially significant in the preliminary study and levels off in later phases. The preliminary study is therefore critical for system boundaries, the formulation of objectives, the selection of the solution path/architectural design, and the compatibility of the solution with its surroundings. Thus, situation analysis is of particular importance during this phase.

During later phases, its focus is on pursuing (monitoring) the development of the expectations for systems design that have been generated by the preliminary study. It also concentrates – if need be – on a deeper procurement and processing of information for each of the successive steps of the PSC.

The curve for possible interventions lies below that of the known problem-solving possibilities (Fig. 6.8), as usually, not all solution possibilities can be applied or executed in a specific case.

*Preliminary studies* concentrate heavily on external information pertaining to the problem field, the solution field and its environment, expectations, and prospects for the future, in addition to opportunities and risks. Information procurement in the *main and detailed studies*, in contrast, are more strongly oriented toward solutions and means.

A solution strategy must be applied principally in the preliminary and main study, where it has an impact on the areas examined in a situation analysis. In a *new systems concept*, it generally makes little sense to engage in detailed structural aspects of the actual condition, as this structure has to be changed fundamentally in any event. Of special interest are the overall effects of the existing system on external factors. In a *systems melioration* (improvement), by contrast, the internal effect mechanisms have to be treated much more closely, because a number of these remain unchanged.

As some problems have already been solved repeatedly in a similar fashion, they are therefore known through one's own experience, through publications, or through the experience of externally consulted experts. For these problems, one can fall back

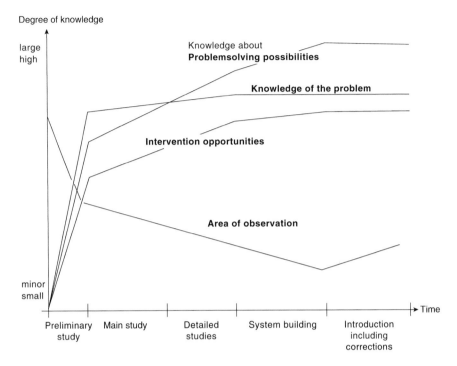

**Fig. 6.8** Change in the degree of knowledge over the course of the development of a system

on checklists, information procurement plans, and other records to determine the type of information that is to be collected and the plan of action. In such cases, it is also conceivable that a problem- and solution-oriented collection of information can be partially combined in the interests of an efficient method.

If the concern is not with routine problems, it is recommended to think about the situation analysis independently without relying on available records. If models are available, these should be critically screened before being adapted to one's own problem. The same applies to information procurement.

Situation analyses exhibit the largest scope during the developmental phases (preliminary, main, and detailed studies). However, if a plan is marked by a high degree of innovation, problems may yet arise in the realization phase (establishment and introduction of the system) as well, which makes it seem reasonable to implement the PSC.

### 6.1.10  Summary

1. A situation analysis consists of a systematic screening and representation of a situation that is intuitively perceived as problematic or rich in opportunities, or of a set of facts that are given in the mandate. Here, the spectrum ranges from a vague discomfort all the way to a specific explanation or definition.

2. A situation analysis does not merely record the actual condition. An unstructured, aimless, and too-detailed involvement in existing or even past circumstances may actually hinder rather than promote an understanding of the problem and the search for solutions.
3. A situation analysis is first of all concerned with compiling accounts about the current condition. Thus, it is intended to discern the differences among the actual condition, prospective developments, and a vague target condition, and thus help to define the problem.
4. The purpose of a situation analysis is:

   - To structure and examine the problematic area (system and environment) with the objective of discerning the "correct" problem, understanding it better, and being able to work on it
   - To demarcate the area of intervention for the necessary measures
   - To create an information base for the subsequent steps of formulating objectives and developing solutions, for reviewing possibly pre-existing problems or tasks, and for revising ideas about a target condition that may already have been established

5. Four typical approaches are applicable in a situation analysis:

   - A *systems-oriented* approach whose attention is directed to a better understanding of the system that is to be developed or changed
   - A *cause-oriented* approach whose immediate focus is on working out in a concrete manner the symptoms and deficiencies of an existing condition. Once these are relatively clear, questions should be asked about their causes. This too should be made an unambiguous matter within the project group. It is only when these prerequisites have been met that it makes sense to speak of appropriate measures capable of rectifying the causes.
   - A *solution-oriented* approach allows one to engage in solutions as early as during the situation analysis. However, this should not be taken in the sense of arriving at specific solutions. The aim is merely to become familiar with the state of technology, with what is possible or impossible, and to create an appropriate problem awareness in those persons who only know the present situation and therefore do not have any notions of a better one in the future.
   - A *time-oriented* approach should focus on developments in the problem field and those in the solution field.

6. A situation analysis should be:

   - Forward-looking and environment-oriented
   - Open with regard to objectives, solutions, and applicable resources
   - Serviceable in character for the successive procedural steps of formulating objectives and developing solutions
   - Capable of distinguishing facts from assumptions and opinions, and comprehensible or sufficiently secure with regard to information sources, information processing, and conclusions

7. A situation analysis contains descriptions and illustrations with sufficient information for:

   – The boundary of system and environment (with respect to the problem field, the solution field, and the area of intervention)
   – The structures and processes, the functioning methods, relationships, element properties
   – The deficiencies, problems, causes, and influencing factors (for example, in the form of a plausible, comprehensible catalog of weaknesses and strengths)
   – The developmental tendencies of relevant influencing factors in the problem and solution field
   – The risks and opportunities (in the form of a catalog similar to the catalog of weaknesses and strengths)
   – The interest in a project. Different interests are often due to the different persons representing them, frequently because of their different functions. Although different interests are permissible, they should be transparent and thus open to discussion and manageable
   – The possibilities for intervention
   – Restrictions that must be observed, freedom for design, and other realities
   – Possible approaches to a solution, also looking at the desired effects, drawbacks, and side effects
   – An assessment of the situation and possible ideas for a solution offered by participants and persons concerned (as a basis for a joint understanding of the problem)

8. Finally, a situation analysis contains a comprehensive representation of the problem and documentation, in addition to references to those aspects in the system and environment that should be subject to monitoring during the further phases of treatment.

### 6.1.11   Self-Check for Knowledge and Understanding: Situation Analysis

1. What is the purpose of a Situation Analysis? Is it equivalent to a review of the current situation?
2. Which views in the situation analysis do you know and consider characteristic and important?
3. How are detected weaknesses or failures, conceivable causes, and measures to resolve the problems mentally and logically interconnected?
4. Is it accepted to discuss solutions as early as in the situation analysis?
5. Is it possible that there are different boundaries one has to observe within a special task?
6. Which techniques and tools do you know about for situation analysis?

7. Does it make sense to use a working hypothesis in the situation analysis? Does this not mean that one works with prejudices?
8. Does the situation analysis have the same significance in all phases of a project?

## 6.2   Formulation of Objectives

### 6.2.1   Purpose and Terminology

A formulation of objectives is initiated by the question, "What is to be achieved or avoided?" The answers one may receive to this question are called objectives; for example, "Cost effectiveness or performance should be increased, hazardous emissions should be reduced," etc.

#### 6.2.1.1   Objectives

Objectives are statements about what a solution should *achieve* or *avoid*.

Formulating objectives is of great importance to the problem-solving process. Objectives are meant to govern the search for a solution; they should not be invented retroactively to justify a solution. To fulfill their purpose, they must be formulated and made known to and accepted by the participants in the problem-solving process.

Objectives are not self-evident; they must first be worked out. It therefore makes sense to apply methods and techniques that support the process of finding and formulating objectives.

The following sections are primarily concerned with the requirements for well-formulated objectives and offering practical advice.

#### 6.2.1.2   Place in the Problem-Solving Cycle

Figure 6.9 shows the place of "formulating objectives" in the PSC and its relation to other steps.

*(a) Relation to situation analysis:*
As ideas about objectives already occur when the impulse for a project is given, objectives often are already contained in a rough formulation of the project mandate. Especially during a situation analysis, important impulses occur; when the participants recognize deficiencies, difficulties, or opportunities, their ideas about possible and desirable changes also become more precise.

In addition, a solution-oriented perspective during situation analysis provides suggestions about role models/paragons or the state of technology that could or should influence the objectives.

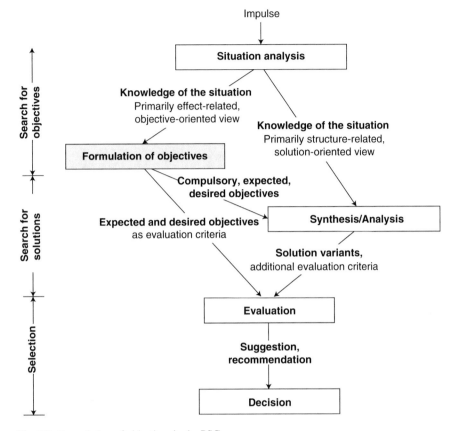

**Fig. 6.9** Formulation of objectives in the PSC

*(b) Relation to the search for solutions:*
In the problem-solving process usually various people are working together. The their activities are advantageously aligned by using explicitly stated and broadly accepted objectives.

Different expectations frequently play a role concerning solutions, and usually different interests exist. It is therefore reasonable to clarify conflicts about the objectives before beginning to search for solutions. Failing to do so only leads to a false peace of mind. Problems arise at the latest during the evaluation, because divergent expectations cannot normally be met all at the same time.

*(c) Relation to the steps of evaluation and decision*
Decisions about possible solution variants have to be made at the end of the PSC. Objectives are the basis for developing assessment standards (criteria) to compare and evaluate different variants. The saying of Marcus Aurelius, "For someone who does not know his harbor, no wind is favorable," means in respect of a project: if I do not know the objectives of a project, I have no standard for assessing solutions.

An important principle of rational decision-making is to prefer those solutions that meet objectives as far as possible. Therefore, in formulating objectives, one should summarize already expressed expectations or objectives and systematically work on those that may be imprecise, unclear, or contradictory; this means to supplement and structure them, to check them for completeness, and examine them for contradictions. The objectives should count for all stakeholders – the project team and the commissioning party – as an obligatory/mandatory basis for further action.

## 6.2.2 The Formulation of Objectives at Different System Levels

As the problem-solving process traverses different concretization levels in accordance with the systems engineering procedural principle "from the general to the detail," objectives have to be worked out and dealt with at different levels. Objectives for partial concepts are derived from each of their higher-level concepts or supplemented by additional considerations. Thereby, the objectives become increasingly more concrete, that is, they contribute increasingly as building blocks of the overall concept. Here, higher-level concepts help to give orientation to a more detailed design.

## 6.2.3 Mental Approaches, Principles, and Guidelines for an Action-Oriented Formulation of Objectives

The requirements described in the following are characterized as *principles* that should be observed when formulating objectives. They can also be regarded as quality criteria for well-formulated objectives. As this view is detached from concrete tasks, only the formal quality can be evaluated.

### 6.2.3.1 Operational Formulation of Objectives

Unclear or imprecise requirements, such as "improve the personnel situation" or "simplify operating procedures," should be regarded merely as impulses for directing one's thinking and subsequent action. They need to be made more concrete, that is, specified more clearly. For what is meant by *personnel situation*, or *improve*, or *simplify*?

One refers to an *operational formulation of objectives* when the formulation is *intelligible* and its *achievement discernible*. It has the following features:

- It names the *target subject* that is ready for change or a new design. To WHAT are the objectives tied? Which subject is to be changed or designed? Frequently, the beginning of a preliminary study merely speaks of a "solution" or a "system."

– It formulates the *qualities* or the *substance of the objectives* pertinent to the object. WHAT is to be achieved or avoided (for example, reduction of pollution, increase in range, etc.)?
– It says something about the *degree* to which these qualities are achieved. HOW MUCH is to be achieved (for example, at least 25%)?
– It is clear about the *temporal aspect*: WHEN do the objectives have to be achieved (for example, within 1 year)?
– It addresses the matter of *location*: WHERE should the desired effect occur or become discernible? (Within the system that is to be influenced/designed, or outside of it, or both?)

The following possibilities exist for making objectives operational:

1. Indication of a clearly *factually stated condition*: only if the solution possesses this quality can a subobjective be counted as valid, for example:

   – The solution should not require any structural changes.
   – The solution should be realized by (date).

2. Indication of the *target direction, without restriction* (open formulation of objectives): Only the target direction is stated but no quantitative criterion, for example:

   – Profitability (return on investment, ROI) should be as high as possible.
   – Amount of investment should be as small as possible.

3. Indication of the *target direction*, with a *restriction* (restrictive formulation): a scale is used (such as monetary units), a target is indicated (as low as possible), and additionally a limit (a restriction) is formulated (the maximum amount tolerated …).

For example, the investment should be as low as possible, but at the most …

Of course, there are cases in which the achievement of the objectives is not readily discernible, much less measurable, unlike the cases above. This sometimes leads to the temptation of omitting the objective altogether. But we think that this is neither necessary nor useful because most of the time it is quite possible to define *replacement measurements* or indicators that ascertain the degree to which objectives are fulfilled.

For example, to specify the objective of a user-friendly solution for a machine, a replacement measure could be specified as the training of a technician for a maximum of 1 day to learn how to use the tool. The time limit would be determined and later assessed by an expert (technician).

We advise that an important objective should not be excluded if it cannot be measured. It gives information about the values of the stakeholders, what seems to be important for them, even if they cannot or do not want to name this precisely.

### 6.2.3.2    Formulation of Desirable or Expressly Undesirable Effects

Objectives may consist of both the *achievement of desirable* and the *avoidance of undesirable effects*.

*Desirable* effects might be, for example, low acquisition and/or operating costs, high profitability, high performance, high flexibility, long lifespan, great user-friendliness and acceptance, easy changeability, etc.

*Undesirable* (negative) effects might be, for example, high acquisition costs, high operating costs, major noise pollution, limited mobility, flexibility or expand-ability, etc.

Neither a maximum achievement of all the desired effects nor a complete avoid-ance of all the undesirable ones is realistic. Instead, an acceptable limit can be set, above or below which a solution should not pass (for example, operating costs a maximum of … \$/€, performance minimum …kilowatt, etc.); see also Sect 6.2.3.10 (The Principle of the Freedom from Contradictions in Subobjectives).

### 6.2.3.3    Difference Between System Objectives and Project Course
###        (Procedural) Objectives

The question "What should be achieved or avoided?" pertains especially to what the new system, when it comes to be utilized, is intended to effect or avoid. Therefore, we speak of *system objectives* or design objectives.

However, it is just as reasonable to formulate important objectives that have an impact on the path to project completion and thus serve as a foundation for estab-lishing a project plan. We refer to these as project course or *procedural objectives*. Formulations of these kinds of objectives are concerned with intermediate goals that must be adhered to (for example, the deadline for completing the main study), resources of a financial nature (for example, the project budget that has to be adhered to) or of a personnel nature (for example, those persons who absolutely must be included/made to participate).

### 6.2.3.4    Structure of a Catalog of Objectives

Because in most cases a target object is required to have several necessary charac-teristics (effects/demands/features), there are usually not just one but several sub-goals involved.

An organized compilation is called a package or a catalog of objectives (the example in Table 6.1 shows an excerpt from a catalog of objectives for an efficiency-improving project – for didactic reasons shown in a more formally correct manner than is usual in practice). A generally valid catalog of objectives, which can be used as a checklist, is found in Table 6.3.

**Table 6.1** Example of a structured catalog of objectives

*Catalog of Objectives*

| Project: | Improving efficiency in purchasing |
|---|---|
| Phase: | Detailed study |
| Target subject* | Ordering process |

| Objective class | Standards | Condition/ restriction | Priority | Remarks |
|---|---|---|---|---|
| *Financial objectives* | | | | |
| Cost-effectiveness | High savings in purchasing | At least 10% | R | |
| Impact on liquidity | Lowest investment possible | 200,000, – maximum | R | |
| *Functionality:* | | | | |
| Performance/functionality | Reduction of ordering time | 2 days maximum | C | |
| Security/reliability | Reduction in error frequency | At least 50% | R | |
| Capacity/stability | Capacity number of orders per day | At least 400/day | R | |
| Expandability | Number of orders per day | Up to 600/day | D | |
| Interface requirements | Linkage to accounts receivable | | C | |
| Ease of maintenance | Time for repair of disorders | 4 h maximum | R | |
| *Personnel objectives* | | | | |
| Required employee qualifications | Time for training current personnel | 1 day maximum | D | |

*C* compulsory objective: condition/restriction must be observed absolutely (killer requirement)
*R* recommended objective: objective is important, but not absolutely necessary
*D* desired objective: adherence desired ("nice to have")
* = subject matter to which the objectives are attached

### 6.2.3.5   Thinking in Terms of Objectives and Means

Action-relevant statements about a desirable target condition may be illustrated by multi-level hierarchical structures. Figure 6.10 shows an example. The mentioned means are possible ways/solutions by which the relevant objectives can be achieved. These means can relate to one another in an "and" or an "or" relation.

From the perspective of two consecutive levels, the statements on the top level indicate objectives, and the statements on the lower level indicate means or measures. This makes it clear that the term *objective* is used relatively in respect of a certain level; thus, a statement is not simply an objective or a means, but it can be both an objective and a means depending on the level from which it is observed. Sometimes, the term *purpose* is used for the top level to characterize three hierarchical levels.

This approach is important with regard to the following:

– A representation of several levels of means and objectives shows causal connections that put subordinate objectives into a larger overall context.

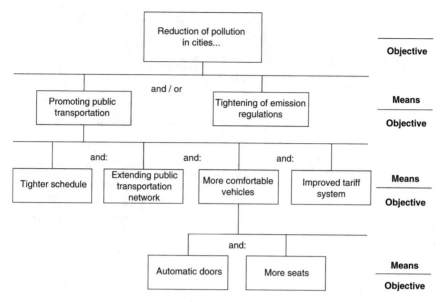

**Fig. 6.10** Hierarchy of objectives and means

- Once a means on a particular level has been determined, its realization becomes an objective. The overall causal connection can be ignored for the time being. This reduces complexity.

To determine the current position in the hierarchy, two simple questions can be asked:

- The question WHY? Or FOR WHAT? Points upward in the objective–means hierarchy (i.e., it points to the objective).
- The question HOW? Or WITH WHAT? Points downward (i.e., to the means or the measure).

This important fact is further elucidated by the phase concept. The relationships in Fig. 6.11 reveal the following.

The target subject of the *preliminary study* appoints the place/area, where improvements should be achieved (here: air pollution). The target characteristics represent the desired effects of the as yet unknown solution (here: reduction of pollution by 50%, without, for reasons of simplicity, indicating which pollutants are meant). In the preliminary study, different variants of solution principles are worked out, such as promoting public transportation or tightening emission regulations.

If the solution principle *promotion of public transportation* is chosen, it subsequently becomes the target subject of the *main study*. The target characteristics are the requirements for the solution that can now be formulated more concretely, precisely, and completely.

| PHASE | Target subject | Target characteristics / Target contents (requirements to be met by target subject) | Means/solution variants (naming the variants) | Remarks |
|---|---|---|---|---|
| PRELIMINARY STUDY | air pollution (solution yet unknown) | Reduction of pollution by 50 % ............ | Promoting public transport or Tightening of emission regulations | The variant promoting public transport was chosen |
| MAIN STUDY | Promoting public transportation | Expansion of capacity by 50 % Maximum additional burden on annual municipal budget ....$ | Rough, overall concept with the components: → Expanding public transportation network → New tariff system → New vehicles | A particular overall concept was chosen. One of its elements is: New vehicles |
| DETAILED STUDIES | New vehicles | More seats Less noise for passengers and environment ... | | |

Fig. 6.11 Thinking in terms of objectives and means in the phase process

The overall concept developed in the main study describes solutions by working out solution components and their interactions. These components now become target subjects of various detailed means. Again, the target characteristics are the requirements for the solutions, for example, new vehicles as one component of the solution of the main study become the target subject of a *detailed study*.

This hierarchy of objectives and means is also suitable as a model for linking statements about target conditions at different planning levels.

### 6.2.3.6   The Principle of Orientation on Facts and Values

This principle, which could also be called the principle of limited objectivity, signifies that objectives are always a combination of facts (nonsubjective) and values (subjective; Fig. 6.12).

Facts may relate to the results of a situation analysis and the detected deficiencies and their apparent causes, to possible threats, and to opportunities, etc. (for example, facts about role models, the state of technology, best practices in one's own or

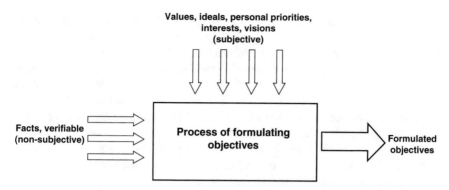

**Fig. 6.12**   Orientation of objectives to facts (nonsubjective) and values (subjective)

other industries). These *facts*, however, are merely guideposts for the formulation of objectives, but are not yet objectives per se. It must still be established to what extent the current situation can be changed, what viable opportunities can be exploited, how much money and time one is willing and able to expend to achieve the objectives, etc. These influences are of course to a significant extent driven by values and therefore subjective.

Because formulating objectives, in terms of combining facts and values, can be a very complex process, a complete analytical penetration is eluded. We therefore have to be content with a minimal requirement: because the process of formulating targets cannot be made sufficiently transparent, it should at least be attempted with regard to the results, the formulated objectives. Observing the principles introduced here does not make discussions arising from different assessments and value judgments pointless; these principles should merely enable more precise communication.

### 6.2.3.7   The Principle of Neutrality Regarding Solutions

Objectives are not meant to describe the solution, the HOW, but the WHAT, i.e., the desired and/or expressly undesirable *effects* of a solution as yet unknown.

The following formulation of objectives accords with this principle: reduction of pollution, no unreasonable limits put on mobility, reduction of noise pollution, and the smallest possible financial burden on the municipal budget.

A formulation of objectives like this allows various solutions, such as promoting public transportation, tax breaks for eco-friendly vehicles, and legal regulations for industry, business, and households to reduce air pollution.

At the same planning stage, the following formulation would not be neutral with regard to a solution. The objective of the project is "to restrict private automobile traffic through tax measures." This would be only one of several possible solution variants. The previously described hierarchy of objectives and means can lead to such an understanding.

A formulation of objectives that is suspected of unduly limiting the area of solution should be critically scrutinized and replaced by a formulation that is focused on the desired effect. The question of WHY or FOR WHAT can help formulate such a solution concept. Such questioning mentally elevates one's reflections to the next higher level of the objectives–means hierarchy (see Sect. 6.2.3.5).

Because it is also possible to question an objective that is more neutral in its formulation ("Why reduce pollution?" Possible answer: "To avoid health problems"), the question arises, when should questioning be discontinued? For this, there are no objective criteria. But a good indication for terminating the questioning process might be when those participating in the process of formulating objectives think that an adequate consensus has been reached and no further substantiation is necessary. In other words, the formulation of objectives has not unduly excluded any solution that could be intuitively accepted.

### 6.2.3.8   The Principle of Completeness for the Content of Objectives

Solutions that are methodically worked out generally are based on multiple types of objectives. A generic catalog of types of objectives is depicted in Table 6.2 and should be taken as a checklist for the classes of possible objectives and their organization.

**Table 6.2**   Generic catalog of objectives

| |
|---|
| (a) Financially relevant objectives |
| Requirements of cost-effectiveness: as a rule, relation between expenditure: profit, ROI |
| Objectives that comprise costs and financially measurable revenues. They are expressed, for example, by key indicators such as: running costs, cost savings, payback periods, etc. |
| Impact on liquidity: for example, amount of investment, financing from cash flow, no outside funds |
| (b) Functionally relevant objectives |
| Performance or functionality of a system: for example, output per time unit |
| Safety |
| Quality |
| Flexibility, for example in respect to |
|   Coverage of short-term load peaks or |
|   Medium- or long-term opportunities for expansion or reduction, adaptation to changing demands, etc. |
| Interface design to be able to link a system with one or more others |
| Service and maintenance aspects |
| Autonomy versus dependency and much more |
| (c) Personnel-relevant objectives |
| All objectives that contain desired or undesired effects on personnel, such as: |
| Ease of operation, ergonomics, working conditions |
| Personnel qualifications |
| Nondependence on personnel |
| (d) Social objectives |
| Objectives directed at the observance of ecological effects (environmental pollution, waste disposal) |
| Objectives relating to the personal acceptance of solutions (user, operator) in addition to |
| Objectives of a general social nature, etc. |

This list shows that the frequently emphasized principle of holistic thinking should naturally also have a bearing on the formulation of objectives. It even finds its essential source there! The principle of the completeness of objectives can also be understood in the sense that all important information and interests have been taken into account or at least deliberated.

With regard to *information*, it especially concerns:

- Information about deficiencies, difficulties, risks in the problem field
- Information about presumed chances and opportunities
- Superior objectives (e.g., superior-ranking concepts, corporation goals)

In respect of *interests*, the following must be kept in mind: in accordance with the rule of fairness and prudence, deliberations about achieving positive effects or avoiding negative ones should include the interests of those persons who are positively or negatively affected by a solution. A solution that favors one-sidedly the interests of certain groups and neglects others is seldom a good solution.

This is not only significant for its ethical aspects, but entails a quite pragmatic component as well: the larger and more influential the group of people whose ideas about objectives are treated marginally or neglected, the greater the risk that they will not accept the later solution, that they will fail to support it, and boycott or even fight against it.

The following count as interest groups (stakeholders) who articulate their wishes directly or indirectly: management, customers (internal/external), staff (involved parties, participants), departments and their staff (users, executors), suppliers, society, the public, the state, etc.

After the step of formulating objectives is completed, an exhaustive catalog relating to the target subject becomes available. The catalogue lists the content that is important for the formulation of the objectives.

### 6.2.3.9   The Principle of Priorities in the Formulation of Objectives

Prioritization should stress the relative importance and discipline with which objectives should be adhered to. This consideration figures prominently again during the assessment stage because by giving different weights to single criteria (derived from objectives), one can invest them with different meanings.

In a formulation of objectives, it usually suffices to use the following three classes of objectives:

1. *Compulsory objectives*: a condition absolutely must be adhered to. A solution that fails to do so is useless. If a variant fails to fulfill only one of these objectives, it has to be discarded.

    For example: it is COMPULSORY for the solution to be achievable without legislative amendments; the annual budget strain MAY NOT exceed …$/€; $CO_2$ content MUST be reduced by at least 25% with regard to the actual condition.
2. *Recommended/main objectives*: Objectives of great significance but not compulsory. Hereby, the planner/developer gets the signal that in developing a solution,

he or she has to pay special attention to these characteristics, though without the objective being absolutely necessary.

For example: a solution without a legislative amendment is recommended. Solutions that do not possess this target characteristic are not considered unsuitable, but they are judged to be inferior to those that do. If a scale is available, a restrictive or an open formulation may be used. If a restriction is implemented, the stated value on the scale provides an important orientation for finding a solution, but adherence to this value is not an indispensable prerequisite for the suitability of a solution.

3. *Desirable/secondary objectives*: Desirable objectives may be formulated in a like manner as recommended objectives. However, adhering to them is less binding (= "nice to have" objectives). Because compulsory objectives severely narrow the field for the solution search, it is advisable to use these sparingly and only when they are clearly justifiable. Compulsory objectives should also be formulated in an operational way, i.e., the formulation should be comprehensible and it should be possible to check the achievement of the objectives.

### 6.2.3.10   The Principle of the Freedom from Contradictions in Subobjectives

The subobjectives contained in a catalog of objectives may have various relationships with one another.

1. Mutual *support*: achievement of subobjective A supports the achievement of subobjective B.

   For example: A, short response times, and B, superior user friendliness. This is an agreeable case.

2. *Independence* (indifference): achievement of subobjective A can be independent of the achievement of subobjective B.

   For example: A, low operating costs, and B, aesthetically pleasing. This is an nonproblematic case.

3. *Competition of objectives* (contrary effect): subobjective A and subobjective B oppose each other. The more A is achieved, the less B is achieved. This often occurs in connection with cost objectives.

   For example: A, low costs, and B, high quality or performance. In this case, a compromise needs to be identified.

4. *Conflict of objectives* (contradiction): subobjective A and subobjective B, either because of a logical condition or the current situation, contradict each other to the extent that they cannot exist simultaneously.

   Example of a *logical contradiction*: A, promoting market X, and B, remove product Z from the product range, even though its main sales are in market X.

Example of a *situational contradiction*: A, local product, and B, functional demands are not currently met by local products; or A, imperative functional demands, and B, unrealistic cost limitations.

Standard situations of conflict or competition can often be defused or even resolved if the search for a solutions is undertaken in an intelligent manner. These are then great innovative thrusts! For example:

– Desired functionality can be fulfilled more cost-efficiently through a different technology or simpler approaches, or for a long time, high quality and low costs were considered to be contradictory demands. However, Japanese production companies have solved this contradiction in part by re-interpreting the concept of quality and by directing attention to the customer's appreciation of quality (in contrast to not scrutinizing demands for quality at all or leaving them to the con-ceptions/declarations of engineers).
– Even for production to be both economical and flexible does not have to pose an insurmountable contradiction.
– Dynamics and consumptions used to be seen as irreconcilable elements in a car. BMW has solved this contradiction with the concept of "efficient dynamics."

The following strategies help to resolve conflicts in a catalog of objectives:

(a) *Setting priorities*, for example, declaring compulsory, recommended, and desired objectives or changing an existing set of priorities (for example, chang-ing a compulsory objective into a recommended or desirable objective).
(b) Introducing *minimum and maximum values*, for example, a minimum perfor-mance of …kW, but as economically as possible; or an upper cost limit within which peak performance is expected.
(c) *Removing/circumventing* the causes of conflicts when there are no possibilities for compromises such as in (a) and (b). Thus, one is left with two options: elevating one's deliberations to the next highest level (thinking in terms of objectives and means), for example, why withdraw from market X? Answer: the risk is too high. New question at a higher level: are there other ways of limiting risks? The second option means deleting the cause of the conflict. Thus, one of the demands is eliminated.

Figure 6.13 shows a matrix of relations among the objectives. The relations do not have to be of a logical nature, but may also arise from personal views (for example, in Fig. 6.13: the objectives "room for five people" and "pleasing aesthetics" may be in opposition only for fans of sport cars, but not for a family man).

In many cases, a contradiction in the objectives can only be recognized when no solution can be found that conforms to all compulsory objectives.

If in such case, an external, off-the-shelf solution had been sought, nothing remains to be done but to correct the objectives retroactively. In the case of self-developed solutions, however, important demands should not be too hastily modi-fied, especially when the objectives are important and there is still time available. It could be that the solutions developed so far are lacking in inspiration.

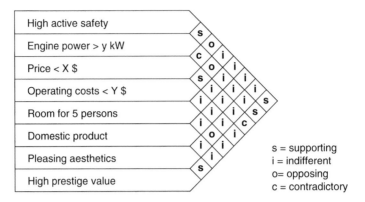

**Fig. 6.13** Paired comparisons for detecting the reciprocal influence of objectives (matrix of relations among objectives)

It is especially difficult to master conflicts of objectives when there are *opposing interests* or even *conflicts of interests*. Although an authoritative word can have a powerfully clarifying effect, it should not be the norm to lead a project team in this way. There is the danger that people will internally reject the project or even openly sabotage it. And considering that the purpose of formulating objectives is to agree upon a common direction of thought, this kind of behavior would be self-defeating.

It is therefore advisable to find a balance among the different interests and to record this in a formulation of objectives that has been jointly agreed upon. This process may take time and is usually not attainable without compromises concerning the objectives. However, such a delay is compensated for by an expedited and more straightforward planning and realization with less sand in the gears and fewer detours.[2]

However, compromises may also have a negative aspect if the solution becomes unattractive and is no longer of any real interest to anyone.

### 6.2.3.11   The Principle of a Manageable Catalog of Objectives

The more attention is given to the principle of completeness and to all important information and interests, the greater the risk that the abundance of requirements not only impairs the manageability of objectives, but may even endanger their realization. That is why mechanisms for selection and reduction must be provided. Because these have already been explained for the most part, they are only summarized here:

-   Structuring a catalog of objectives (see Sects. 6.2.3.4 and 6.2.3.8) creates an overview and order.
-   The principle of setting priorities helps to distinguish the important from the less important, for example, compulsory, recommended, and desirable objectives (see Sect. 6.2.3.9).

---

[2]A good guide to transacting negotiations by consciously distinguishing between interests and positions is Fisher et al. (1991). Getting to Yes: Negotiating Agreement Without Giving In.

- The principle of the freedom from contradictions can help to eliminate objectives.
- The principle that it must be clear whether objectives have been met (operational formulation of objectives) makes it possible to eliminate possibly trivial demands, such as those that say the same thing with different words (see Sect. 6.2.3.1).

Beyond that, one should of course always ask oneself if the mandated effects are truly relevant, that is to say, are all the objectives to avoid certain negative consequences truly that important, or do they not rather stem from an aversion to making changes and/or a fear of conflicts?

Thereby, we want to emphasize once more that formulating objectives is often based on social interactions that are concerned with factors such as mutual trust, aptitude for learning, adaptability, and openness to others, but no less with qualities such as a determination to set limits, authority, persuasive power, and much more.

### 6.2.4 Methods for Formulating Objectives

The formulation of objectives allows the use of methods that are already known from other contexts:

- Creative methods such as brainstorming, brain-writing (card technique)
- Key indicators and key indicator systems
- Standardized target catalogs, etc.

### 6.2.5 The Process of Formulating Objectives

The procedural recommendations below indicate, without prescribing a required sequence, the steps that the reflections presented so far are to follow.

1. Naming the *target subject*: at the beginning of a preliminary study, one should focus on the target subject that has to be developed (see Sect. 6.2.3.5) and avoid describing a specific solution (neutrality regarding solutions).
2. Compiling *target characteristics*: here, it is helpful to ask: what difficulties/ opportunities did the situation analysis reveal, and should they be considered target qualities that characterize the objectives? Independent of these, what further target qualities should be formulated? In doing so, what examples or models can guide us? Would it be reasonable to take this direction of thought? What potential drawbacks might be included in the catalog of objectives under the formulation "what is to be avoided"? What are the objectives for the project course?
3. Design of a meaningful *classification of the catalog of objectives* by means of the question: what are the essential topics or key aspects?

4. Producing a first *catalog of objectives*, classifying and supplementing existing target ideas. Complementing the classification if important target ideas cannot be filed.
5. *Systematic analysis of the catalog of objectives*: checking whether the above-mentioned principles have been adhered to, asking especially: is there neutrality regarding solutions, are the contents of the objectives, information, and interests complete, can the achievement of the objectives (their scope) and their operability be identified, have priorities been set (compulsory, recommended, and desirable objectives), and is there freedom from contradictions (resolution of target conflicts)?
6. *Supplementing, restructuring, and tightening* the catalog of objectives.
7. The *approval* of the catalog of objectives as a collectively accepted working hypothesis. A record must be kept of possible causal connections, such as recourse to the results and insights from the situation analysis, of thinking in terms of objectives and means, of subjective value judgments, etc.

All participants and persons concerned with the formulation of objectives should keep in mind that situations, opportunities, and value judgments, or their own opinions about all these, do not and should not constitute the final word. If reasonably argued suggestions for change come up and maybe even prove to be compelling, these will have to be subject to a renewed discussion, as unpleasant as this might be in a particular case.

However, a later change of objectives, for whatever reason, should not happen solely in the heads of individual persons, but should be discussed, justified, and jointly agreed upon or rejected – both by the team and by the commissioning party.

### 6.2.6   Restrictions

The reflections presented here are to be interpreted with common sense and they should not be regarded as binding regulations. The scope of their significance depends upon many factors, for example, in which phase of the project one finds oneself, how extensive and risky the project is, how large the proportion of work accomplished by outside partners is, which objectives are contractually agreed upon and no longer changeable, and much more.

The correct formulation of objectives in the preliminary and main study phases is usually of greater importance for the overall success of a project than the formulation in the detailed study; it is the earlier phases of the project that set the basic course for its success.

A well-thought-out formulation of objectives is especially significant for large and high-risk projects, as any error could entail a considerable expense.

If the project is worked on "in-house" to a large degree, the project staff often know what is an issue and what is important. It will also be easier to make subsequent changes, if need be, although this is not without risk and can certainly be seen negatively as well (hence one should tend towards nonbinding agreements and amendments). However, if large parts of the project are worked on "off-site," an exact formulation of objectives in terms of a functional specifications document is indispensable in most cases.

## 6.2.7 Summary

1. Formulating objectives is an essential thought and work stage in the problem-solving process. Objectives are to be developed as important *control parameters* for the solution search. At the same time, this step should serve to detect and resolve *conflicts and incongruities* concerning the expectations of several people.
2. To start the formulation of objectives, the following two questions are useful:

   - What is it hoped a solution will achieve, or avoid? (*system or design objectives*)
   - What is to be observed on the way to the solution? (*procedural objectives*)

   Both types of objectives are important.

3. A *hierarchy of objectives and means* shows not only that solutions are a means of achieving objectives, but also that realizing these solutions can itself be declared an objective for further efforts.
4. To fulfill its purpose of being a guidepost for the solution search, the formulation should observe several important *principles*:

   - Objectives are not only justifiable by *facts*, they also include *value judgments*.
   - Objectives should be deliberately formulated to be *neutral with regard to solutions*, i.e., to be effect-oriented, to avoid excluding solution ideas because of some bias.
   - Objectives should be formulated in a *problem- and action-oriented* fashion.
   - A formulation of objectives should list *all the important effects*, and possibly the features of a potential solution toward which participants in the formulation process do not wish to adopt a value-neutral attitude (principle of completeness).
   - Objectives should be formulated as *precisely* as possible, using identifiable features and, if possible, measuring scales.
   - A formulation of objectives should express *priorities* (compulsory, recommended, and desirable objectives).
   - In a formulation of objectives, *contradictions* and opposing interests that could block solutions should be made transparent and clarified (principle of the freedom from contradictions).
   - A catalog of objectives should be *surveyed* and *managed*.

5. Several methods and tools are available for formulating objectives, such as a catalog of objectives, polarity profiles, etc.

### *6.2.8   Self-Check for Knowledge and Understanding: Formulation of Objectives*

1. What is a goal, an objective?
2. How are the steps "formulation of objectives," "situation analysis," "synthesis/ analysis," and "evaluation" connected, linked together?
3. Why do compulsory objectives not play a role in the "evaluation" step?
4. What is meant when one demands that objectives have to be formulated in an operational way?
5. Why does it generally make sense to distinguish between system objectives and project course (procedural) objectives?
6. Can certain demands be objectives in addition to means?
7. Does it make sense to use the term "objective" (here: objective as the opposite of subjective) goal/target/aim/?
8. What is meant by demanding that objectives have to be solution-neutral?
9. How can we give different importance to objectives?
10. What can one do if goals oppose or contradict each other? Give examples.

## 6.3   Search for Solutions: Synthesis/Analysis

The sequence of steps in the PSC consists of synthesis followed by analysis; solutions are developed (synthesis) and critically tested (analysis).

### *6.3.1   Purpose and Terminology*

Synthesis/analysis builds on the results of the situation analysis and the formulation of objectives (Fig. 6.14):

*The situation analysis* yields knowledge about the problem, and potentially also insights into the solution and ideas about possibilities for intervention.

The step of *formulating objectives* supplies information about objectives understood as effects or qualities that are expected from the solutions or about standards (criteria) with which to assess the suitability of solutions (compulsory, recommended, or desirable objectives).

The search for solutions consists, on the one hand, of the *(concept or solution) synthesis* that, as a constructive, creative step, has the purpose of finding, conceiving, designing, and constructing solutions; it consists, on the other hand, of the *(concept or solution) analysis* that, as a critical step, has the purpose of examining solutions critically in a systematic manner to improve, modify or, if need be, reject them.

These steps should result in solution concepts that, although considered to be basically suitable, nonetheless differ with regard to quality or degrees of preference.

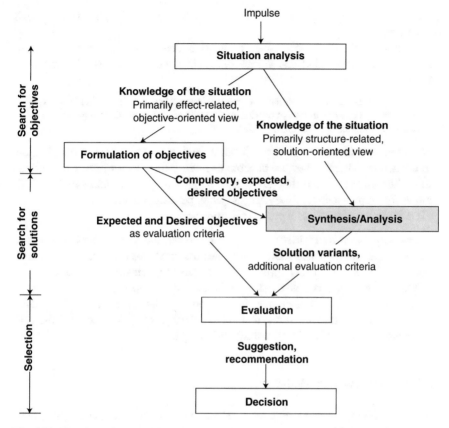

**Fig. 6.14** The place of synthesis/analysis in the PSC

Determining or working this out in more detail is the purpose of the ensuing evaluation and decision.

The process of searching and finding solutions is marked by the mutual interaction of the procedural steps of synthesis and analysis, which are dealt with later.

A synthesis is a highly creative act. It consists of "intuiting" the whole solution concept, of discerning or "finding" the requisite solution elements, and of intellectually assembling and combining these elements into a viable whole.

## 6.3.2 Synthesis of Solutions

### 6.3.2.1 The Importance of Creativity

The search for solutions to a problem or a task presupposes expertise in the problem and solution field. Knowledge about specific methods and tools is also helpful. However, the determining factor for finding new and suitable solution approaches is the creativity of the designers, developers, constructors, etc., whether as individuals or as a team.

*Creativity* is the ability of people to generate ideas, concepts, compositions, or products of whatever kind that are new in their essential features and previously unknown to the creator.

> "Creativity can comprise the formation of new patterns and combinations taken from experience, the transmission of known relationships to new situations in addition to the discovery of new relationships" (Drevdahl 1956).

Creative people are characterized by an inventive curiosity, critical imagination, and deductive thought. They are capable of freeing themselves from conventional and traditional views, and they know how to look at problem and solution fields from different perspectives. They are, as a rule, persistent optimists.

For the practical work of a project this means:

– Creativity exists in a variety of shapes. However, the above list of qualities of creative people indicates some approaches that systematically promote creative action. This can be further stimulated by means of → creative techniques.
– Creative, ingenious people must be included in the project work.
– Producing a stimulating climate that supports creativity (distance from daily events, reduced time pressures, relaxed atmosphere, etc.). Appreciating the new, accepting it first before immediately criticizing it.

### 6.3.2.2  Working with Models

Treating complex problems requires an abstraction of concrete facts. Abstractions in the problem field can lead from descriptive, graphic illustrations of reality to quantifiable, graphic, structural models, and even to mathematical abstractions. Then, abstraction is again used in the solution field to design solutions.

Figure 6.15 depicts this process: through abstraction a representation of the problem is produced. Based on this representation, a solution model is developed from which, through a course of progressive concretization, a solution concept results. The process is successfully concluded when a solution concept is finally available that appears to solve the problem in a satisfactory manner.

The transition from abstract model formulations of this sort to a concrete solution takes place over several steps of a progressive concretization, whereby the principle of variant creation and reduction can and should be implemented at each step. One should develop variants of the most different kinds at each level of concretization and elaborate particularly promising solutions by repeatedly improving, refining, and perfecting them.

### 6.3.2.3  General Design Principles

In working out solution concepts, one should take into account the three *design principles* listed below by way of example:

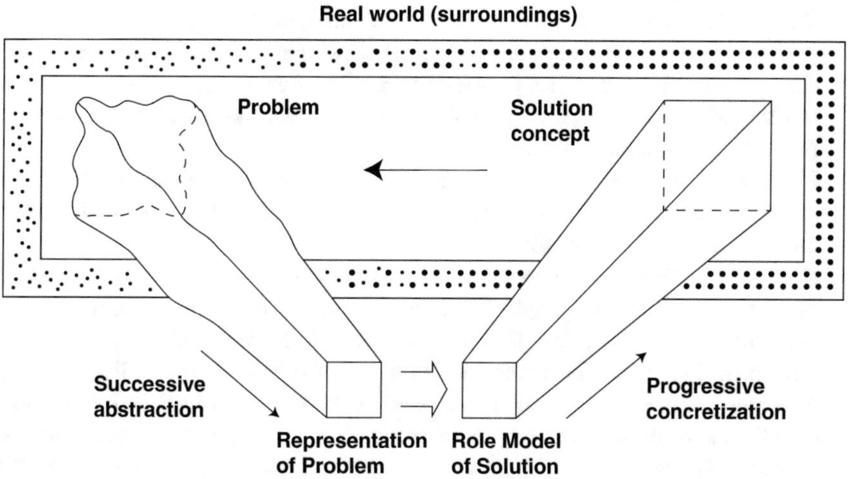

**Real world (surroundings)**

**Fig. 6.15** Abstraction and concretization

1. *Minimizing constraints:* means preferring solutions that allow the most room for further developments, such as changes, detailing, and later exchange of solution components (modules).
2. *Minimizing interfaces:* pertains to the structuring of a system and means creating a minimum number of interfaces. Any interfaces between solution components (for example, building groups) or organizational units, system aspects, should be simply and clearly defined. Here, it must be kept in mind that the structure of complex systems is often not identical with traditional technical or manufacturing constraints.
3. *Modular structuring:* means that solution components are designed to be reused several times and in other systems as well, or that customary components or current organization regulations can be used.

These three principles cohere with each other insofar as modular solutions with hierarchically structured components usually have a positive effect on minimizing constraints.

Here, one could refer, in addition, to the "*Fundamental Principles of Good System Design*" found in the literature. T. Bahill, for example, lists 34 of these[3]:

- In your work, give higher priority to the *more risky objects* or entities of a design: "Work on high risk entities first" (Bahill and Botta 2008) and "Do the hard parts first" (Rechtin 1991). This has several advantages: there is less risk of creating unnecessary costs. Unresolvable design concepts can be aborted earlier. Bahill also thinks that thereby the overall expenditure for changes will be less: because risky objects will probably have to be changed retroactively, this always requires a change in other objects as well.

---

[3] Bahill, A.T. and Botta, R. (2008): Fundamental Principles of Good System Design.
Bahill, T: http://www.sie.arizona.edu/sysengr/slides/

– Be content at first with *satisfactory designs* and *do not optimize* them *during an early stage* of development: one cannot do so with any finality because after any change in a design, one has to optimize it anew. Furthermore: if you optimize, test the criteria by which it is done. Are they in fact those criteria that will be influential for a later application?
– Use *open* standards whenever possible: The exchange of data, material, and energy, the inter-operability of systems, and the insertion or exchange of concept modules are thereby simplified.
– It is recommended whenever possible to incorporate *reserves in the design* (design margins): these might be security factors, budget reserves, tolerances, performance capacity, etc. This allows concepts to be better adapted to new demands. However, incurring additional costs represents a disadvantage. Nonetheless, one should think about where new demands might be heading and how the capacity to meet them could be installed in a concept. For example: the Boeing 747 was equipped with the capacity to handle heavier payloads. The engineers purposefully gave the aircraft larger wings and tailpiece than was first deemed necessary.
– *Design for testability*: as early as the planning stage one should think about how to test, as simply and reliably as possible, the important features that are demanded by the solutions.

### 6.3.2.4  Creating and Reducing Variants

The principle of creating and reducing variants was explained earlier in connection with the action model (Part I, Sect. 2.1.2). It will be taken up again here and dealt with in more depth.

*1. Thinking in variants*
An essential characteristic of methodical action is not to be content with the first solution that meets the compulsory objectives. One should rather try to obtain as comprehensive an overview as possible of the solution possibilities that are conceivable at a certain level of observation or concretization (functional, scientific/technical, or structural solutions), and then work on a selected part of the solution spectrum.

The cause of this course of action is the wish for an optimal solution. Practical problems usually do not have an absolute standard for determining the quality of a solution. This quality can be known only through comparison with other solutions.

Therefore, when a solution has been found that fulfills all compulsory objectives, this does not mean it is the best possible solution. It is merely a solution that suffices for the core demands. The call to search for variants is meant to reveal whatever potential exists for improvement, to stimulate the pursuit of other solution approaches, and thus to inspire confidence in the quality of the solution found, in addition to a sense of security about the continuing development of the system. ("A [single] solution is no solution" and "not the first-best but the best solution counts.")

Even when the first solution found proves ultimately to be the best, there is still opportunity to take up partial approaches or good solution elements from other variants and thus gain an overall better result. Here, one should make an effort to look for a wide variety of basically different approaches to the solution, which means not creating alternatives simply by varying the details. This is a step that can take place at the next lowest level of concretization.

*2. Avoiding unreal variants*

Variants (alternatives) worked out in a synthesis/analysis sequence should have the same logical status and represent real alternatives. An example serves to illustrate this. If the insufficient capacity of a manufacturing plant is the problem, proper alternatives would be those of a fundamental type, for example, expansion of in-house manufacturing, outsourcing orders, phasing out less profitable product lines, etc.

One should not mix in subvariants of the alternatives named above, such as expansion of in-house manufacturing at the present location, expansion of in-house manufacturing in a neighboring foreign country, placing orders domestically, etc. One should instead think about some basics first: expansion of in-house manufacturing (yes/no), or external delivery, or streamlining the product range. This can also be referred to as different architectural designs of a solution.

*3. Reducing the solution spectrum*

Although a diversity of variants in the solution search is temporarily desired, one should not, for reasons of time and expenditure, carry them too far along the subsequent levels of concretization. The diversity must be reduced earlier.

The variants that, in contrast to others, do not meet compulsory or recommended goals, or that meet them less satisfactorily, can be eliminated without a formal analysis, that is, simply on the basis of a greatly reduced number of criteria. Others cannot be so easily assessed and are therefore carried over to the evaluation and decision steps. If necessary, such decisions to reduce diversity should be brought about jointly with the commissioning party.

There is no doubt that experience and intuition play an important role in the selection of a strategy for pursuing solutions over one or more concretization levels. The same goes for an open and relaxed working atmosphere in the team.

### 6.3.2.5   Interaction of Synthesis and Analysis in the Course of System Development

Here two questions are of interest. First, how does the innovative character change in the course of system development? Second, how many concretization levels should there be within the PSC?

**1. Decreasing innovative character**

It is useful to distinguish between innovation processes and routine processes. The first possess a relatively high novelty value for those participating in the process, given that at present the participants have only relatively vague notions about the plan of action and the later outcome. In routine processes, on the other hand, one can rely on known action and behavioral patterns whose outcome can be estimated.

In the course of a system development, the amount of routine will and must increase, and the amount of innovation decrease. If this is not the case, the project will become more difficult to manage. A simple example may serve as illustration. The success of a building project will be notably compromised if the architect intends to use new materials for the supporting structure that have not been used before and that require special (i.e., clearly above-average) skills for their workmanship and installation. However, the more he or she turns to practiced and familiar routines, the lower the risk of failure.

Reducing the amount of innovation has an effect on the synthesis/analysis. To the extent that recourse is taken to known and proven solution components, less expenditure will be required for the synthesis and, in this connection, for the analysis relating to a single concept as well.

This is not equivalent, however, to a reduction of planning efforts in the detailed studies phase. Because of the large number of detailed concepts that have to be worked out and brought into agreement, the total expenditure during this phase may actually be greater than that for the preliminary and main studies.

### 2. Several concretization levels in one PSC

We have already referred several times to the principle of increasing concretization during the course of system development. However, this transition from one concretization level to the next does not necessarily have to be connected to the transition to the next phase as well. Each functional variant may contain several PSCs, and each PSC can certainly have several concretization levels.

This thought is illustrated in Fig. 6.16. Two functional variants (A and B) on the first concretization level are considered for a certain solution approach.

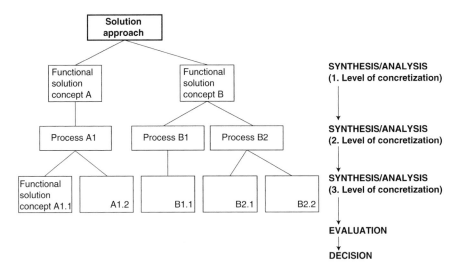

**Fig. 6.16** Multiple synthesis/analysis steps with increasing concretization in the PSC

Procedural solution possibilities are developed at the second concretization level for these variants.

The following third concretization level with the technical conception then allows an evaluation and decision to be made, and thus the transition to the next step in the PSC.

## 6.3.3   Strategies for Finding Solutions (Synthesis)

Habitual, familiar routines seldom lead to better solutions, to innovative products and processes, to awareness of novel opportunities and to the handling of as yet unknown risks.

To be sure, many mature solutions stem precisely from the routine, professional improvement of details, which are the subject of much fine-tuning. Such a strategy of piece-by-piece (incremental) improvement may indeed be useful for a limited time. However, at some point it becomes unproductive, necessitating some basic rethinking of the solution concept, that is, new architectural designs infused with a pronounced pioneering character. For situations like these, one cannot fall back on habitual routines. There are, however, various kinds of search procedures and heuristic methods for opening up the solution field and for finding solutions in an efficient manner. The reflections below are intended to provide some impulses and suggestions.

### 6.3.3.1   Demarcation of the Solution Field

The field in which the solution of the problem or the accomplishment of the task is possible is the sphere of action for designers, planners, and constructors. This field is defined by *three categories of influencing factors* (according to Rittel):

- Restrictions and limitations that arise from the situation and upon which no influence could or should be exerted by the designers, planners, and constructors or the commissioning group or the body responsible for setting objectives (so-called *context variables*)
- Function and performance volumes that were prescribed as compulsory, recommended or desirable objectives (so-called *performance variables*)
- Design parameters (so-called *design variables*) characterizing the actual free space available for creating solutions

Context variables are worked out in a situation analysis, performance variables are established in a formulation of objectives, and design parameters, even when they sometimes remain partially undefined, characterize the actual free space for solutions. The use and exploitation of this free space depends largely on the distinct abilities of the designers, planners, and constructors.

An example from industrial construction should help to illustrate this. A property's usage for a manufacturing facility is invested with *restrictions and limitations* (= *context variables*) such as the building code with zone plans, boundaries, and spaces between buildings, gable heights, building lines, and also the trade inspectorate's regulations on worker protection, industrial emissions, and safety precautions, etc.

*Function and performance volumes* (= *performance variables*) pertain to, for example, specifications for machines and facilities, operational processes and mass flow rates, operational side functions, number of work places, transportation connections, etc. They also include requirements for expandability, convertibility, upgradability, or demolition, and ones for aesthetic considerations and aspects of corporate identity, plus for the use of certain materials and their maintenance.

The specific *room for design* (= *design variable*) comprises the manner in which grounds, space, and nature are used and organized, the allocation of operational functions and processes to the spatial potential, the distribution of construction volume among single cubes and their inner and outer design, etc.

This can also be called *systems architecting*.

### 6.3.3.2  Different Starting-Points for the Search for Solutions

The following reflections are related to the system improvement described earlier. Here, a distinction is made between the "from-the-outside-to-the-inside" strategy, which corresponds to the "from the general to the detail" process, and the "from-the-inside-to-the-outside" strategy, which can make immediate improvements at the core of the problem.

*"From-the-Outside-to-the-Inside" Strategy*
This strategy starts with the desired effects of a solution and its relationships to the surroundings to work out the requisite components of the solution.

As in this case, decisions about solution principles and concepts must often be made on the basis of incomplete insights into problem relationships and solution opportunities, a certain confidence in the feasibility of future solutions is indispensable. Such confidence may rest on experience, trust in the skills of the problem solvers, optimism, the courage to engage in limited risks, and the ability to make subsequent corrections, etc., or it may be bolstered by the fact that different options are kept open for as long as possible. This strategy carries the relatively high assurance that a solution will be suitable to the surroundings. However, if the feasibility of a solution was not judged realistically enough, the consequences could have a negative impact on expenditure for development and, in an extreme case, lead to a termination of development.

This strategy is suitable for the conception of new systems. It corresponds to the procedural principle "from the general (the whole) to the detail" and it can be improved by elements of the "from-the-inside-to-the-outside" strategy described below.

*"From-the-Inside-to-the-Outside" Strategy*

In this strategy, the idea of an overall concept embedded in the surroundings is relinquished for the time being. One begins by assembling known and available solution components, designs them according to functional and performance requirements, and tries to adapt them retroactively to the conditions of the surroundings. Although this strategy leads more quickly to a finished solution, its utility, quality, and the duration of its serviceability are largely open questions.

This strategy is in principle opposed to the methodology of systems engineering and should not be used for the conception of new systems. It is useful for a stepwise improvement of an existing and known solution, as the alteration of single components usually does not have too much influence on the impact that a system as a whole has on the outside. It can be of service when a solution has to be improved under time pressure or possibly even repaired (stopgap solution), or when a total reworking or new designing of a system would take too long or is already planned for a future point in time.

The "from-the-inside-to-the-outside" strategy is especially advantageous for changing systems in the social domain through stepwise or piece-by-piece improvements (K. Popper's incremental "piecemeal engineering").[4] In addition, it may be necessary or expedient to plan a partial step and then to complete it immediately. The effects of this step can be taken into account when the next step is planned and realized. Because one refrains from making large-scale changes, there is a lower risk of a system's compatibility being irrevocably impaired. On the other hand, this strategy could also lead to a patchwork in the sense that a high level of expenditure is invested in conserving an existing, outdated *architecture*.

*Procedure with Changing Starting Points*

The suitability of "piecemeal engineering" is called into question when a comprehensive change proves to be necessary. To establish meaningful relationships among the partial steps, it is in this case useful to develop at least a rough framework that can be used as a coordinating instrument. This is accompanied by the influx of elements of the "from-the-outside-to-the-inside" strategy.

For the new conception of a system, one should principally choose the "from-the-outside-to-the-inside" strategy. However, it can be augmented by the advantages of the "from-the-inside-to-the-outside" strategy. In this regard, see the remarks on agile systems engineering (Sect. 2.2.2).

Critical system elements that are especially important and whose detailed configuration is expected to be difficult, should be given priority in the developmental process. This has already been noted in Sect. 6.3.2.3 (Bahill).

---

[4] Popper, K. (1957): The Poverty of Historicism.

### 6.3.3.3   Systematic Search Strategies

A nonsystematic search for solutions would be, for example, the *trial and error* method (and further searching), which could also be called a search process of a *test* nature.

Several systematic search strategies are sketched out below; the solution field for these is indicated briefly in a linear or a cyclical fashion.

*(a) Linear search strategies*
A linear process means a sequential search for a solution without planning for the possibility or necessity of taking recourse to earlier decision steps.

*Routine Process*
This process is distinguished by one procedural step following the next one in a routine manner. It assumes that solutions are easy to find and that no special selection problems arise, or that the planner is an experienced professional (Fig. 6.17).

This procedural principle can be applied within the framework of systems engineering to partial problems that show a relatively small degree of innovation. This is often the case when considerable progress with system development has already been made (particularly detailed studies or, as the case may be, system building) and its routine implementation has become increasingly significant. Here, conscious efforts to apply creative techniques do not play any essential role.

*Non-optimizing Search Strategy*
As solutions here are not apparent at each stage, they must first be searched for. However, the search is stopped as soon as one has found a solution that is deemed functional, and one passes immediately to the next level of concretization (Fig. 6.18).

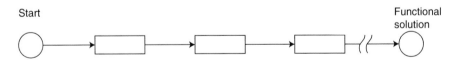

**Fig. 6.17**   Routine process

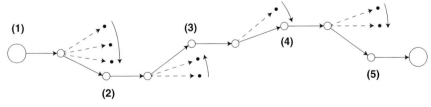

• = **non-functional solution**        ○ = **functional solution**

**Fig. 6.18**   Non-optimizing search strategy

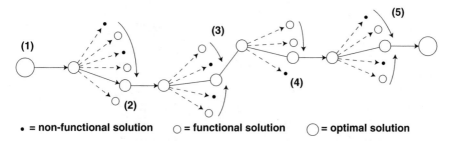

**Fig. 6.19**  Single-step optimizing search strategy

In a plan to remedy a lack of space in a production facility (1), the decision to expand (2) would accordingly be met with an additional building (3), to be located in the north of the grounds (4) as a two-story building (5).

With this strategy, too, one dispenses with searching for alternatives to a functional solution variant. It should only be used for relatively unimportant aspects that do not have an essential impact on the concept and where there are no innovative plans in the background. This is the case toward the end of the solution search when conventional solution components are selected or in situations that are relatively close to the stage of completion.

*Single-Step Optimizing Search Strategy*
This strategy is characterized by the creation of variants at all stages. On the basis of criteria to be established, an appropriate variant for further work is selected, one that promises the best chance of success (Fig. 6.19). Such a strategy is also called myopic (short-sighted). It pertains to a localized search method that does not guarantee, or usually makes it unlikely, that an overall optimal solution will be found.

In the above-mentioned plan to remedy the lack of space in a production facility (1), the decision to re-organize and change the set-up (2), in place of a decision to expand (as above), would be met by converting the procurement of external parts to just-in-time delivery for production (3) instead of, for example, reducing the storage area in the production floor by building a second story for the new storage are. The resulting demand for improved delivery (4) is not met separately, but rather optimized for both delivery and shipping.

In the context of systems engineering, this process is recommended above all for the preliminary and main studies.

*Multi-Step Optimizing Search Strategy*
The linear strategies elaborated thus far assume that it is expedient in each case to plan only the next step, then make a decision, and only then plan the step beyond that, etc.

The multi-step optimizing strategy departs from this idea. It is characterized by a range of variants that are not confined to single steps but instead fan out over several steps. A decision is made only after certainty has been reached about the suitability of an idea (➔ fault tree analysis, ➔ decision tree process); Fig. 6.20.

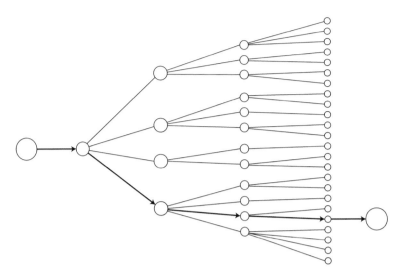

**Fig. 6.20**  Multi-step optimizing strategy

Because of the expenditure associated with this strategy (variants explosion), it should be used with caution. An important reason for its use would be that the risk connected with determining a certain solution principle is relatively high and that the additional expenditure required for developing alternative solutions at lower levels appears bearable in proportion to the risk.

*(b) Cyclical search strategies*
Applying a linear strategy is usually possible over only a few steps. Often, one reaches a point where the path taken cannot be continued, or only with modifications. Then the path must be retraced to an earlier decision and the process starting from there must be repeated in modified form (Fig. 6.21).

We can illustrate this process with the example of a parcel of land that is examined for the construction of a manufacturing facility. The search begins with the first allocation of operational units (A). A plausible location (a) is examined more closely (A') but abandoned because of difficulties with the rail connection (b). A recourse understood as a cycle becomes necessary. A new search starts with a second allocation of operational units (B); the concretization efforts (B') yield a solution (d) for the possibility (c).

### 6.3.3.4   Mathematical Methods of Operations Research[5]

The methods listed in the encyclopedia under the heading ➜ Operations Research can also be understood as evaluation techniques. To apply these techniques, however, it is necessary to develop quantitative models for the scenarios that are to be evaluated.

---

[5] See, for example: Hillier, F.S.; Lieberman, G.J. (2004): Introduction to Operations Research

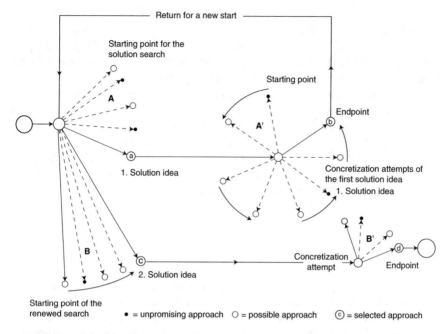

**Fig. 6.21** Cyclical search strategy (according to Rittel)

If a mathematical solution algorithm can find the optimal solution in a reasonable amount of time, it takes over the role of the evaluation step. In such a case it will aid in detecting the best possible solution. Thus, a comparison of variants, such as in the methods discussed so far, becomes unnecessary. Examples of these efficient algorithms are → linear optimization (optimal location, optimal production program, etc.), and nonlinear optimization.

Because it is difficult to render practical problems and their solution algorithms in adequate models, these approaches can usually be implemented only in partial areas.

### 6.3.3.5 Heuristics as a Strategy for Finding Solutions

Search processes can be improved and supported by intuitive work that uses methodical components. These components are based on a deliberate transfer of analogies, similarities, or even oppositions (among others → analogy method, → bionics, → synectics), they are logically/rationally applicable (among others → morphology, → problem-solving tree, → simulation) or mathematically deducible (among others, linear programming, branch and bound, combinatorics).

Systematic heuristics and the inventive algorithm attempt to cover the entire synthesis/analysis process: systematic heuristics incorporate into a "construction kit" proven methods of a solution search in such areas as development and construction in mechanical engineering. In an application-directed context it contains methods for finding and specifying solution-oriented formulations of tasks, for planning procedures, and for seeking and assessing solutions. It also comprises methods for determining information needs and for evaluating experience.

The inventive algorithm (➜ TRIZ according to Altshuller) goes through a specified series of working steps similar to the PSC. *Tables* have been developed for the operative stage of finding solutions in technical systems. These are based on the recognition that, on the one hand, groups of requisite characteristics and effects can be attributed to certain scientific phenomena, forces, etc., and that, on the other hand, there are only a limited number of procedural principles to resolve technical contradictions, i.e., between a target and an actual condition.

## 6.3.4  Analysis of Solutions

It was pointed out at the beginning of Chap. 6.3 that *synthesis* is to be understood as a *constructive/creative* step and *analysis* as a *destructive/critical* step. The latter examines solutions systematically to find approaches to improve them or arguments for their elimination. Here, the express intent is not (yet) a comparative evaluation of variants, but rather a critical screening of the suitability of each variant by itself.

### 6.3.4.1  Intuitive Versus Systematic Analysis

This critical screening is possible:

– In the sense of an *intuitive* coupling. The very moment an idea appears for a solution or solution element (synthesis), it is usually accompanied by a critical interaction (analysis) concerning the manner and extent of its suitability.
– In the sense of a *formal* sequence. Important planning results are systematically and critically analyzed by means of concrete queries.

The first-named screening, the intuitive analysis, consists of barely predictable responses to creative action. Intuitive analysis has a positive and a negative component. It is positive because it can have a fruitful effect and lead to improvements or wholly new ideas. It is negative because it is liable to disqualify solution concepts as unsuitable or impractical too soon.

In the second-named screening, analysis is an independent, formal process that should be structured with regard to the content and be prepared in an organized fashion. This is especially useful when important planning results are at hand or important decisions have to be made.

Below, we deal only with the formal, systematic analysis.

### 6.3.4.2   Contents of Systematic Analysis

The contents of analysis can be summarized with the following six areas of questioning, which, depending on the type of project and its current phase, have varying significance:

- Analysis of *formal aspects* – suitability for appraisal, fulfillment of compulsory objectives
- Analysis of *integration ability* – effect-oriented approach, looking outward
- Analysis of *functions and processes* – looking inward
- Analysis of *operational efficiency* – usability and operational fitness, ease of service, security and reliability
- Analysis of *prerequisites* and *conditions*
- Analysis of *consequences*

**1. Analysis of formal aspects**
Here, two points are important: the suitability of a solution to be appraised and the fulfillment of objectives.

*Suitability for appraisal* means that well-founded assertions can be made about the functioning of a solution in the context of given specifications (function and performance volumes, in addition to limiting conditions and restraints). Their degree of detailing depends, of course, on the project phase in each case. No further detailing of the concept should be immediately undertaken. Rather, a concept should be tested to see if in its momentary stage of concretization is complete to the extent that an assessment of the solution can be made. Only variants that are assessable can be subsequently evaluated.

Then, one must assess if the given compulsory objectives have been met or can be met in the further course of development and to what extent the observed variant can or cannot fulfill essential recommended objectives. Here, the starting point is the catalog of objectives that was worked out in the procedural step of formulating objectives, but which may have changed and expanded because of altered situations, insights, or value judgments. Variants that violate valid compulsory objectives must be rejected or, as far as deemed useful, reworked in a further step of synthesis. If this cannot be accomplished with any variant, the compulsory objective must be accordingly modified or eliminated.

**2. Analysis of integrability**
The concern here is to screen the solution that has been worked out with regard to its ability to be integrated and its relationship with the environment. The effect-oriented approach, the "look outward," stands to the fore, whereby special attention is paid to the effects of a solution from the perspective of its environment and, as the case may be, of a system at a higher level. Regarding the effect-oriented approach, the coherence of input and output relations should also be examined. This may be carried out by balancing the relations between output and input or checking if the resources of the system are at all plausible for the desired output in light of the given input.

Priority is given to the interfaces between system and surroundings, the respective input and output with its provenance and usage, in addition to its reception and transmission. Increasingly, it is both reasonable and necessary to check for opportunities for later changes and adaptations, but also for removing components (for example, recycling of old cars, electrical appliances, demolition of high-rises in inner cities and of nuclear power plants, etc.). One might also check whether the modules of a product family could be used.

### 3. Analysis of functions and processes (looking inward)

In this step, one should look inward: input, which may consist of information, material, energy, and more, in addition to input carriers, are tracked through the system from one element to the next. The mental tracking of these processes is finished when the conversion of input into output has been completed, the balance of their transfer is correct, and the output is ready for being taken over into the next system. The preparation of a complete output list and the comparison with the inputs serve to create a detailed quantity balance, by which unwanted input disposals or unplanned output creators can be detected. If outputs exist that cannot be tracked to their inputs, or if inputs disappear without a resulting output, this can be taken as evidence that the analyzed concept was not sufficiently thought through. It can also mean, that additional functions and processes must be planned, or even if these functions and processes were originally provided in the concept, they nonetheless have not been part of a critical analysis.

The internally directed approach gives rise to process illustrations that are important for subsequent steps.

### 4. Analysis of operational fitness

The focus here is on usability and operational fitness, ease of service and maintenance, security, reliability, etc.

In pursuit of this question of analysis, one should attempt to place oneself successively in the position of the operator and user, the suppliers and clients in the broadest sense (information, material, energy, etc.) and the service personnel, and to consider the planning results from their perspective. Further aspects that should be observed here are service and maintenance, system care, and updates. These issues become especially significant in phases that are close to realization.

It is expedient, moreover, to take account not only of normal cases but also of overload or partial load states and extreme input or output states to evaluate the behavior of the system and the intervention possibilities under these conditions.

In the analysis of reliability and security, one observes element after element, determines their functions, and considers:

- How great the likelihood is that the elements will become inoperative (initially often qualitative considerations such as "great," "small," etc., will suffice)
- The manner in which they fail to function or may react incorrectly
- The consequences of a breakdown or malfunction
- The relationships that may be disturbed or interrupted

- How this could be avoided
- What to do if this could not be avoided (emergency organization)
- And much more

Starting from the responses to these questions, suitable measures have to be worked out to avoid breakdowns or, should this be judged to be too elaborate or impossible, to remedy or control them (redundancy, breakdown organization).

Useful tools for considerations of reliability and security[6] are analyses of failure types and their effects. These can be expanded to include analyses of the significance of failure types and failure effects. Also to be considered are path analyses and fault tree analyses (see ➜ security analysis, ➜ reliability analysis).

### 5. Analysis of prerequisites and conditions
Prerequisites and conditions for the realization and functioning of the solution have to be worked out very clearly. In particular, those prerequisites must be determined that are indispensable for the functioning of the observed solution and which, if not complied with, would jeopardize its realization.

Examples of such prerequisites would be:

- The availability of neighboring solutions that are essentially connected with the observed solution (up- or down-stream production operations)
- The creation of personnel and infrastructure prerequisites that are not directly an object of the observed project but of other, parallel projects, such as personnel training, energy supply, etc.

Particular attention should be paid to the designing of such prerequisites in the further course of system development and realization.

### 6. Analysis of consequences
Associated with the selection of a certain concept or the later operation of a system are positive and negative consequences of financial, personnel, organizational, and other kinds. Of special concern are measures to avoid, limit, or alleviate negative consequences. Also to be considered are questions regarding whether a solution can actually be produced and implemented and whether it will be accepted by the concerned parties. With regard to the inclusion, and thus to the analysis of production requirements, a partial-parallel development, as expounded in the concept of "simultaneous/concurrent engineering," has definite advantages (see Part I, Sect. 2.2.1.8).

In present times, a systematic, methodically supported risk analysis is required at the latest during the "analysis of consequences." Ever more complex end-products, greater time pressures, etc., increase the risk that project goals are not met. Questions such as the following have to be clarified: what damages could ensue if a solution fails to function, is not available on time, exceeds cost limits substantially, or does not fulfill the required specifications? How can this be determined in a timely fashion within the context of project monitoring? What are the indicators (risk cockpit)?

---

[6] Nancy (2012): Engineering a Safer World.

How does monitoring function? On the methodology of this risk analysis, see T. Pfletschinger (2008).[7]

Important consequences can yield additional criteria for the subsequent evaluation of variants. Such consequences may lead to limitations or supplements for compulsory objectives, or changes of compulsory objectives into recommended or desirable objectives. Also, system adaptation, modification, expansion, and possibly reduction, or even dismantling, can be considered.

## 6.3.5   Methods and Tools for Synthesis/Analysis

There are many methods and tools available for creating and further developing solution approaches, for generating variants and modeling situations, for analyzing solutions with regard to their different requirements, and for simulating and illustrating system behaviors.

They can be divided into three groups, which are described in Part VI: creative techniques, modeling and illustrative techniques, and analytical techniques.

## 6.3.6   The Procedure of Synthesis/Analysis

The different approaches described earlier, understood as synthesis strategies and analysis principles, may be characterized as general thought principles that follow the steps sketched out below. Here, it can be appealing to first approach the question directly without any systematics. Only when this attempt has been carried out and no ideas are forthcoming, or only one, is it advisable to try a systematic path.

The process can be roughly divided into six steps that tend to describe a sequence:

1. Analysis of the design task
2. Generating and compiling solution ideas
3. Systematic ordering of solution ideas that appear suitable
4. Working out solution concepts
5. Systematic analysis of solution concepts
6. Reworking or following up on solutions

The sequence of tasks is not strictly linear. In practice, it is often interrupted and the steps are run through cyclically and thus repeatedly for a variety of reasons. Moreover, the treatment of individual steps entails different focal points in accordance with each situation. Persons work on an alternating basis in teams or singly.

**Step 1. Analysis of the design task**
Here, it is a good idea to recall the results of the situation analysis and the formulation of objectives and to clarify the "design or construction question" for which an

---

[7] See Pfletschinger, T. (2008): Risiko-Management.

answer must subsequently be found. It may also be necessary to reach agreement about a search strategy (routine process, optimizing search strategy, etc., according to Sect. 6.3.3.3).

**Step 2. Generating and compiling solution ideas**
Thinking alone cannot completely penetrate or describe the creative process – a process, moreover, dependent on an individual's skills and predilections. Synthesis is first of all a creative matter that should not be constrained by detailed instructions. However, thought-provoking impulses are useful when the creative process has come to a standstill or a solution has already been found and it is time to look for supplements or alternatives. The remarks below pertain primarily to such situations that promote the creative process.

*Working hypotheses*: at the beginning of the search for solutions, one should attempt to formulate several solution principles that differ widely from each other. Such principles should not, however, be understood as rigid guidelines, but rather as working hypotheses that are to be modified as insights increase. They must evolve from expertise and knowledge of the situation.

*Preferring important objectives*: the more difficult a problem is or the more comprehensive a system of objectives, the more likely it is to make an unconscious or arbitrary selection among the different objectives during the solution search. Thus, one will concentrate on a single or just a few objectives and look solely at these while searching for solutions. Therefore, it is essential that the selection of objectives follows an orderly sequence. This can be accomplished by deliberately concentrating on compulsory objectives in the search for solutions. Although one cannot thereby forcibly bring about solution ideas, one's willingness to accept suitable ideas may be increased. Solution elements for other, less important objectives can be incorporated later.

If useful solutions have already been found and the focus is on developing alternatives, it may now be expedient to concentrate primarily on recommended or desirable objectives, although they may actually be considered less important, and to search for solutions with especially these objectives in view. Even though this proposition, insofar as it is successful, may not create true "rivals" to the original solutions, it could lead to additional solution approaches to improve the original ones.

*Hypothetical dissolution of limitations*: in situation analysis and in the formulation of objectives, restrictions and compulsory objectives are established, often of necessity and without closer knowledge of concrete solution possibilities. This can later prove to overly limit the search for solutions. Here, it may be appropriate to undertake an initially hypothetical, stepwise removal of particularly obstructive limitations, thereby conceivably opening perspectives upon new solution dimensions. Should this result in new and promising opportunities, it might be worth considering whether or not the restrictions accepted earlier or the formulated compulsory objectives could be modified (feedback on the formulation of objectives, possibly additional situation analyses, consultation with the commissioning party, substantiating the advantages associated with removal). But this does not mean to suggest that – waiving all limitations – one should always first develop "ideal concepts," which are later brought down to earth through the introduction of constraints and compulsory objectives (see "ideals concept" in Part I, Sect. 2.1.4.3).

The following questions can open up new dimensions of thought:

– Which constraints or compulsory objectives are perceived as especially limiting on the search for solutions?
– Which additional solutions could make their elimination possible?
– What other consequences might elimination have? (Additional situation analyses may be needed to answer this question.)
– Would it be expedient to remove limitations on the basis of considerations like these?

*Resources for stimulating solutions*: another avenue for finding solutions might be to take up and, if need be, augment the catalog of resources, which was prepared in the situation analysis, and the intervention possibilities listed there, and then to try to find solutions on the basis of alternative means. The resources should serve as an impulse to find solutions. This procedure is employed to advantage in connection with a solution-neutral formulation of objectives, as it hampers a premature adherence to a particular resource.

*Solution base for generating variants*: sometimes one finds solutions without being aware of the principle or basic idea on which they rest. Carving out this base can lead to new ideas. To find the base it could be useful to bring in a third party that, unencumbered by details, might more easily be able to detect the base or lead the designer to it by posing questions.

*Result*: the approaches named above can be employed for both overall and partial solutions and principally at every level of concretization. They claim neither completeness nor all-around applicability and may be summarized as follows:

– Application of intuitive approaches before systematic ones.
– Application of the principle of increasing concretization in creating models.
– Formation of working hypotheses.
– A mental orientation toward particularly important objectives creates better conditions for good solutions.
– A mental orientation toward less important objectives can yield solution approaches for improving good solutions.
– Challenging certain restrictions or compulsory objectives can open up new solution dimensions.
– A catalog of resources can serve as an impulse for seeking solutions.
– Carving out the basic idea on which a certain solution is founded can stimulate the search for alternative basic ideas.

### Step 3. Systematic ordering of ideas

Appropriate solution ideas should be ordered incrementally and repeatedly. This has the virtue of enabling the recognition and improvement of whatever appears to be relevant, useful, realizable or promising. Such ideas can also be augmented with further opportunities and, where appropriate, diversified (renewed variant creation).

Here, intuitive analysis, which was addressed earlier, plays an important role. We have already referred to the danger of prematurely excluding seemingly strange ideas.

### Step 4. Working out (designing) solution concepts/architectural designs

The focus here is on the concrete conversion of ideas into solution concepts at the level of concretization that is in accordance with the phase of development currently being worked on. This is an important creative leap, requiring conceptual skills, i.e., a rather critical measure of ideas and experiences.

### Step 5. Systematic analysis of solution concepts

Each of the reasonably attractive and apparently useful solution concepts must now be systematically analyzed. For this, the analysis principles explained in Sect. 6.3.4 should be applied.

### Step 6. Revising solution concepts

Insofar as analysis yields concrete indicators of flaws, deficiencies, or improvements, the working out of solution concepts (step 4) must be taken up again. If the results are satisfactory, the solution just analyzed can be deemed suitable for the next step. This may entail revisions or further work on the next level of concretization, or evaluation of solution concepts. Evaluation can begin when possible solution variants can be surveyed adequately and when important course settings must be decided.

### Suggestions for Carrying out the Synthesis/Analysis

The *search for ideas* and the development phase require creative team members. An environment that fosters creativity is essential for teamwork. The target is to find a wide variety of solution principles and basic ideas. Even ideas that at first appear strange or purely novel should not be dismissed. Keeping a catalog of ideas makes it possible to recapture lost ideas. How consequentially the above recommendations are carried out depends on the situation, the mind-set of those involved, and also on time constraints.

A systematic *analysis* requires critical team members (analysts, doubters, destructors, the curious, caricaturists). However, criticism that is too vehement and discouraging can create tensions among personnel. Criticism should therefore be as positive as possible. The simple suggestion that one does not know how to solve the problem, but that the present solution needs improvement, can serve to alleviate tension.

On the whole, this step should be informed by the recognition that it is still relatively easy to change solution concepts (in the sense of removing substantial flaws or making significant improvements) while solutions are still in the draft stage. After detailed planning is complete and the realization process has begun, repair work is usually much more elaborate, if at all possible.

### Documentation

This refers to the documentation of the solutions themselves, in addition to the data that have led to the solutions or are required to understand them.

*Documentation of solutions*: solutions should be documented, with a view to the evaluation, when the search for solutions is complete or, as need be, at the end of preceding stages. Thereby, a distinction is made among at least four categories of recipients, each with different requirements:

– The *commissioning party* must understand the solutions sufficiently well to make a reasonable selection from the different variants.
– The *system developer* must be able to detail and implement the chosen solution.
– The prospective *users*, *operators*, and *maintainers* must have an understanding of the type and the conditions of their activities.
– The effects of realized solutions and/or of their utilization must be assessable by the *persons impacted*.

The first and the last categories of recipients are also often interested in having information about solution approaches that have been eliminated and the reasons for their elimination.

As a rule, one and the same description cannot cover all the requirements.

*Documentation of data*: given the variety of purposes, one should be able to resort to propositions that are the results of the synthesis/analysis and thus lay the groundwork for subsequent steps in the PSC. These propositions must be documented, including their rationales, to enable other planners, commissioning parties, participants, and persons concerned to reenact the working out of solutions. They facilitate the work of the planner himself when re-opening a case. Documentation is the basis for monitoring the development of essential prerequisites and for the operability and other functions of the solutions. Personnel changes in the team are easier to manage when the work results of team members have been recorded.

A written record of situations shows where gaps may be detected. It promotes uniformity in the vocabulary that is used and thus serves the means of communication. Therefore, both essential data about the process and important interim results of the work should be recorded in writing and kept in an orderly file.

### 6.3.7   Summary of the Search for Solutions

1. The search for solutions consists of the steps of synthesis/analysis and forms the creative core of the PSC. Its purpose is to produce solution variants whose suitability has been checked and that can be systematically compared in the next step (evaluation).
2. The search for solutions is prepared through situation analysis and a formulation of objectives, but it is also limited by these.

   – A situation analysis results in knowledge of the situation (problem field and insights into the solution field).

- A formulation of objectives contains a structured summary of the demands and requirements agreed upon with the commissioning party. It thus also contains insights regarding evaluation criteria.

3. The search for solutions consists of a creative/constructive step and a critical/analytical step. The constructive step is called synthesis, the critical step analysis.
4. A synthesis has three functions:

   - Intuiting a whole, a solution concept, a possible systems architecture
   - Recognizing or working out the requisite solution elements
   - Conceptually assembling and joining these elements into a model that makes a suitable whole

   Creativity plays a major role here.
5. In analysis, a distinction has to be made between an intuitive and a formal, systematic analysis:

   - An intuitive analysis is spontaneous and unplanned; it is the spontaneous, critical reaction to a creative action and can offer both advantages (improvement recommendations) and disadvantages (for example, a possibly premature rejection of solutions because of prejudices).
   - A formal analysis should be implemented when important planning results are available in the form of solutions that must be critically examined before their further development. It derives systematically from analysis principles that in each case give prominence to a different, but important, point of view (evaluative capacity, compulsory objectives, integration, functions and processes, operability, prerequisites and conditions, consequences). Here, it may prove useful to consult or commission other persons for the task.

6. Models play a role in the design of solutions and in their critical analysis (for example, through simulation).
7. The principle of variant creation and reduction is an important methodological component.
8. Innovative features should be successively reduced during the course of development. Routine processes should play an increasing role in the work on partial solutions and in their realization.
9. There are several strategies for the solution search:

   - The "from-the-inside-to-the-outside" strategy is of special significance for system melioration.
   - The "from-the-outside-to-the-inside" strategy offers more opportunities when fundamental changes have to be made.
   - Mixed strategies are often expedient.

10. A series of methods, tools, and techniques (for example, creativity techniques, modeling and illustrative techniques, analysis techniques) support the search for and selection of solutions.
11. A systematic procedure in the search for solutions can take the place of or follow an impulsive frontal approach.

### 6.3.8  Self-Check for Knowledge and Understanding: Search for Solutions: Synthesis/Analysis

1. What does synthesis mean? What does analysis mean?
2. What is creativity and what is its role in the search for solutions?
3. Explain Fig. 6.15 relating to the idea of abstraction and concretization
4. Does the innovative character increase or decrease during the course of a project?
5. Which typical search strategies for solutions do you know?
6. Give some examples of methods, tools, and techniques for the search of solutions.
7. What is the logic in differentiating between an intuitive and a formal analysis?
8. What are the contents of a systematic analysis of solutions?

## 6.4  Evaluation and Decision

We now turn to the selection process, which involves the evaluation and decision steps. Their place in the PSC is illustrated in Fig. 6.22.

Roughly described, the evaluation step serves to prepare the decision. For that, three conditions must be fulfilled:

1. Distinguishably different *solution variants* must be available, from which one may and should choose.
2. *Evaluation criteria* are required for signifying which qualities or effects are considered essential.
3. There must be the *capability* to appraise and to rank the variants that are to be evaluated according to the degree to which they fulfill the criteria.

*Solution variants* are worked out in the synthesis. In the analysis, they are examined critically with regard to their adherence to the latest compulsory objectives, their proper functioning (process logic, integration ability, security and reliability, completeness), their comparability, their fulfillment of necessary requirements, and their expected consequences. Unsuitable variants have to be reworked or removed. They no longer make it to the evaluation.

Operationally formulated subobjectives, which were designated as recommended or desirable in the formulation of objectives, are particularly suitable *evaluation criteria*. The list of criteria (criteria plan) that is thus created must be supplemented, however, by additional criteria, which arise only once the solution concepts (synthesis/analysis) are known. This is not an unproblematic claim as now criteria or subobjectives are introduced into the evaluation, which were not considered necessary in the formulation of objectives. The addition of criteria, though, should not be an arbitrary process but rather an expression of a learning process, offering the chance of reaching a better solution on the basis of target conceptions that have been supplemented or modified.

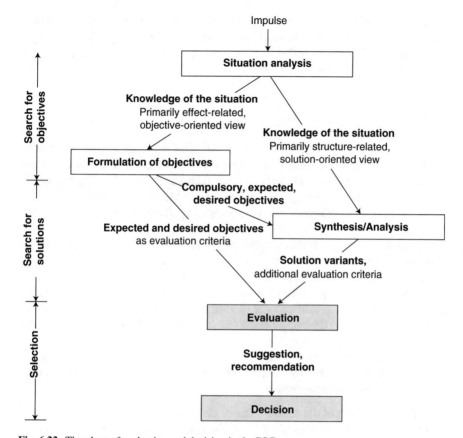

**Fig. 6.22**  The place of evaluation and decision in the PSC

The *capability* of assessing variants comprises both a knowledge of the situation and expertise about the effects, qualities, and operating conditions of variants; it also means being capable of rendering arguments and making judgments.

Decision follows evaluation. Depending on the phase of the project, it involves the resolution to work out the selected solution variant in detail or to begin with the realization or, possibly, to terminate the project.

## 6.4.1  Purpose, Terms, Fundamentals

Before we deal with the procedures and processes in more detail, we shift our perspective somewhat and take up the problem of decisions in general: although decisions are particularly important at the end of the PSC, they are also necessary in the context of the target search, the solution search (synthesis/analysis), and project management.

A decision-making situation exists when one can or must choose among several alternatives for action. The choice is largely determined by the expected effects of certain actions. Therefore, before making a decision, one should become informed about the consequences of one's choice.

### 6.4.1.1   Different Types of Decisions

A decision-making situation represents a barrier in the course of action. It is overcome by the resolution to execute future actions with reference to the choice made. Sometimes this barrier is not even perceived and a choice is made unconsciously. It is only when the barrier is subjectively perceived and exceeds a certain level that the acting parties become aware of the decision-making situation and make a deliberate decision.

Inasmuch as decision-making situations are often identical or similar, it is useful to develop decision-making routines or rules that facilitate a habitual and thus efficient execution of the evaluative and decision-making process (for example, choice of suppliers, batch size policies, rules on discounts, handling of applications in public administration, etc.). When no decision-making routines are available, there are two ways to master decision-making situations: improvisation or methodical support.

*Improvised decisions* are made, often under time pressure, on the basis of only a few criteria and a not very profound analysis of the starting situation and the consequences. They are justifiable when:

– The consequences of the decision are relatively unimportant
– The course of action initiated by the decision can be relatively easily influenced later on
– The quality differences among the present action courses are very great, showing that one solution has a clear advantage over another, which of course simplifies decision-making
– Or when the opposite is the case and it is immaterial which variant is chosen because the differences in quality are insignificant

With an improvised decision, one must keep in mind that its value depends greatly on the experience that the decision-maker has acquired in similar situations.

If the four conditions above cannot be met, the decision-making process must be *methodically supported*, whereby it will be expedient to differentiate between interim and final decisions:

In the context of systems engineering, *interim decisions* are those that are made during the synthesis/analysis. Even if not often expressly referred to as a decision, the pursuit of a certain solution idea (synthesis) or its critical examination (analysis) still has a determining influence on the further course of action, so that this choice, whether conscious or not, is of a decision-making nature.

*Final decisions* are often of a formal nature, meaning that the expressions of will of several people should and must be procured before continuing the course of action. These decisions are always made at the end of major planning activities, particularly after the synthesis/analysis. They may pertain to, for example, alternative solution principles, *architectural variants*, alternative overall or detailed concepts, alternative procedures or resource allocations, etc. A certain special status is assigned to so-called target decisions, which have to be made when there is a conflict regarding objectives (see Sect. 6.2.3.10).

Thus, the following attribution tends to be valid: final decisions should preferably be methodically supported; interim decisions can also be improvised. But if the import of interim decisions is far-reaching, it would additionally be advisable here to use procedures that make a methodically supported decision possible.

Furthermore, we want to make a deliberate distinction between improvisation and intuition: *improvisation* refers to the process or the (minor) formal and temporal expenditure that is carried out in preparing the decision. *Intuition* refers to the unconscious assessment of factors that influence the decision, that is, the consequences of the decision, which are based on subjective factors such as experience, gut feeling, "having a good nose," and the skill to evaluate complex situations even when only a little information is available. It plays a part in both improvised and methodically supported decisions.

### 6.4.1.2   Methodically Supported Decisions

In a methodically supported decision it is assumed that the quality of the decision is augmented by knowledge about the consequences of the choice. Therefore, sufficient information must be procured. In addition, formal processes are utilized, allowing information relevant to the decision to be processed in such a way that a recommended decision can be derived logically and often even mathematically.

So-called → evaluation methods and → economic feasibility calculations are relevant for decision-making problems that come up in the planning of large projects. Beyond that, there is an array of other methodical approaches for working on partial aspects, relating in particular to illustrating the problems and structures of decision-making (→ decision tree, → polarity profile). Also to be mentioned are the approaches of → operations research to detecting optimal solutions mathematically.

### 6.4.1.3   The Process of Preparing a Decision and the Resolution

The logical process depicted below and in Fig. 6.23 shows how to prepare and complete decision-making situations that are handled methodically.[8]

*Step 1*: analysis of the decision-making situation. Which decision is to be made and why? From among which possibilities should one choose?

---

[8] Note: this process should not considered bureaucratic – it can be done in several hours, depending on the complexity of the topic and the general mood of the team.

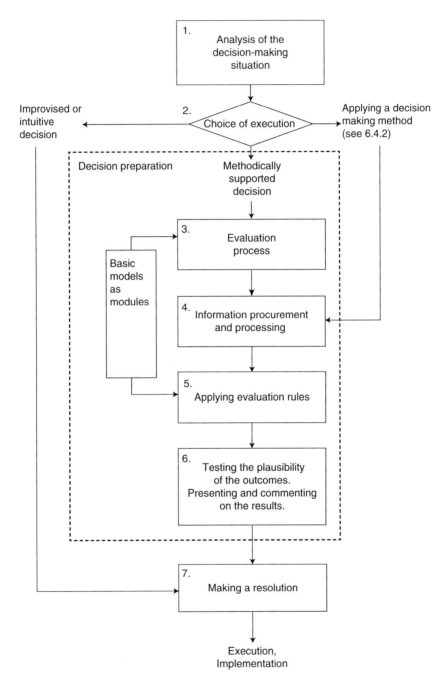

**Fig. 6.23** The process of preparing a decision and the resolution

*Step 2*: selecting the method of execution. One chooses a methodically supported process especially when:

–  The decision is considered to be important and significant, particularly when it has lasting consequences (setting the future course)
–  No obvious favorite can be identified among the different variants with regard to essential effects or qualities
–  The persons responsible for the decision have clearly different opinions, perceptions, or expectations (for example, in respect of the criteria that have to be observed, their significance, etc.). A methodically supported process thereby also helps to raise awareness of different value judgments and to incorporate even quite divergent criteria in the decision
–  The decision does not have to be made under extreme time pressure, thus allowing sufficient time for its preparation. This aspect, however, can also be influenced by good planning and good project management

*Step 3*: establishing the evaluation process, for which various prototypes are available as modules:

–  The models and methods of traditional ➔ economic feasibility and investment calculations assume that the essential characteristics and effects of solutions can be expressed in monetary units. However, many decision-making situations cannot be adequately portrayed by this method.
–  In contrast, the processes of value-benefit analysis or cost-effectiveness analysis, which are outlined in the following section, allow a broad spectrum of different types of criteria.

*Step 4*: *information procurement* and *processing* serves in the preparation of solution variants and decision-making situations for their entrance into the evaluation process. If the information procured and processed in analysis is not sufficient, additional analyses must be carried out.

*Step 5*: the implementation of the model should result in a *decision recommendation*.

*Step 6*: this recommendation is to be tested for plausibility and presented to the decision-making body together with the eligible alternatives.

*Step 7*: even in the case of well-prepared decisions, the *resolution* may diverge from the recommended decision. For this, there can be a whole series of vaguely acceptable reasons, for example, the intuition of the decision-makers who do not accept the results of the decision preparation, or circumstances, situations, and value judgments that have changed in the meantime. Therefore, additional clarifications, new evaluation steps, etc., may be necessary.

As can be seen from Fig. 6.23, the resolution is at the same time the completion of the *improvised decision*. The difference between the two types of decisions is that no discernible decision-making methods are applied in improvised decisions.

## *6.4.2   Evaluation Methods*

The methods represented here are limited to those few that we consider to be characteristic of or universally applicable to an evaluation in the comparison of variants. These are the balance of arguments, the value-benefit analysis (point rating, scoring model), and the cost-effectiveness analysis.

### 6.4.2.1   The Balance of Arguments

The basic idea of this very simple procedure is to list the advantages and disadvantages of single variants in the form of verbal arguments (for, against). Thus, this method creates a kind of survey of the decision-making situation, even though it is not especially efficient and transparent and should actually not be used in systems engineering. We describe it because it often suffices for simple decision-making, but above all because it can serve to illustrate the characteristics of the processes to come.

As an example, we may use the evaluation and selection of an apartment for a young family with two children (see Table 6.3). This problem has the advantage of being generally easy to understand. Following the systems engineering action model, we should of course adhere to the PSC and derive objectives, understood as the demands made on the apartment, from their intended purpose and from the situation of the family. From that, and with consideration of the housing market, we may extrapolate assessment criteria for an evaluation. As the primary concern here is illustrating a method, we skip all these steps and refer to the two examples in Part IV, Chaps 8 and 9.

The advantages and disadvantages of the balance of arguments are obvious.

Advantage: there is a certain ordering of arguments (better than merely spinning thoughts in one's head).

**Table 6.3**  Balance of arguments (apartment example)

|  | Advantages | Disadvantages |
|---|---|---|
| Apartment A | Short distance to school and work<br>Top floor, open view<br>Good insulation<br>Friends of parents nearby<br>Big apartment, good floor plan | Relatively noisy (street noise)<br>Overhead heating<br>Neighbors not so nice<br>The most expensive apartment |
| Apartment B | Attractive surroundings<br>Good shopping opportunities<br>Connection to district heating<br>Very nice neighbors<br>Grandmother nearby<br>Low rent | Inconvenient route to work<br>No acquaintances nearby<br>Smallest apartment, no room to expand |
| Apartment C | Good shopping opportunities<br>Little street noise<br>Largest apartment<br>Low rent | Unattractive neighborhood<br>Long distance to school and work<br>Poor insulation<br>Inadequate floor plan |

Disadvantages: there is no uniform assessment standard. Not all arguments are used for all variants. It is not made clear what is important and what is less important. With more than two variants it is not clear what is compared with what: advantages or disadvantages compared with which variants?

The balance of arguments is therefore suitable only for relatively simple decisions, namely those that are already intuitively apparent, but should still be tested, though with little expenditure. It does offer opportunities for expansion, as is illustrated below.

### 6.4.2.2 The Evaluation Matrix as a Basis for the Comparison of Variants

The evaluation matrix presented in Table 6.4 is at the core of a series of evaluation methods (point rating/multi-criteria method/value-benefit analysis) which, under different names, express the same thing. It serves as a good illustration of the

**Table 6.4** Evaluation matrix

| | | Variants | | | | | |
|---|---|---|---|---|---|---|---|
| | | $V_1$ | | $V_2$ | | $V_3$ | |
| Objectives | Weight $\sum = 100$ | s | w*s | s | w*s | s | w*s |
| $O_1$ | $(w_1)$ | $(s_{11})$ | $(w_1{}^*s_{11})$ | $(s_{12})$ | $(w_1{}^*s_{12})$ | $(s_{13})$ | $(w_1{}^*s_{13})$ |
| | 50 | 5 | 250 | 8 | 400 | 6 | 300 |
| $O_2$ | $(w_2)$ | $(s_{21})$ | $(w_2{}^*s_{21})$ | $(s_{22})$ | $(w_2{}^*s_{22})$ | $(s_{23})$ | $(w_2{}^*s_{23})$ |
| | 40 | 3 | 120 | 2 | 80 | 3 | 120 |
| $O_3$ | $(w_3)$ | $(s_{31})$ | $(w_3{}^*s_{31})$ | $(s_{32})$ | $(w_3{}^*s_{32})$ | $(s_{33})$ | $(w_3{}^*s_{33})$ |
| | 10 | 10 | 100 | 7 | 70 | 8 | 80 |
| Overall fulfillment of objectives | | | | | | | |
| $FO_1$ | | | 470 | | | | |
| $FO_2$ | | | | | 550 | | |
| | $FO_3$ | | | | | | 500 |

$V_1$, $V_2$ ...: *Variants* available for selection

$O_1$, $O_2$ ...: Measures (criteria, subobjectives) by which variants are assessed. They represent the *objectives*

$s_{11}$, $s_{12}$...: Assessment *scores* that express to what extent the variants fulfill each of the subobjectives. For this, different scoring scales are available: scales of 0-5, 0-6, 0-10, and others.

The present illustration is based on a 0–10 scale: a variant that is very good in respect to a certain criterion receives the score 10, one that is barely sufficient = score 1, and those that are insufficient or do not fulfill criteria (which, however, may not be compulsory objectives) receive a score of 0; the remaining scores can be used in an incremental gradation

$w_1$, $w_2$...: *Weight*. The significance of one subobjective in relation to other subobjectives is expressed by weight. Recommended is the conventional measure that the total weight = 100 (thinking in percentages)

$w_1{}^*s_{11}$, $w_2{}^*s_{21}$ ...: *Weighted achievement of subobjectives* by the solution variants (the weight of the subobjective times the score of the respective variant)

$FO_1$, $FO_2$ ...: Values per variant, representing the overall *fulfillment of objectives*. They result from the sum of the weighted achievements of the subobjectives:

$FO_1 = w_1{}^*s_{11} + w_2{}^*s_{21} + w_3{}^*s_{31}$, etc.

$FO_2 = w_2{}^*s_{12} + w_2{}^*s_{22} + w_3{}^*s_{32}$, etc.

problems of evaluation (1). To be evaluated are various solution variants, which are considered viable in principle, but have markedly different advantages and disadvantages (2). One must take into account very diverse subobjectives or criteria (such as performance, investment amounts, operating costs, personnel, flexibility, etc.) when evaluating these variants. The central problem of evaluation is aggregating the fulfillment of objectives, which is measured by different standards, into one key indicator.

A valid evaluation rule is that the variant that fulfills the objectives to the greatest extent is the best (maximum $FO_1$).

### 6.4.2.3  Value-Benefit Analysis (Synonym: Scoring Model)

A value-benefit analysis is an application-oriented interpretation of the evaluation matrix shown above. The only expansion illustrated here is that the criteria are consolidated structurally, i.e., in perceptibly logical groups. This method is applied to the example of an apartment selection in Table 6.5. Of course, the method is widely applicable to other and even more complex evaluation situations, such as the evaluation of manufacturing processes and concept variants/architectural designs for products; it can also be used for evaluating candidates when selecting job applicants, and much more.

The advantages of a value-benefit analysis over the balance of arguments are that:

– *Criteria* have to be established for assessing variants
– The process demands that *all* variants are to be measured by the *same standards* (criteria)
– The criteria can be assigned different degrees of importance (*weight*), which must be rendered transparent
– The criteria are *subdivided*, allowing partial results to be used for assessing plausibility

The evaluation of the quality of a solution with respect to a certain criterion can be indicated by giving a score on a scale from, for example, 0 to 10. Score 0 means very bad or not existing, score 10 means excellent. The scale may be inverted according to the respective criterion (highest performance and lowest cost may both get score 10).

We look more closely at procedural issues, special questions, and partial problems in Sect. 6.4.3.

### 6.4.2.4  Cost-Effectiveness Analysis

This process, illustrated in Fig. 6.24, differs from a value-benefit analysis in that the cost criteria are at first considered separately from the other criteria. In the upper left of the figure, an *efficiency value* is calculated for each variant in the same manner as in a value-benefit analysis. Cost criteria are evaluated separately in the upper right of

**Table 6.5** Value-benefit analysis (apartment example)

| | Weighting (w) | | Variants | | | | | |
| | | | A | | B | | C | |
| Criteria | Group | single | s | w*s | s | w*s | s | w*s |
|---|---|---|---|---|---|---|---|---|
| 1. Location of building | 20 | | | | | | | |
| Attractivity of the neighborhood | | 6 | 6 | 36 | 10 | 60 | 2 | 12 |
| Way to school | | 3 | 10 | 30 | 5 | 15 | 1 | 3 |
| Shopping opportunities | | 3 | 4 | 12 | 8 | 24 | 8 | 24 |
| | | 3 | 8 | 24 | 4 | 12 | 4 | 12 |
| Street noise | | 5 | 3 | 15 | 6 | 30 | 8 | 40 |
| Subtotal 1 | | | | 117 | | 141 | | 91 |
| 2. Building | 15 | | | | | | | |
| Appearance | | 3 | 4 | 12 | 6 | 18 | 4 | 12 |
| Location of the apartment | | 6 | 10 | 60 | 8 | 48 | 6 | 36 |
| Insulation (noise, heat) | | 3 | 8 | 24 | 6 | 18 | 2 | 6 |
| Heating | | 3 | 4 | 12 | 10 | 30 | 6 | 18 |
| Subtotal 2 | | | | 108 | | 114 | | 72 |
| 3. Social environment | 15 | | | | | | | |
| Neighborhood | | 10 | 4 | 40 | 10 | 100 | 6 | 60 |
| Friends of parents | | 2 | 8 | 16 | 4 | 8 | 4 | 8 |
| Friends of children | | 3 | 6 | 18 | 8 | 24 | 6 | 18 |
| Subtotal 3 | | | | 74 | | 132 | | 86 |
| 4. Apartment | 30 | | | | | | | |
| Size | | 20 | 8 | 160 | 4 | 80 | 10 | 200 |
| Floor plan | | 10 | 8 | 80 | 6 | 60 | 4 | 40 |
| Subtotal 4 | | | | 240 | | 140 | | 240 |
| 5. Costs | 20 | | | | | | | |
| Required investment | | 2 | 8 | 16 | 8 | 16 | 1 | 2 |
| Rent and operating costs | | 18 | 3 | 54 | 8 | 144 | 7 | 126 |
| Subtotal 5 | | | | 70 | | 160 | | 128 |
| Total | **100** | **100** | | **609** | | **687** | | **617** |

s score, (scale 0–10) w weighting

the figure. The values deriving from the cost criteria do not have to be weighed, but are simply added together, as they are measured in the same dimension. But they have to be apportioned to the same period of time, for example, cost per year. For acquisition costs, this is accomplished through yearly depreciation and the interest over the expected period of usage. The result is the total cost per variant and per year.

In contrast to a value-benefit analysis, efficiency values and cost assessment are not added together, but their ratio to one another is established through *division*. The result, in the form of a cost-effectiveness index, is a value that expresses the cost of one point on the effectiveness scale. The applicable rule for decision-making is to prefer that variant that shows the lowest value, here $V_3$.

An inconvenient feature of the cost-effectiveness analysis is that, when used as the sole evaluation criterion, it does not differentiate enough. In the case of two variants, one of which has double the efficiency value at double the costs compared

| Efficiency criteria | Weight w | Variants | | | | | |
| --- | --- | --- | --- | --- | --- | --- | --- |
| | | V₁ | | V₂ | | V₃ | |
| | | s | w*s | s | w*s | s | w*s |
| E1 Capacity | 40 | 8 | 320 | 7 | 280 | 6 | 240 |
| E2 User friendliness | 25 | 5 | 125 | 7 | 175 | 6 | 150 |
| E3 Service friendliness | 10 | 5 | 50 | 10 | 100 | 10 | 100 |
| E4 Environmental compatibility | 25 | 3 | 75 | 6 | 150 | 8 | 200 |
| **Efficiency value E** | **100** | **570** | | **705** | | **690** | |

| Cost criteria ($ in thousand / per year) | Variants | | |
| --- | --- | --- | --- |
| | V₁ | V₂ | V₃ |
| C1 Personnel costs | 140 | 180 | 120 |
| C2 Depreciation and interest | 160 | 135 | 120 |
| C3 Energy and maintenance | 260 | 265 | 190 |
| Total costs C ($ in thousands) | **560** | **580** | **430** |

| Variants | V₁ | V₂ | V₃ |
| --- | --- | --- | --- |
| Total costs C | 560 000 | 580 000 | 430 000 |
| Efficiency value E | 570 | 705 | 690 |
| **Cost per efficiency point (C/E)** | **982** | **822** | **623** |

**Fig. 6.24**  Cost-effectiveness analysis

with the other, the same ratio numbers result. Here, compulsory objectives, for example, those that limit the total costs or the efficiency value or some of its essential components, could help in the selection and facilitate a decision.

The results can also be illustrated graphically (Fig. 6.25) by positioning each variant (V) in accordance with the values of C (costs) and E (effectiveness). High efficiency at low cost appears in the positive area and vice versa. However, it should be noted that the arrangement of the scales (selection of the zero point on the scale) could easily lead to a distorted, i.e., an inflated picture of the results; moreover, the differences are not necessarily as obvious as in the graphic illustration. Therefore, if the results risk being wrongly interpreted because of a distorted picture, the scales should be entered at the coordinate axis beginning at 0.

The cost-effectiveness approach is especially suitable for those evaluation and decision-making situations where costs play an important role and where their listing in separate accounts is appreciated.

### 6.4.2.5  Other Evaluation Methods

Other evaluation methods, for example, all methods of economic feasibility or investment calculations, if applied by themselves, are usually not conducive to good decisions among the different variants, as they must assume that the decision can be

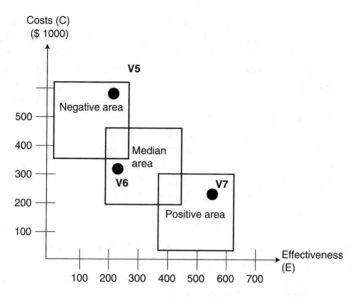

**Fig. 6.25**   Cost-effectiveness, graphic illustration (variants V5 to V7 are fictional)

reduced exclusively to an economic computation or that the evaluated variants are of equal value with regard to other features such as performance, quality, longevity, operability, styling, etc.

These methods signify different things in the context of evaluating additional variants: for example, whether or not the variants satisfy internal corporate ROI policies, how long the amortization periods are, and much more. Thus, a reasonable combination with other methods is preferable to an insistence on exclusivity (→ economic feasibility calculation, → cost-benefit analysis). A particularly interesting method is the real options approach, described in Part I, Sect. 2.2.5.

### 6.4.3   Process of Evaluation

The following guidelines pertain to the process in general, against the mental background of a cost-benefit analysis:

1. Establish the *participating group* in the evaluation: who should understand and back up the realization of results? Who is important? As opinion-shapers, as supporters, as opponents? (The latter should be allowed to weigh up pros and cons in a businesslike atmosphere. To exclude them could be counterproductive.)
2. Choose a *shorthand description* for each of the variants to be evaluated. The chosen description should clearly characterize the respective variant and be comprehensible to everyone participating in the evaluation process. It would make sense to recall the basic characteristics of the variants once more before evaluating them.

3. Finally, establish a *criteria plan*. This plan, as has been mentioned, may consist of subobjectives already secured in the formulation of objectives and those that only first emerged from the solution search. Thus, it is admissible both to take up new criteria and to omit originally determined criteria, insofar as these have shown themselves to be premature or irrelevant. However, in the latter case, it should be checked whether it was not precisely these criteria that led to the exclusion of certain variants. Such criteria would then have to be "rehabilitated," i.e., taken up again in all fairness.
4. Establish the *significance* of each subobjective (weighting).
5. Determine to what extent *subobjectives have been accomplished* (assigning scores).
6. Calculate *overall usefulness* (score times weight, adding up).
7. *Plausibility test*: are the results plausible or do they contradict an intuitive expectation? If so, why?
8. *Sensibility analysis*: does the evaluation result change if there is a variation in the assignment of weights or scores within a reasonable framework?
9. *Analysis of risks and potential problems*.
10. Determine, if necessary, the *economic feasibility* of the overall solution (important after the preliminary and main studies).

Below, we deal with special questions and subproblems that appear in the application of the value-benefit or the cost-effectiveness analysis. The order follows the procedural steps listed above.

### 6.4.3.1  Establishing the Participating Group

It is certainly useful to consult one or more representatives of the eventual decision-making body in the evaluation. There are several advantages to this:

– Decision-making is clarified, particularly with regard to the significance (weighting) of criteria.
– The decision-makers must interact in more depth with the qualities or effects of single variants.
– The decision-making process becomes more transparent on the whole, because, after all, the results of the evaluation tables require a variety of interpretations.
– Results worked out in common are, as a rule, more sustainable than those that simply receive recognition and consent.

Here, an assignment of tasks could be conducted in such a way that primarily the decision-makers determine the criteria and their weighting, whereas the members of the project team assume priority in the assessment of variants (rating).

### 6.4.3.2  Establishing Criteria

In formulating a criteria plan, one must be especially careful not to include certain qualities or effects of solutions multiple times. An example may serve to illustrate this: if operating costs and investment amounts are both to be considered when determining the fulfillment of objectives, one must take care that operating costs also include investment costs, usually through depreciation.

Exactly what a certain criterion has to evaluate should be clear. If, as in the present case, investment costs are already in fact included in the operating costs via depreciation and interest, two possibilities are conceivable: either investment costs are eliminated from the criteria catalog altogether, or their evaluation is limited to a consideration of the difficulties of capital procurement, of aspects of liquidity, risks, etc. In a case of elimination, one should check whether the cost side by and large has the significance ascribed to it or if the weighting of operational costs should be increased.

Another problem might be that individual features can have different effects. The engine performance of a car can be positive with regard to acceleration, but negative with regard to consumption or insurance rates. The weight of a vehicle can have a positive effect on comfort and security, but a negative one on consumption. Thus, horsepower and weight would not be the actual evaluation criteria, but merely the *indicators* that offer reference points for how well or how badly a particular variant rates in respect of consumption, security, or comfort (Table 6.6).

This consideration is a relatively simple matter in the assessment of known or existing solutions. It becomes difficult when solutions exist only in drafts and therefore lack any operating experience. For example: in an industrial plant, noise pollution of the environment is an important criterion. Now, two possibilities are conceivable: the engine unit, as the main noise-maker, is isolated, or a noise barrier is erected around the whole plant. It is difficult to reliably estimate the quantitative effects of either of these measures in the planning stages. Here, additional analyses or even experimental arrangements of prototypes might be necessary. Advice from experts, conclusions drawn from analogous models, etc., could prove helpful.

### 6.4.3.3   Treatment of Compulsory Objectives

Here, three subquestions are of interest:

(a)  What role do compulsory objectives play within the context of an evaluation?
(b)  May compulsory objectives be questioned?

**Table 6.6**  Derivation of criteria from characteristics or features

| *Characteristics and/or features* lead to → | *Criteria* understood as positive effects (to be achieved) or negative effects (to be avoided) |
|---|---|
| Horsepower | + Acceleration |
|  | + Active security |
|  | - Operating costs (insurance, fuel consumption) |
|  | - ... |
| Vehicle weight | + Comfort |
|  | + Passive security |
|  | - Operating costs, ecology (fuel consumption) |
|  | - ... |
|  | ... |

(c)  Under what conditions is it reasonable to formulate later, additional compulsory
     objectives?

*(a) The role of compulsory objectives*
It is helpful to distinguish between two categories of compulsory objectives: those
whose achievement can be answered simply by *yes* or *no*, and others whose achievement can take markedly different forms. In the first category, variants that do not
meet the objectives can be eliminated. However, this category allows no differentiation among the remaining variants and is irrelevant for the further evaluation. If the
compulsory objective is "electric current switchable from 240 to 120 V," the variants
that do not fulfill this condition are not admitted to the evaluation. The compulsory
objective is no longer necessary.

The second category includes those objectives that represent restrictions, but, beyond
that, remain noteworthy. If the compulsory objective is "operating costs per year 50,000
maximum," all variants that go beyond this limit would have to be excluded. However,
operating costs could continue to serve as an evaluation criterion. In such cases, though,
one should not undertake a high weighting, seeing that expensive variants have already
been excluded and only the difference from 50,000 has to be assessed.

*(b) Questioning compulsory objectives*
Another consideration in the context of compulsory objectives has already been
referred to. Sometimes, there are existing solution ideas or variants that would be
very advantageous in many areas of interest. However, they are not allowed in an
evaluation if they violate a compulsory objective. If the retraction or change of the
compulsory objective obviously entails major advantages, this question should not
be declared off-limits for formal reasons. Of course, agreement with the commissioning party must be sought.

*(c) Introducing compulsory objectives at a later point*
The question whether or not to introduce an additional compulsory objective during
an evaluation process comes up when a variant is judged to be very inferior in
respect of a fairly important criterion, but on the whole is rated favorably.
Accordingly, a restriction leading to the elimination of this variant would in this
case be warranted if it can be reasonably justified. As in the preceding case, it concerns a correction of an objective in terms of a learning process and would in any
event have to be discussed with the commissioning party/decision-making body.

### 6.4.3.4  Number of Subobjectives

The question regarding the "right" number of subobjectives that should be taken into
account is difficult to answer, as it depends very much on the type of existing evaluation problem. Generally, it may be said that the more comprehensive a criteria plan:

–  The more nuance can be given to assessing different variants
–  The more difficult it is, on the other hand, to determine the relative significance
   of a subobjective (weight allocation)
–  The more expenditure is involved in the evaluation

A growing number of subobjectives do not necessarily produce more objective evaluation results, because the tables become harder to survey and the opportunities for a deliberate manipulation may even increase. A workable and proven dimension for many evaluation situations is an approximately 20–25 subobjectives (the rationale: the total situation can still be well represented on an A4-sized sheet of paper).

### 6.4.3.5 Weighting of Subobjectives

#### Basics
Both the establishment of a criteria plan and the allocation of weights frequently involve juggling and gauging so many different views and values that an agreement cannot be effected immediately and several iterations are necessary. To avoid unnecessary repetitions of the evaluation process and to achieve the best possible consensus with the decision-making body with regard to a recommended decision, it is expedient, as already mentioned, to ascertain what the decision-making body thinks about establishing subobjectives and allocating weights, and even to involve it actively in the process.

#### Limitation of Weight Reserves
Experience has shown that limited reserves are more carefully managed than unlimited ones. This also applies to weight reserves. Limiting weight reserves (for example, to a weight total of 100 or 1000) results in a more attentive allocation of weights and is furthermore practical for any later changes of weights (weight shifts, new intake, or elimination of subobjectives).

If the total weight were unlimited, which technically would be quite possible and, in the case of later changes, even facilitate matters considerably, then a newly received subobjective could be assigned a certain weight without any changes having to be made in the weights of the other subobjectives. Because the total weight would increase to the same extent, all other subobjectives would thereby have to participate proportionally in an equal devaluation (= inflation). It would not be necessary to check the effects on all the other criteria. The reliability of accuracy and efforts may suffer as a consequence.

Therefore, it is recommended, in both the reception of new and in the elimination of existing criteria, to scale the total weight to a fixed sum (for example, 100). Thus, one is forced to examine the overall relations. Although this undoubtedly has the drawback of having to recalculate the evaluation scheme, it is made less difficult by the increasing the use of PCs and the implementation of spreadsheet programs.

#### Process of Allocating Weights
In allocating weights to criteria, it is expedient to proceed from the general to the particular. With respect to the target structure shown in Fig. 6.26, this means to undertake first a rough distribution of weights at the level where the classes of objectives are found (financial, functional, and social objectives). Within these classes, a detailed allocation is made among the subclasses of objectives, and from there a detailed allocation at the operational level.

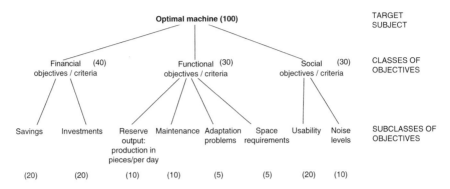

**Fig. 6.26** Allocation of weights according to the principle of node weighing

This process has the following advantages:

– A rough distribution allows the weights of whole groups to be put into proportion with each other.
– The process is made more transparent: all the weight reserves are no longer available for single objectives, but only deliberately limited reserves.
– The agreement process is simplified: later weight shifts often remain limited to an area of a particular class of objectives.
– The significance attributed overall to a class of objectives no longer depends on the more or less arbitrary number of criteria belonging to that class.

If the allocation of weights is begun at the lowest level, there is a tendency to over-emphasize classes of objectives that involve many criteria, even if each single criterion is given only relatively little weight.

Note: we have reservations about the detection of criteria weights by *comparing* single criteria *in pairs* (for example, in the ➜ analytical hierarchy process method), as is sometimes propagated. This method is elaborate, requires later calculation processes, and its result is not readily transparent or comprehensible for the decision-making committees.

### 6.4.3.6  Determining the Achievement of Subobjectives

Scores are used to express to what degree a variant fulfills a certain subobjective. Using a 0–10 scale is practical. Variants that rate excellently with regard to a certain criterion receive a score of 10; those that are utterly insufficient a score of 0 (for example, when they do not exhibit a feature or effect demanded by the objective – which, however, may not have the nature of a compulsory objective). Average achievement receives a score of 5. The remaining numbers are used to indicate gradation. Of course, many other scales are also conceivable and applied in practice.

Below, we deal with several special issues.

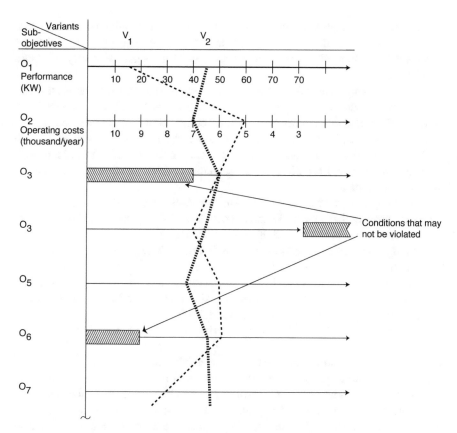

**Fig. 6.27**   Graphic illustration of the fulfillment of objectives

### Objective Measures as Starting Points

People often establish scores solely by means of a rough assessment, without a special determination process. Such a method of determination is not satisfactory; it can be improved if the fulfillment of subobjectives can be established on the basis of measured values. This aspect was referred to earlier in connection with the demand for an operational formulation of objectives.

Here, one assumes that measuring scales should help initially to illustrate the fulfillment of subobjectives (for example, Euro/$, kW, mph, amount; Fig. 6.27). Conversion into scores does not take place until the next step.

### Transformation of Characteristics on a Scoring Scale

The particular question here is whether or not the assignment of scores should exhaust all the space available to it, in other words, should the variant that is best with regard to a certain subobjective be assigned a score of 10 in any event, and the worst a score of 1 (or even 0)?

This question is equivalent to asking about the orientation point that should serve as the basis of the evaluation: do the variants that are to be evaluated furnish the reference point themselves (relative standard) or is it sought externally (absolute standard)?

A relative standard is easier to manage because only the best and the worst of the existing variants are sought; these are assigned a score of 10 or 1 (or 0), on the basis of which the remaining results may be interpolated. However, it has some significant disadvantages, for example, when variants that have been introduced into the evaluation at a later point make it necessary to change the reference base. Moreover, it distorts the proportions when the measurement results of different variants lie very close together.

For example, in a case where different machines have to be evaluated, if their operating costs lie between 5000 and 5500, it would surely not make sense to give the best variant a score of 10 and the worst a score of 1.

To solve this problem, the following possibilities are conceivable: one eliminates this criterion if the difference among variants is only minor, or one assigns a score of 5 to a median variant (which does not actually have to exist) and takes into account up and down deviations by means of additions and deductions. In the upper case, reasonable scores would range between 4 and 6.

A graphic illustration of this curve yields a so-called utility function.

### Utility Function as an Instrument for Determining Scores

The recommendation to plot the course of a utility function on a graph is intended to make the value judgments transparent, for oneself and for others, and is thus open to discussion. If there is uncertainty about the assignment of scores or the conversion of measurement scales into scoring scales, the attempt should be made to manage the problem by means of graphs. Figure 6.28 illustrates an example of the course of different utility functions.

The upper-left graph shows a linear course with an upper (200 kW) and a lower (100 kW) barrier. Variants with less than 100 and more than 200 kW are not allowed; no intersection point is possible in the diagram.

The upper-right graph shows the course of a progressive utility reduction, signifying that less noise pollution in the upper area yields a higher score than in the lower area.

The converse is true in the bottom-right graph, in which the additional utility of higher performance finds increasingly less appreciation. Although a performance of more than 200 kW is acceptable, it does not entail an increase in points, as the maximum score of 10 is already assigned, starting from 200 kW.

The bottom-left graph shows the optimal date for which both a deviation that goes beyond it in addition to one that falls below it are acceptable, though both lead to deductions in scores.

The attempt to illustrate value judgments in a graphic form is especially useful for the additional reason that it diverts attention from concrete solutions for the moment and compels one to focus on one's own value judgments.

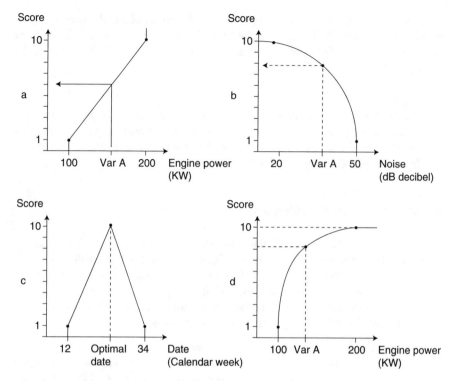

**Fig. 6.28** Example of the course of utility functions

When composing such a utility function, it is advisable to start with extreme values and to ask oneself: when would I begin to rate a performance or a price as very good with a score of 10? When would it become unacceptable (a score of 0), or barely acceptable (score of 1)? Thus, one would have found two supporting pillars between which any desired function may be interpolated. With this instrument, the project team can work out a transparent and comprehensible evaluation scheme, which reduces coincidences, errors, and arbitrariness in the assignment of scores.

**Various Scales**
These kinds of utility functions can be depicted, however, only when the effects of their characteristics can be measured and represented in so-called *cardinal scales*, for example, price, costs, kW, $m^2$, square feet, sec, etc.

However, there are also effects or features of solutions where this is not possible, and only nominal (verbal) descriptions for the quality of solutions are available. For example: confidence in the manufacturer: very great, great, less great, etc.; design: excellent, very good, pleasing, takes getting used to, etc.; flexibility, for example, in the use of other source materials: high, medium, low. Here one speaks of *nominal scales*, which can be converted into a scoring scale.

*Ordinal scales* establish a ranking of variants in respect of a certain criterion, but which do not additionally quantify the gaps. In these cases, variants are ordered according to rank (first, second, third, etc.), from which the scores are to be determined.

**Scaling Matrix**

As an instrument for transforming cardinal, nominal, and ordinal scales into a scoring scale, one can produce a scaling matrix (Table 6.7). It also contains criteria that can be accessed in cardinal measuring. In this case, a scaling matrix replaces a graphically depicted utility function.

### 6.4.3.7  Plausibility Considerations

Once the results of an evaluation have become available after the first calculation, they should undergo a plausibility test. This can be completed relatively quickly if the decision situation is obvious, i.e., if the intuitively expected results harmonize with those that have been calculated and an all-round favorite has emerged.

If this is not the case and the calculated result comes as a surprise that cannot be readily accepted, the following measures are recommended:

– Checking the *evaluation chart* for *calculation errors*.
– Checking the *criteria plan* for *completeness*: often an intuitive expectation includes subobjectives of which one is not aware and which for some reason are not contained in the criteria plan. To become aware of these, one should look for the reasons why another variant is considered better or worse than what the evaluation results indicate. Here, it may be necessary to supplement the criteria plan and to renew the evaluation process.
– Checking *utility equivalents*: this step is based on the thought that establishing a partial utility (score times weight) may lead to constellations that do not appear justifiable.

  The following example serves to illustrate this: a scaling matrix developed in connection with the acquisition of a machine would have yielded the results shown in Table 6.8. The formula allows one to determine how the assessment of operating costs relates to the assessment of performance.

  Concretely: if we choose the more economical variant ($7000/year), what is the equivalent sacrifice in performance?

  Result: a difference of $1000 in operating costs is of equal value to us as a difference in performance of 4 kW.

  Now the question can be asked: does this result really tell what we want to express with regard to our values?

  If no: in what direction should corrections be made? Are an additional 4 kW of more or less value to us than $1000 per year operating costs? Depending on the outcome, we would have to change either the utility function or the weighting of the performance criterion.

**Table 6.7** Scaling matrix for determining scores

| Criteria (examples) | 0 | 1 | 2 | 3 | 4 | 5 | 6 | 7 | 8 | 9 | 10 | Scoring scale |
|---|---|---|---|---|---|---|---|---|---|---|---|---|
| | Very bad | | Bad | | Medium | | | Good | | Very good | | General nominal scale |
| Confidence in manufacturer | Not in the least | | Low | | Medium | | | High | | Very high | | Nominal scale |
| Design | Totally unsatisfactory | | Barely acceptable | | Acceptable | | Pleasing | | Very good | | Excellent | Nominal scale |
| Design | 6 | | 5 | | 4 | | 3 | | 2 | | 1 | Ordinal scale |
| Area required (square feet) | >26 | 26 | 25 | 24 | 23 | 22 | 20 | 19–15 | | 14–10 | | Cardinal scale |

**Table 6.8** Plausibility test

| Criteria | Weight w | Variant A Quantitative value | s | w × s | Variant B Quantitative value | s | w × s |
|---|---|---|---|---|---|---|---|
| Engine power (kW) | 30 | 100 kW | 4 | 120 | 110 kW | 5 | 150 |
| Operating costs per year ($/€) | 12 | 8000 | 4 | 48 | 7000 | 5 | 60 |

Formula: $\dfrac{\text{diff.}(\text{Engine power})}{\text{diff.}(w^{*}s)(\text{performances})} \times \text{diff.}(w \times s)(\text{operating costs}) = \dfrac{110-100}{150-120} \times (60-48) = 4$

- *Sensibility analysis*: checking allocations of weights and scores (possibly changing the course of the utility function) within an admissible framework (see Sect. 6.4.3.8). This is especially appropriate and also necessary if the evaluation process has led to differences of opinion or problems of agreement about the selection and/or weighting of criteria or the assigning of scores to variants, or if it has caused general uncertainties about the assessment.

It would be incorrect to interpret all these measures to mean that the evaluation chart should be manipulated until the intuitively expected result can be mathematically confirmed. Their purpose is simply to enable us to identify unconscious value judgments or possible global judgments, to test if these are warranted and, if they are suitable, to introduce them into the evaluation process.

Also, this often entails a necessary correction of intuitive perceptions, especially when these are based on an unconscious overestimate of certain subobjectives whose significance, however, can now be considered in an ordered form.

Therefore, a *systematic method* and *intuition* should not be rivals, but should rather be seen as two approaches that supplement, control, and, if need be, correct each other.

### 6.4.3.8   Sensibility Analyses

A sensibility analysis (also called sensitivity analysis) helps to determine if preferred variants change when the prerequisites that have resulted in this preference also change. As mentioned earlier, this can pertain to the criteria employed and their weights, and to the allocation of scores.

The following possibilities are conceivable:

– Checking the *interim results*: thus, in the evaluation example in Table 6.5, the subtotals would be scrutinized regarding their plausibility: do we truly believe that variant B is the best with regard to the criteria groups 1, 2, and 3?
– Checking the *weights*:

  Should not criteria group 4 (the apartment itself) carry a higher weight than 30? Are the other criteria truly that important?

  Is the distribution of weights between size and floor plan (20:10) a reasonable one?

  Is street noise not underrated at 5 points?

  How does the result change if we change the weights within an admissible framework?

– Where does a variant receive notably many or few points (w * s)? Which variant? Is this in fact warranted? How does the result change if the scores are changed within an admissible framework?

Performing a sensibility analysis is quite easy with the help of a spreadsheet program. However, the analysis should be limited to truly essential variants, because otherwise one may easily lose an overview of what was changed and why.

It is often the case that although the numbers may change, the results in the decisive categories remain stable. It may even be the case that differences, and thus the head-start of a favorite, increase. This alleviates the decision-making problem. But matters become difficult when the results of a sensibility analysis lead to a change in ranking.

Besides performing further clarifications and examinations, which are popular patent remedies in such cases, one can choose a different approach and turn the question around: if the results lie close together, it may not even be that important which variant is preferred. The risk of a wrong decision is minor. We let the chairman decide, or someone else who has the courage to bear the responsibility for this decision. Or: what risk does this or that variant entail? If we have not agreed about the advantages in the first place, we may perhaps come to a quicker agreement about the risk and suggest a variant that involves less risk. This is taken up in the next section.

**Table 6.9**  Matrix for risk analysis

| Risks | Variant A | | | Variant B | | |
|---|---|---|---|---|---|---|
| | S | P | P × S | S | P | P × S |
| Delivery problems of subsuppliers | 4 | 3 | 12 | 2 | 4 | 8 |
| risk of delays | 1 | 6 | 6 | 3 | 2 | 6 |
| **Total risk evaluation** | | | **18** | | | **14** |

Inspired by the Kepner–Tregoe method, for example, Rummler and Brache)
S degree of severity if risk occurs
P probability of occurrence

### 6.4.3.9   Analysis of Risks and Potential Problems

It is recommended to carry out an additional special risk analysis in those cases where various risks or resulting problems in connection with different solution variants are to be expected. Here, one could again apply a process that uses an evaluation matrix. In this case, the criteria by which each variant is to be judged are the possible risks or the potential problems. For each variant and risk one would evaluate the *probability* that a risk event might occur and, should it occur, its *severity*.

A scale from 0 to 10 can be used to score *probability of occurrence of a risk (P)*. 0 means that the risk will most probably not occur; 5 means that the risk is moderate; and 10 means that the risk has the highest probability of occurring.

The *severity (S)* of a risk that is to be expected if a risk occurs, can likewise be expressed on a scale of 0 to 10. 0 means practically no effect; 5 means moderate severity; And 10 means great effect = catastrophic effect. Note: the scope does not necessarily depend exclusively on the type of risk, as it can also be influenced by a variant that is fairly resilient vis-à-vis risks.

Now, for each risk and variant, the value P * S is calculated and the total for each solution variant is generated, so that the total value can be viewed as a key indicator for the total risk occurrence of a solution variant (Table 6.9).

However, one must check whether a risk has not already been factored into an evaluation of utility, as this would falsify the result.

See also → risk analysis, → security analysis.

## 6.4.4   On Objectivity in Evaluation Processes

In connection with the performance of plausibility and sensibility analyses, it may be asked, what is it about these methods that is actually objective? The answer is short and clear: almost nothing. Let us look at each of the procedural steps.

At least partially *subjective* and thus contestable are:

- The selection of the variants being evaluated variants. They are, after all, worked out according to partially subjective goals

- The selection of subobjectives for the criteria plan
- The weighting of subobjectives
- The allocation of scores (when do we assign a score of 10, or 1 or 0?).

*Objective* are merely a few calculation operations:

- Determining subutilities (weights × scores)
- Determining total utility (sum of the subutilities for each variant)

If the doors are so obviously open to manipulation, one has to ask why such processes are being propagated in the first place. Two important reasons can be given:

> There is no method or tool that excludes subjectivity, and the reason is that no solution has an objectively determinable value. There is only value in a particular context, and this value is essentially determined by the subjective appraisal of the problem situation, the goals that are pursued, the subjective influences that play a part in the solution search, and finally the likewise subjective influences on the assessment of the solution.
>
> The method shown is an excellent instrument for rendering the decision-making situation transparent. The completed evaluation chart is at the same time the *rationale* for a recommended decision. It contains the criteria, their significance (weight), the subutility calculated in the process, and the summed total utility. It thus compels one to think about value judgments and to structure them; thus, it helps to avoid purely intuitive and quite arbitrary decisions. Nonetheless, it allows, as we have seen, room for intuition.

Given this backdrop of undisputed subjectivity, it seems only consistent to keep the process as simple as possible and comprehensible for the decision-making bodies – and not to complicate it unnecessarily, thereby possibly even creating a specious "scientific" objectivity.

### 6.4.5   Preparing a Decision: The Economic Feasibility Calculation as a Supplement

As mentioned earlier, evaluation techniques are not meant to examine the meaningfulness of a solution for its own sake. Instead, they allow a *comparison*, which is oriented to the objectives, with other variants.

Particularly in the course of the early phases of a problem-solving process, the question often arises whether an enterprise is at all economically profitable, or if it would be preferable to terminate the project.

For these kinds of questions an ➔ economic feasibility calculation, for example, would be appropriate. With this technique, those variants that are favored in the value-benefit analysis can be comprehensively evaluated with regard to their economic effect. Thus, the objectives applied in the variant comparison could also serve as the starting point for this calculation. Additional references may be found in the encyclopedia (Chap. 16) under the key-word ➜ "cost-benefit calculation."

### 6.4.6 Documentation of the Evaluation Step

To see the evaluation step as a sustainable basis for the decision-making process, it is useful to properly document all considerations and results and to present these to the decision-making entity. In addition, to an account of the evaluated variants and the evaluation schemes, documentation includes, in particular:

- The rationale for the criteria and their weighting
- The shape of the utility functions, or the allocation of the variants in the scaling matrix
- The rationale for the scores (based on which characteristics and indicators). This may be provided, for example, in the form of footnotes attached to the evaluation scheme
- The plausibility and sensibility analyses
- The recommendation and its summary rationale
- Possible flaws or risk factors and the measures to control them
- The valid compulsory objectives and the variants that were eliminated because they failed to meet those compulsory objectives at the beginning of the evaluation step
- The economic viability of the favored variants

Proper documentation should also prevent solutions that were treated intensively and seriously, but were finally excluded through well-considered arguments from being brought into the discussion again without any essential change in the prerequisites or the solution idea.

### 6.4.7 Decision

Building on the results of the evaluation, one should now select the variant that is to be described in further detail or brought to realization. The greater the participation of representatives of the decision-making body in the evaluation process (possibly even in the solution search), the fewer difficulties they have during the decision-making phase. First, they will be more familiar with the facts underlying the particular solutions; second, they will have had more opportunities to contribute their value judgments, intuitive views, and expectations to the process of the solution search and evaluation.

   With regard to the retroactive effects that the decision phase may have on the evaluation and the solution search, we refer the reader to Part II, Sect. 3.3 (on repetitive cycles). In respect of the collaboration of the commissioning party/decision-making body and the planning group in the search for objectives, the search for solutions, and the selection, see Part II, Sect. 3.4.

## 6.4.8   Summary and Rounding Off

1. The purpose of an evaluation is not to assess the suitability of a particular solution, that is, the task of (solution) analysis; it should rather determine which of the several qualified variants is the best, second-best, etc.
2. Important decisions, which have far-reaching consequences and for which several persons are responsible, should be supported methodically.
3. For this, there are different evaluation methods, for example, the value-benefit analysis and the cost-efficiency calculation.
4. The inclusion of individual representatives of the decision-making body in the evaluation process generally makes for more sustainable decisions.
5. Various instruments, such as the graphic representation of utility functions or the creation of scaling matrices, support the evaluation process by making it more transparent.
6. Allocating weights to the criteria is a highly subjective evaluation task that should be borne primarily by the decision-makers.
7. Likewise, this applies to establishing the principal shape of the utility functions.
8. Assigning scores, i.e., assessing how well or how badly particular variants rate according to certain criteria generally requires in-depth expertise and knowledge of the situation and the solutions. It should be carried out primarily by specialists from the project group.
9. Results of an evaluation step should be tested by plausibility and sensibility deliberations.
10. No evaluation is objective in the sense of delivering a solution that can consistently be proven to be the best. Each evaluation is based on a variety of subjective value judgments and appraisals of a situation.
11. Intuition and methodology (systematization) are not opposites, but different approaches that should support and monitor each other.
12. A final, comprehensive testing of the meaningfulness of a decision recommendation, with special attention paid to economic feasibility, should help to achieve more security, that is, to identify flaws and risk factors that will have to be observed more carefully after the decision.
13. The real options approach (see Part I, Sect. 2.2.5) allows a mathematical analysis of options that was not previously available from traditional business economics. This approach, moreover, facilitates planning systems that emphasize the flexibility of processes and results (agile systems engineering).

14. The results of an evaluation and the accompanying deliberations should be properly documented so that they may be transparent and replicable.
15. An evaluation does not replace the volitional act of a decision. It does, however, render decision-making transparent, as it compels the participants in the decision to think thoroughly about their standards and about the functions and effects of different variants, and to structure these accordingly.

### 6.4.9 Self-Check for Knowledge and Understanding: Evaluation and Decision

1. Which steps in the PLC are the main information suppliers for the evaluation
2. What is the essential difference between analysis and evaluation?
3. Are improvised (intuitive) decisions acceptable or should decisions always be methodically supported?
4. Which methods of comparative assessment do you know?
5. Why does it make sense to limit the sum of weights of the criteria (100 or 1000)?
6. Why does it make sense to represent the course of a utility function on a graph?
7. Why is it recommended to conduct reasonability analyses or sensitivity analyses in the evaluation process?
8. Are the methods for comparative assessment "objective"?
9. Why use evaluation methods that are not "objective"?
10. Is it possible for the decision-making body to deviate from the proposal of the project group?

## 6.5 Special Cases and Situational Interpretation

We have frequently pointed out that the methodology of systems engineering is not a panacea for all types of problems. Rather, it provides general recommendations for action that require intelligent and situation-related interpretation. By this, we mean the ability to relate to a situation and to single out those methodological elements that are found to be reasonable and that one is prepared to implement and carry out, and of course, conversely, to omit or modify those elements where this is not the case.

Below, we consider some *cases of application* that we regard as typical in respect of the requirement of a *situational interpretation*.

– Modifications of the "living object" (Sect. 6.5.1)
– Improvement, melioration plan (Sect. 6.5.2)
– Plans of limited scope (Sect. 6.5.3)
– Plans of unusually large scope (Sect. 6.5.4)
– Programs (Sect. 6.5.5)
– Staggered deployment (Sect. 6.5.6)

However, we also consider special situations to which developers, designers, planners, and implementers are subject, such as:

- The relative inexperience of the participants (Sect. 6.5.7)
- Keeping options open (Part I, Sect. 2.2.5)
- Entering disorganized problem-solving processes (Sect. 6.5.8)
- Shut-downs and terminations (Sect. 6.5.9)

### 6.5.1  Modification of a "Living Object"

Often, an enterprise is not involved in new planning (as a "wide open field"), but rather in a local change or an expansion, or even a re-dimensioning or improvement of an existing solution. An additional consideration is that the old system should continue to run as undisturbed as possible during the modification. The special requirements resulting from these scenarios are:

- Good situation analyses that relay a well-founded understanding of present functions, processes, and relationships, as these present conditions are usually also influenced and impaired, even when they are not the object of the renovation.
- Scrupulous realization planning that fulfills two requirements: each realization step should represent an advance toward the new solution and create a stable condition that allows a continuous, even if sometimes limited, operation.

This can often be done only with great difficulties or additional efforts. Such as alleviating steps or interim measures that do not advance the solution by themselves but are necessary in the interest of maintaining operational functions (for example, provisional traffic or transport routes and working places, relocations of functional units, and much more).

The project management team is especially important for accurately planning and directing processes, and for relaying timely information to all participants and concerned parties and coordinating them.

### 6.5.2  Improvement (Melioration) Projects

It is a characteristic of improvement (melioration[9]) that essential circumstances of the actual condition are not to be changed, but only improved in certain areas.

This is often the case because an existing system may be functioning, but not optimally. Dissatisfaction, for example, with functions or performances, is the reason for engaging in improvement plans. When improving an actual condition, it is obvious where one has to look. By this, we mean that one does not fundamentally

---

[9] Derived from the Latin comparative form: good = bonus, better = melior, best = optimus

call into question the existing functions, processes, or material inputs, but instead tries to improve these partially and gradually according to the formulated objectives.

Although this obvious procedure, which was perfected in Japan through ➜ *Kaizen* (improvement), may be faster and less elaborate, its application is not without limitations. On the one hand, there is the risk of treating localized symptoms while overlooking that issues lying outside this local area must be included to solve a problem overall. On the other hand, every concept becomes outdated at some time and a new conception becomes necessary.

Although a system melioration presupposes a basic definition of the problem, it is itself the result of a decision about which individual problems are to be tackled:

- There exist a number of individual problems that may be only loosely related to one another.
- The overall solution results in the sum of individual solutions that often, though, cannot be selected independently of one another.

Many tasks in both the economy and the public sector constitute melioration plans. Their peculiarities pose special challenges to the process. These are dealt with below on the basis of the steps involved in the PSC.

**Situation Analysis**
Frequently, the starting situation is not clear and an inquiry into the details is necessary, as the basic state of affairs is not up for dispute.

It is advisable to first run through the steps of the PSC quickly, to gain an overview, to carve out the core points of the subproblems, to detect the relevant causes, and to formulate possible solution approaches in the sense of working hypotheses. Then one can proceed more specifically with the analyses required for defining particular problems. (This process is a modified implementation of the "from the general to the particular" principle.)

**Formulation of Objectives**
Often, only quite vague objectives can be formulated for the melioration plan as a whole. These objectives are barely helpful in the search for solutions to single problems (and in the evaluation of variants). Sometimes, the boundary between objectives and solution ideas is not clear at the outset.

However, even when the triggers for melioration plans and objectives can be articulated relatively clearly, there is often much uncertainty whether or not they are at all realistic and attainable with the planned resources. The reason for this is that it takes vast experience to appraise how individual improvement measures bring about the total result.

**Solution Search**
Although a systems melioration is concerned with solutions for several individual problems, it is advisable to try to formulate common themes and principles for the

solutions instead of an overall concept. This increases the chance of finding single solutions that harmonize with each other.

Various solution variants are designed and assessed for each individual problem. Insofar as problem and solution fields are independent of each other, one can choose the best single variant in each case. However, this is often not the case; then, the integration of concepts takes on special significance. In such a situation, a procedure roughly as follows is reasonable:

- For each individual problem, several independently applicable solution variants are developed.
- The individual problems are then interpreted as parameters and their solution variants as expressions of a ➔ morphological matrix.
- Through a combination of parameters, useful overall solutions in the morphological matrix can be constructed and evaluated (with elimination of incompatible or clearly less suitable pairs).

**Evaluation**

In melioration projects, the actual current condition should be included in the evaluation. This not only allows the single-solution variants and components to be compared relative to one another, but it also makes evident the complete extent of the effects of any changes. It could be the case that such an overview forces one to the conclusion that certain parts of the improvement recommendation are not worth the activities to realize them.

## 6.5.3   Initiatives of Limited Scope

The systems engineering action model as a whole is directed toward the completion of complex projects. For plans of limited scope, simplifications are of course admissible and reasonable. Such plans, for the most part, do not require an actual project organization; because they are usually easy to survey, the circle of participating persons is small (conceivably, only one person works on the problem). Hierarchies are limited to a few, often to only one or two levels, and the tasks at both levels are managed by the same team. In such cases, aspects of the phase model and the PSC are usefully combined into a simple action plan. As the plan is easy to survey, it is possible to continue to introduce modifications, which become increasingly necessary during the course of the work owing to the growth in knowledge.

At the beginning of the work, however, it must be made clear whether or not such limits imposed on the scope and depth of the process are appropriate, to avoid working out a solution that relates to a problem that is imperfectly understood.

### 6.5.4 Initiatives of Unusually Large Scope

Plans for solving social challenges of the future, for example, traffic or energy concepts, often have an unusually large scope. Generally, they are directed toward goals on the horizon that are far-off in time with regard to the applicability of their results, they go through many stages in which their technical, economic, and ecological feasibility and social acceptance are tested, and they often depend on other major plans. Their characteristic features are:

– A large circle of participants and concerned parties whose intentions and preferences undergo frequent changes
– Examining and planning teams that shift at each step and in each phase of development
– High innovative content and degrees of uncertainty
– Constant changes with regard to appraisal of the development of the system and its environment, and of the resultant system demands, etc.

The only thing constant in these plans are unforeseeable changes. To control these changes, a many-tiered, broadly conceived "preliminary study phase" with extensive examination and study programs, which cover the expanded problem and solution fields, is necessary. Also required is strong project organization from the outset that emphasizes the areas of configuration and interface management and documentation as well.

The considerations described below also apply to some extent to these kinds of reflections.

### 6.5.5 Programs

Many plans that have a similar goal are referred to as programs, often located in the public or semi-public sector. The goal of research and development plans may be to exploit a chance (for example, solar energy production, application of micro-electronics in mechanical engineering) or to avert a danger (for example, reduction of atmospheric pollution, humanization of the workplace). The goal of a concentrated package of measures might be to contribute to reducing individual motorized traffic in inner cities, lowering costs in healthcare, or cutting overheads in a company.

Each plan can be developed in principle independently of other plans and in accordance with the systems engineering action model. Given the relatively large latitudes that research and development plans should possess, especially during the early phases, it is highly probable that there will be unrecognized, mutual influences in the problem and solution fields belonging to other plans or the latent appearance of both positive and negative side effects in neighboring fields. In imposing measures and sanctions, there is the risk of an unwanted accumulation of consequences, and of the weakening of the intended effects.

Therefore, it is expedient to define and initiate the plans of a program as interactive components and to monitor and direct their execution in the form of program management and controlling. This also includes, for example, regular joint discussions about progress, results, and further intentions.

## 6.5.6  Staggered Implementation

In complex systems, notably in business and organizational domains, is not possible to construct and introduce a system with one stroke, because resources are often limited or the associated risk is frequently too great. Therefore, system building and system introduction often take place in several stages. However, if these stages extend over a longer period of time, the system environment and/or the expectations made of the system run the risk of changing so greatly in the interim that the developed concepts are no longer the best or even suitable (see also Sects. 2.2.4, 3.2.3.5, and 3.2.3.6).

One can counter this circumstance, at least in part, with a process that uses the following guiding ideas:

–  An overall concept is developed in the main study that contains the functioning method and the realization principle of the most important subsystems or system aspects, but is directed to a stage-by-stage realization.
–  Detailed concepts are worked out in a staggered manner and brought to realization as soon as each is finalized.
–  Before going on to the next stage, the overall concept is checked and, if need be, adapted, thereby allowing an influx of both the insights gained in earlier detailed studies and in their realization and of new environmental factors and demands.

This process has a number of advantages:

–  It is more considerate of the usually limited capacity of qualified personnel and of the receptivity of the users.
–  The experience that the systems designer and the system user have gained from the establishment, introduction, and usage of the system can be better exploited.
–  Risks can be kept lower and there is also greater flexibility in adapting objectives to new situations.
–  The plan brings earlier application benefits.
–  If central system parts are realized first, the degrees of freedom are reduced as desired, in addition to the uncertainties in the design of further system parts.

A process of this kind represents a differentiation in the phase model. The sequence of phases after the main study is now valid only for individual stages, for which reason different subsystems may find themselves at very different stages at the same time.

A successful implementation of these guiding ideas presupposes that there has been a skillful selection of the sequence according to which the different system parts are worked on. Thus, the viewpoints on establishing priorities are frequently in opposition:

- Demands from the standpoint of the system:

  Logical sequence for utilization
  Conceptual significance of system aspects

- Initial training of the system team
- Urgency of resolving certain flaws (or exploiting opportunities)
- Simple things first
- Desire for quick financial benefit

It may be the case that certain parts must be introduced into a provisional version and modified or replaced later. Project management becomes more elaborate and the commissioning party must collaborate more intensely for the release of the stages.

These processes are also useful in the situation "Modifications of the living object."

## 6.5.7 Relative Inexperience of Participants Because of Pioneering Situations

Pioneering situations may occur for the planners/developers and for the commissioning party, and also for both sides simultaneously.

*Pioneering situations arise for developers and planners* not only when the solution to the problem is not known, but also when the solution and the path to it are generally known (based on the state of science, of technology), but their own implementation experience is lacking. The pioneering character may pertain to the system, to new components for established functions (for example, in control engineering), to new materials and the method of their processing, to potentially applicable production techniques, and to methods of finding and designing solutions. In such cases, it appears expedient to use experience learned elsewhere: one may study the solutions realized there or secure external know-how by consulting experts.

*Pioneering situations for the commissioning party* occur when the commissioning party or the persons or committees appointed by it do not or insufficiently understand the complexity of the problems, the expansion of solution fields, the implications of solution variants or the necessity of a formal project organization, and the mechanisms of completing a project. These situations can come to light when demands are constantly changing, conflicts about goals are unresolved, decisions are postponed, and much more. Here, it may be quite helpful to hold joint meetings at the start of the project and discuss not only the problems, expectations, and limitations of the solutions, but also the logic and the methodology of the process model.

If the *pioneering situation* is very pronounced *for both sides*, it is necessary to engage in intensive risk reduction.

This includes the following possibilities:

- Dividing plans into smaller segments, learning in smaller dimensions
- Consulting experienced outside sources

- Deliberately segmenting phases with the option of a phase-out or a correction
- Imposing prototypical approaches during the early phases, making experimental adjustments, creating manageable pilot situations, with the option for a later decision not only to withdraw or make corrections, but also to expand plans.

### 6.5.8   Entering Disorganized Problem-Solving Processes

Sometimes, the necessity for target-oriented cooperation within an organizational framework and the implementation of common procedural instruments is not recognized before false results have appeared in the development of a solution, deadlines and costs have been exceeded, and the achievement of project objectives has been rendered dubious. What can reasonably be done in such cases?

Starting at point zero, with a preliminary study and clarifications regarding the most useful frame of action, is usually excluded because of the expenditure involved. Another prohibitive factor is the fear of recriminations, associated with an impairment of the working climate, which would be detrimental to the possibility of bringing the project to a successful conclusion. Instead, one should introduce the systems engineering action model as a common doctrine in association with a review of the status quo, i.e., the achievements to date, the remaining activities, and the time and resources required for them. This can be achieved, for instance, by carrying out a short and intensive review of the situation, following the logic of the PSC:

- Search for objectives: what situation are we confronting? What objectives are we pursuing?
- Search for solutions: what solution approaches are available? Do they meet the objectives given above?
- Selection: what decision(s) must be made?

Special attention must be given to checking the plausibility of the overall concept being pursued, to incorporating the detailed studies into the overall concept, and to defining the interfaces.

The organizational framework for the cooperation of all participants should be newly defined or redefined, whereby the informal channels existing so far should be broadly exploited.

As part of the course planning for the fulfillment of system specifications, i.e., for the provision of performance and quality, one should also consider:

- Attainable stages of completion when budgets and schedules are adhered to
- Adherence to schedule with an incomplete, yet operational system, including the required expenditure
- Trimming specifications (target reduction)

These considerations should help in finding an acceptable basis for the continuation of the project or its further realization. They can also be tied to a new agreement about the project mandate and its specifications.

### 6.5.9   Shut-Downs and Terminations

The termination of the operational phase of a system, for example, after it has become obsolete, can also be seen as a plan, which must be undertaken in harmony with the demands not only of the environment or the employed infrastructure but also of the system parts that continue to be operated.

Toward the end of the operational phase of a system, deliberations are made that either have as their goal the conversion of an obsolete system while maintaining, reusing, or further using essential components, or they strive for the recreation of the original state (what we called a "wide open field"), which may entail disposal problems, as in the case of shutting down nuclear power or chemical plants.

Seen under the auspices of methodical planning, these are new and separate plans. Then again, with regard to an obsolete system, it should have been the task of the original planners to theoretically anticipate, i.e., to include in their planning deliberations, the demands that would arise from its conversion or the requirements for recreating a "wide open field," so as not to pose unnecessary hindrances to any new plans.

### 6.5.10   Self-Check for Knowledge and Understanding: Special Cases and Situational Interpretation

- Do you know any other special cases you would find worth discussing? The authors would like to know your views.

## Literature

### Literature on Situation Analysis

Bryman A. (2008): Social Research Methods
Cadle, J; Paul, D. and Turner, P. (2010): Business Analysis Techniques
Checkland, P. (1999): Systems Thinking
Hall, A. D. (1962): A Methodology for Systems Engineering
Pearl, J. (2000): Causality
Vester, F. (2007): The Art of Interconnected Thinking
Wilson, B. (1990): Systems. Concepts, Methodologies and Applications

For details of the bibliographic references, see "bibliography" at the end of this book.

# Literature on the Formulation of Objectives

Christensen, C. (2003): The Innovator's Dilemma.
Dörner, D. (1997): The Logic Of Failure
Drucker. P. (1955): The practice of management
Fisher, R.; Ury, W. and Patton, B. (1991). Getting to Yes
Hale, R.; Whitlam, P. (1995): Target setting and goal achievement
Hall, A. D. (1962): A Methodology for Systems Engineering
Keeney, R.L. and Gregory, R.S. (2005). Selecting attributes to measure the achievement of objectives.
Kotter, J.P.; Cohen, D.S. (2002). The Heart of Change.
Lindblom, Ch.E. (1959): The Science of Muddling Through.
McKeown, M. (2016): The Strategy Book.
Mintzberg, Henry and Quinn, J.B. (1988): The Strategy Process.
Quinn, R.E.; Faerman, S.E. (2010): Becoming a Master Manager
Rouillard, L.A. (2002): Goals and goal setting: Achieving Measured Objectives.
Weihrich, H. (1985): Management excellence: productivity through MBO

For details of the bibliographic references, see "bibliography" at the end of this book.

# Literature on the Search for Solutions: Synthesis/Analysis

Alexander, C. (1964): Notes on the Synthesis of form.
Altshuller, G. (1984): Creativity as an Exact Science
Altshuller, G. (1999): The Innovation Algorithm: TRIZ
Anthony, S. D., Johnson, M. W., Sinfield, J. V., Altman, E. J. (2008): The Innovator's Guide to Growth
Bahill, T.; Botta, R. (2008): Fundamental Principles of Good System Design
Berkun, S. (2007): The Myths of Innovation
Brown, T. (2009): Change by Design.
Chesbrough, H.W. (2006): Open Business Models.
Chesbrough, H.W. (2005): Open Innovation.
Christensen, C. (2003): The Innovator's Dilemma.
Christensen, C.; Raynor, M.E. (2003): The Innovator's Solution.
Christensen, C.M.; Grossman, J.H.; Hwang (2009): The Innovator's Prescription.
Christensen, C.M.; Johnson, C.W.; Horn, M.B. (2008): Disrupting Class.
Colgan, St. (2009): Joined-Up Thinking
Conklin, J. (2006): Dialogue Mapping
De Bono, E. (1970): The Use of Lateral Thinking.
Dörner, D. (1980): Heuristics and Cognition in Complex Systems.
Drevdahl, J. E. (1956): Factors of importance for creativity.
Eppinger, S., Ulrich, K. (2003): Product Design and Development.
Eversheim, W. (2008): Innovation Management for Technical Products.
Fey, V.; Rivin, E. (2005): Innovation on Demand.
Forbes, P. (2005): The gecko's foot. Bio-inspiration - Engineered from nature.
Friend, J.; Hickling A. (1987): Planning under Pressure
Gerardin, L. (1968): Bionics.
Gigerenzer, G. (2008): Gut Feelings.

Guindon, R. (1990): Designing the Design Process.
Hillier, F. S; Lieberman, G. J. (2004): Introduction to Operations Research.
von Hippel, E. (1994): The Sources of Innovation.
von Hippel, E. (2006): Democratizing Innovation.
Horn, R.E. (2001): Knowledge Mapping for Complex Social Messes
Leveson, N. (2012): Engineering a Safer World.
Mann, D. (2002): Hands on Systematic Innovation.
Markman, A.B.; Kristin L. Wood, K.L. (2009): Tools for Innovation.
Marteka, V. (1965): Bionics.
Michalewicz, Z.; Fogel, D.B. (2000): How To Solve It.
Moeller, M.; Stolla, C.; Doujak, A. (2008): Strategic Innovation.
N. N. Authors: Harvard Business School Press (2009): Innovator's Toolkit.
Oak, A. (2011): What can talk tell us about design?
Oak, A. (2012) You can argue it two ways
Osborn, A. F. (1957): Applied Imagination.
Pahl, G.; Beitz, W.; Feldhusen, J.; Grote, K.-H. (2007): Engineering Design.
Pfletschinger, Th. (2008): Risiko-Management
Rantanen, K.; Domb, E. (2007): Simplified TRIZ.
Rechtin, E. (1991): Systems Architecting.
Rittel, H. (1972): On the Planning Crisis.
Rittel, H. W., Webber, M. M. (1973): Dilemmas in a General Theory of Planning.
Rossmann T.; Tropea C.; Vincent, J. (2007): Bionics.
Schantin, D. (2004): *Makromodellierung von Geschäftsprozessen.*
Sherwood, D. (1998): Unlock Your Mind.
Silverstein, D.; Samuel, Ph.; DeCarlo, N. (2008): The Innovator's Toolkit.
Tidd, J.; Bessant, J. (2009): Managing Innovation
Ulwick, A. (2005): What Customers Want.
VanGundy, A.B. (2007): Getting to Innovation.
Verganti, R. (2009): Design Driven Innovation.
Zwicky, F. (1969): Discovery, Invention, Research.

For details of the bibliographic references, see "bibliography" at the end of this book.

# Literature on Evaluation and Decision

Beer, S. (1966): Decision and Control.
Kahneman, D.; Tversky, A. (2000): Choice, Values, Frames.
Keeney, R.L.; Raiffa, H. (1976): Decisions with Multiple Objectives
Keeney, R.L.; Gregory, R.S. (2005). Selecting attributes to measure the achievement of objectives
Kepner, Ch.H.; Tregoe, B.B. (1965). The Rational Manager
Krapohl, D. (2013): A Structured Methodology for Group Decision Making
Saaty, T. L. (2001): Decision Making for Leaders – Analytic Hierarchy Process
Schuyler, J. R. (2001): Risk and Decision Analysis in Projects.
Torgerson, W. S. (1968): Theory and Methods of Scaling
Triantaphyllou, E. (2000). Multi-Criteria Decision Making: A Comparative Study

For details of the bibliographic references, see "bibliography" at the end of this book.

## Literature on Special Cases and Situational Interpretation

Bahill, A. T. and Gissing B. (1998): The Systems Engineering Process
de Weck, O. L., de Neufville R.; Chaize M. (2004): Staged Deployment of Communications
    Satellite Constellations.
Haberfellner, R.; de Weck, O. (2005): Agile SYSTEMS ENGINEERING versus AGILE
    SYSTEMS engineering

For details of the bibliographic references, see "bibliography" at the end of this book.

.

# Part IV
# Case Studies

# Chapter 7
# The Systems Engineering Basics in Our Systems Engineering Concept

In the following pages, the application and interpretation of the methodology are presented using three case studies. The first case study, "Home Construction," is simple and is presented in all phases. The second, "Airport," is substantially more complex. We only examine the preliminary study phase. The third, more recent case deals with a very interesting urban planning (overriding design of city areas with different owners) and within it a very attractive Science Tower (in terms of function and technology in addition to its architecture).

All case studies have real projects as their background. But of course the methodology is in the foreground, and the case studies are merely the vehicles for conveying and demonstrating the methodology.

Before we enter into the case studies, we summarize the core of our systems engineering concept.

## 7.1 Basic 1: Application of the Systems Approach

Systems thinking is current and is applied when answers to the following questions are sought and it is possible to provide the following information:

How do we define the *system* that we are observing? Which system are we dealing with?

- What are the *elements* that make up the system and that we consider important?
- How do we *delimit* our system from its environment? Which elements or factors are important to us but lie outside the system that we must/want to design?
- What is the *superordinate* system?

---

The original version of this chapter was revised. The correction to this chapter is available at https://doi.org/10.1007/978-3-030-13431-0_17

© Springer Nature Switzerland AG 2019, Corrected Publication 2021
R. Haberfellner et al., *Systems Engineering*,
https://doi.org/10.1007/978-3-030-13431-0_7

– Which *systems aspects* are important to us? Through what lens do we look at the system? Which other aspects should we also refer to?
– Which *relationships* exist among the elements that we consider essential?
– How do they relate to the *environment*?
– At which *system level* should we intervene?

> Should we aim for a higher level (= a more comprehensive view)? Are longer-range changes necessary or practicable?
> Is the selected level OK?
> Do we need to lower the level of observation, reduce the scope of the design?

Result should be a bubble chart as an orientation map for the team (elements or influencing factors in the bubbles, relationships in the arrows).

## 7.2   Basic 2: Application of a Recognizable and Accepted Process Model

The systems engineering process model presented, with its four modules (Part I, Chap. 2), is well-engineered and has proven its worth in many projects. But depending on the case of application, other action models may be more advantageous and more appropriate (e.g., simultaneous engineering, agile methods, etc.). The important thing is to discuss the type of procedure and reach an agreement at the beginning of work.

In practice, a systematic process is more accessible if we consider the following questions and find answers to them:

Do the four modules of the systems engineering action model make sense in the specific problem context? Do we think that applying them will be expedient, or do they seem to be entirely or partially unnecessary formalities? Why? What other possibilities of the process do we have in mind? Why?

1. **The *"from the general to the detail"* approach (top–down approach)**

   – Before we analyze the situation in detail, we should structure the problem field roughly (bubble approach, bubble charting) – see Part I, Chap. 1, Systems Thinking.
   – When working on solutions: first make an overall concept or a rough draft before dealing with details; rough objective before detailed subobjectives; fundamental decisions/determinations before detailed ones.[1]

2. **Developing *variants/alternatives* and cutting back/selecting**

   – When we think we have found a solution: what alternatives might there be? Are other possibilities conceivable? In short: we should not settle on the first solution that we discuss or that the commissioning party mentions.

---

[1] Anecdote: an older actor who is already quite hard of hearing drives the prompter to despair. She hisses the appropriate text passage so loudly that it can be heard clearly in the front rows, whereupon the actor snarls to the prompter's box, "No details! Which play?"

- Having to decide between different solutions – as a desired effect – makes you reconsider the problem that should be solved.
- Alternatives make it easier to discern the characteristics (feasibility or suitability) of solutions and may help to improve existing favorites.
- Thinking in variants/alternatives applies at all detailing levels. The resulting variety of variants must be managed, and this occurs through a stepwise choosing/selection of variants that are treated in the next detailing step, as it is impossible to treat and evaluate all variants on the detailing levels. One must concentrate on those that have the greatest promise of success.
- The selection and choice require not only preparatory work in the project group, but also a declaration of will by the commissioning party. This is a desirable learning process and helps this entity to gradually become more familiar with the problem and the possible solutions.
- The selection of a specific solution does not have to be definitive. With very innovative projects it may happen that a selected approach – despite the highest qualifications of the parties involved – subsequently turns out to be neither practicable nor suitable. A systematic process cannot eliminate that entirely, but it facilitates the search for other solutions. In addition, if you consider only one solution and no alternative, there is even less protection against heading down a dead-end street.

3. **Dividing a rather large task into** *project phases* separated from one another (as a management-oriented module of the action model).

   - This *macro-logic* is to be taken as a mental framework that can reduce the risk of undesirable developments and formally provides for including the commissioning party in the interim decisions. The commissioning party and the project group are thus forced to clarify, exchange, and adjust their ideas on objectives and values.
   - At the transition from one phase to the next there should be a formal pause: one should consider, "Where are we? In what direction can we and should we continue?"
   - The phase model is intended to focus attention on the results of a phase that are to be delivered (which questions will we have to answer at the end of the next phase?).
   - Terminating a project during an early phase is not to be regarded as an undesirable development; rather, in a specific case it may be the best alternative.

   *If you prefer an agile approach it is absolutely ok. You should know why and how.*

4. **The** *problem-solving cycle should be understood as micro-logic*, in the sense of a logical sequence of steps to be applied to every project phase.

   - The main points for application are of course the development phases (preliminary study, main study, detailed studies), for this is where most problems will occur that require systematic handling. In the phases approaching building and implementation, routine processes should gain more ground.

– The problem-solving cycle consists of three steps:

(a) *Objective search:* where are we? Where do we want to go? Why? – consisting of situation analysis and formulation of objectives.

*Result:* a series of accepted objectives (desires, challenges, requirements) that should help to steer the search for a solution and help with the subsequent evaluation.

(b) *Solution search:* What possibilities/routes are there, which ones are essentially conceivable for getting there? – consisting of *a synthesis* of solutions (creative step) and an analysis of solutions (critical step).

*Result*: at least two solutions that are theoretically suitable/acceptable because they meet the compulsory objectives formulated in step (a) – albeit not to the same extent.

(c) *Selection:* which one is the best variant, or the one that promises the greatest success? – consisting of *evaluation* as a comprehensible, comparative *assessment* of the existing variants, and a *decision*, as the determination of the variant that should be further detailed or implemented.

*Result:* the client and project group agree on the direction of the continuation of the process.

## 7.3   Basic 3: Application of Methods and Techniques

Systems engineering must be open to all types of methods, techniques, and tools (MTTs) regardless of where they come from. It is important that they support the individual steps during the course of a project. But MTTs generally do not produce ready-made results, but rather results that must be interpreted.

These MTTs are *not* a generic component of the Hall/BWI systems engineering approach. They should be selected and applied depending on the project situation and specific requirements (e.g., customary to the industry or to the commissioning party, or required by the commissioning party, useful with this type of project, etc.).

An assignment of the methods and techniques (M&Ts) to the individual steps of the problem-solving cycle could look like the following:

(a) M&Ts for *situation analysis*:

(i) Procurement, collection, investigation, analysis, assessment, representation, presentation of the results of the situation analysis that correspond to the understanding of the actual condition, the state of the art, the presumed development, and other factors. This category includes such aspects as the methods of inquiry (oral, in writing), observation, desk research (assessment of existing documents, quantitative and/or qualitative type), etc.

(ii) Analysis of the preceding development and/or any others that allow assessment of the future development (trend extrapolation, prediction techniques, Delphi method, etc.)

(iii) Methods for the analysis of flaws and causes, etc.

(b) Methods for ascertaining or developing the *objectives/challenges* and for structuring objectives:

(i) Catalog of objectives, objective tree

(c) M&Ts for *finding* and *analyzing* solutions.

(i) Creativity, simulation, and optimization techniques, etc.

(ii) Techniques for the analysis of solutions, validation and verification, failure mode and effects analysis, risk analysis, etc.

(d) M&Ts for supporting *evaluation* and *selection*.

(i) Value-benefit, cost-benefit analyses, profitability, ROI, and investment calculations, etc.

## 7.4   Self-Check of Knowledge and Understanding: Systems Engineering Basics of our Concept

– Do you agree on the basics summarized in this chapter? Or are there too few, too many or the wrong ones?
– The authors would like to receive feedback.

# Chapter 8
# Case Study 1: Private House Building: Additional Domicile

The following case study is based on a project that one of the authors carried out using the principles presented here – not strictly and formally, but hopefully recognizable in the logic.

## 8.1 Initial Situation

A family, parents with two children (aged 10 and 9), moved from abroad a year ago. Both parents work (father full-time, mother part-time).

They live in a downtown apartment that has the following *advantages*:

- It is adequately large, and has high, airy rooms.
- It is in a nice neighborhood: quiet, not much traffic, and near the center of town.
- It is conveniently located: both parents and children can walk to work or to school.
- After cultural or other functions, it is possible to take the streetcar or even walk home.

*Disadvantages of the Apartment*
It is on the fourth floor and has neither a balcony nor a green area; thus, there is no space for the children to play and run around, no inducements for outdoor activities (hanging out clothes, sunning, playing, barbecuing, picnics, etc.).

*Wishes, Intentions, Objectives*
Visits to weekend houses owned by colleagues and acquaintances generated a desire for an additional place to live outside of the city. The aim: a place to spend free time,

The original version of this chapter was revised. The correction to this chapter is available at
https://doi.org/10.1007/978-3-030-13431-0_17

weekends, vacations, and perhaps as a backup for later. However, in contrast to colleagues, neither of the parents is skilled at or interested in do-it-yourself projects.

Additionally, the apartment in town should be retained.

*Preliminary Considerations*

The husband is not convinced that they should actually implement the wishes. In his view, the role of the visitor is quite pleasant – and for the hosts as well. If everyone were to buy a property, then they would all end up sitting on the property that they built at great expense, and have to maintain at great expense or their own efforts. At the same time, or for that very reason, the desire to visit other people declines. We could fill this market niche and take on the role of professional visitors, as most people who stay in their nice weekend homes would be happy for someone to visit and admire their efforts.

No doubt they will want to show off their beautiful garden, the outstanding views, the tranquility, the pool, the new barbeque, and so on. But his wife does not accept these arguments and feels that when the weather is nice, she will not want to sit around waiting for an invitation.

The husband does accept this argument, albeit hesitantly. As a committed systems engineer, he is sure that he will be able to manage this undertaking with the tools at his disposal. He feels that the philosophy of systems engineering – particularly *systems thinking, project phases*, and the *problem-solving cycle (PSC)* – could help everyone involved:

– *Systems thinking:* what is the system that we want to consider or create (does it change from phase to phase)?
– *Project phases* (macro-logic)

  How to reduce the complexity by breaking down the project into phases?
  What decisions would need to be made at the end of the individual phases?
  What activities would be necessary and how could they be allocated to the
    phases?

– Application of the *PSC* (micro-logic) within each phase.

## 8.2   Preliminary Study

Here the preliminary study serves as a clarification phase and as the basis for the selection of a solution principle or for a possible termination of the task.

**Desired result:**  answers to the following and subsequent questions:

– What do we actually want? What should the intended use of our additional domicile be? A weekend house, a fully-fledged second home…?
– Where should it be located?
– How large should it be?
– Can we or do we want to afford it? …

**Approach:**  use of the PSC in the preliminary study

*Situation Analysis*

- What *system* do we consider in the preliminary study? What influences come into play (see Fig. 8.1)?
- Clarification of *needs and expectations:*

   Requirements: opportunity to spend the free time actively in fresh air and a
      healthy natural environment, hiking, jogging, cross-country skiing, etc.
   Good road and transport connections, easily reached by public transportation
      (for the children as well), etc.
   Certain infrastructure available: doctor, shops, restaurants, etc.

(Answer: an unknown additional housing need that we have identified as an additional residence, of whatever type, and subject to the influences shown above).

*Formulation of Objectives*

These specific objectives and requirements have resulted from a series of discussions:

- Intended purpose:

   Initially: weekends and vacations.
   Subsequently: the possibility to turn it into a second home. When the children
      no longer want to live at home they can leave. The parents can leave the
      apartment in town to the children – or vice versa.

- Clean air, minimal fog in winter.

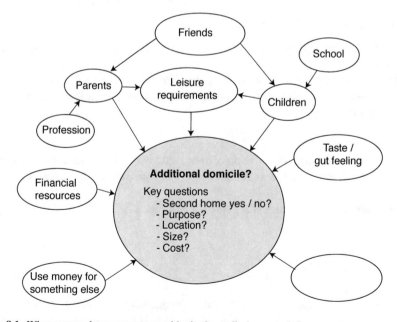

**Fig. 8.1**  What system do we want to consider in the preliminary study?

**Table 8.1** Evaluation of the locations based on the intended purpose (second home)

| Variants ➜ | Radolfshof | Semper | Einöd |
|---|---|---|---|
| Criteria | | | |
| Distance from city | 15 miles/25 km | 19 miles/32 km | 24 miles/40 km |
| Score | 1 | 3 | 4 |
| Public transportation | Bus (frequent) | Bus (infrequent) | Non-existent |
| Score | 2 | 3 | 5 |
| Altitude (fog) | 2700 ft (900 m) | 1800 ft (600 m) | 1500 ft (500 m) |
| Score | 1 | 3 | 4 |
| Infrastructure | Very good | Very good | Scant |
| Score | 1 | 1 | 4 |
| Location of property | OK | OK | Gorgeous |
| Score | 2 | 2 | 1 |
| Result (rank) | 1 | 2 | 4 |

Scores: 1 = very good; 5 = poor
Decision: Radolfshof location

- Good infrastructure (shopping, restaurants, doctors, etc.). Of increasing importance as parents grow older.
- Good transportation connections:

  No more than 30 min drive to the city
  Accessible via public transport: environmental considerations, unsecure availability of fuel in the future? More freedom for the children to pursue their weekend activities in the city without constantly needing taxis?

For the search for a solution this means:

- We want to be at a higher altitude – over 1500 ft (457.2 m) – fresh air, less fog in winter.
- Distance from the city: no more than approximately 15 miles (25 km).
- We lean toward a village setting, not miles from anywhere (infrastructure, public transport).

*Solution Search*
Possible *locations* that meet the objectives listed above are: Radolfshof, Semper, and Einöd. All locations were visited and available properties were viewed.

**Evaluation and Decision:**  – see Table 8.1 and Fig. 8.2.

## 8.3   Main Study

Three sites were inspected in Radolfshof. The one above the village was instantly the first choice. To verify the decision, an architect was consulted. In his view, any type of house could be built on the site. The property was purchased because:

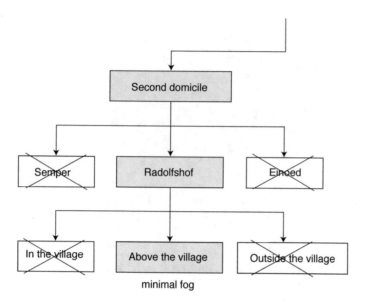

**Fig. 8.2** Summing up the top–down process in the preliminary study

- It was very nicely situated: designated construction-free zone to the south, only meadows and forest.
- Clear advantage with respect to fog (2700 ft/900 m)
- Comparatively inexpensive:

  Even if the entire project is terminated, the property can be regarded as an investment for later.
  No risk associated with the purchase, either technical (appropriate property) or financial.

The *decisions for use and location* determines the solution principle (= system architecture).

*Desired Result of the Main Study*
Work out ideas for space and functions, type (style); planning and illustration of floor layout, front and side views, based on the intended use (weekend house, subsequently extendable as a second home, etc.) and with the help of an architect.

- Specific result: building application plan for local building authorities, which also serves as the basis for detailed planning.
- More precise cost predictions: areas, volumes revealed in the design, ballpark prices per square and cubic meter/foot are known to the architect.

**Approach:** use of the *PSC* in the main study.

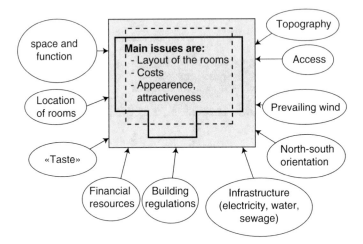

**Fig. 8.3** What system do we have in mind in the main study? (Answer: a house to be designed for a specific plot of land)

*Situation Analysis*

–  What *system* do we keep in mind in the main study? What influences affect it (Fig. 8.3)?
–  What does the topography of the land look like for the selected property? This is in turn important for the design, layout (north–south), slope of the land, access, prevailing winds, connections for electricity, water, telephone, etc.
–  What are to be the main uses of the house (space and function)?
–  What should the house look like? Building style, etc.

*Formulation of Objectives*

–  Satisfy space and functional requirements (which rooms, how large?)
–  Style of building that fits into the landscape, unpretentious
–  Good, well-utilized floor plan
–  Energy-efficient, quick to bring up to temperature (weekends)
–  Good-quality construction, environmentally friendly
–  Building costs not to exceed …$

Variants: architect's drafts (Fig. 8.4).

–  "Orthodox Church" (unsolicited variant submitted by the architects)
–  Style of a West Styrian farmhouse

**Decision:**  for the West Styrian variant.

–  Planning on scale of 1:50
–  Application for and granting of planning permission
–  Begin detailed design (essential for requests for quotations)

**Fig. 8.4**  Summary of the top–down process in the main study

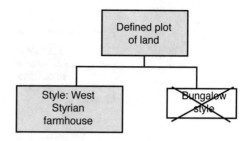

## 8.4   Detailed Studies

Detailing of designs and plans are derived from the building application plan:

- Planning and soliciting quotations for building works, excavations, skeletal structure, and interior construction
- Carpenter: roof trusses, balconies, wood siding, etc.
- Joiner: doors, windows, stairs, etc.
- Roofing, plumbing
- Heating design and installation
- Electrical design and installation
- Sanitary facilities
- Tiling

*Desired Result from Each Detailed Plan*
Basis for quotations (through the builder, architect) and contracting (preparation of contracts, guaranties, penalties for missed deadlines, etc.)

**Approach:**  use of the PSC in every detailed study.

*Situation Analysis*

- What *system* do we have in mind in this detailed study? What influences affect it? (Fig. 8.5).
- The starting point is always the overall concept determined in the building application plan, which is now developed in greater detail.

*Formulation of Objectives*
Defined building specifications from the architect as the basis from which bids can be tendered.

*Solution Search*
Bids from subcontractors.

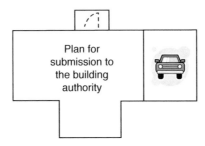

**Fig. 8.5** What *system* do we have in mind in the detailed studies? (Answer: detailed design from architect, carpenter, roofer, etc., based on the overall concept building application plan, construction permit)

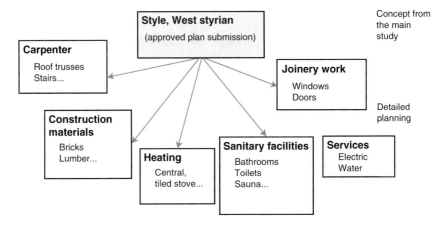

**Fig. 8.6** Summary of the derivation of detailed plans from the overall concept (approved plan submission)

*Assessment and Selection*
Choice of the desired contractor: on the basis of bids (proposed version, bid price, bid comparisons, negotiations, visits to completed jobs/reference buildings, etc.)

*Decision*
Awarding the contract, contract agreement, etc. (Fig. 8.6).

## 8.5   System Building and Implementation

This phase, along with the implementation and application of the system, is not shown here because they do not offer much with regard to methodology.

In Fig. 8.7, the successful completion of the project can be seen. The result turned out very satisfactory with respect to the system architecture and the system design.

**Fig. 8.7** Additional domicile after completion. (Photo: Haberfellner)

The specified deadlines were not met in every case: several craftsmen did not adhere to contractually established delivery dates. The deduction of penalties for missed deadlines resulted in an underrun of the total budget.

## 8.6 Concluding Remarks

The original purpose of the build (children and parents have the option to live either in the city or in the new house) proved to be irrelevant. Both children have moved out of the house and only occasionally come back for visits.

**Summing Up: What Was to be Achieved?**

– Incremental implementation of a project that was initially rather unclear
– Showing that the focus and the system can change during the course of the project
– The phase model creates increasingly specific situations and reduces complexity
– Exit possibilities are intentionally provided for
– The objectives direct the solution search
– Thinking in variants

- The objectives enable decision-making
- The decisions taken channeled further progression
- Use of the logic of the PSC in every phase, but with different (increasingly concrete) content
- …

## 8.7   Self-Check of Knowledge and Understanding: Additional Domicile

1. Can you find the abovementioned basics in the case study? see (Chap. 7).

# Chapter 9
# Case Study 2: Airport Planning

The following example illustrates a more complex case: design of an entire airport.

This case study is based on a real project that was successfully realized. For instructional reasons we have simplified the situation as much as possible to highlight the thinking principles with regard to the methodical aspects rather than others (e.g., technical ones). In addition, various discussions and decision-making processes were considerably more complicated and emotional than they are presented here.

This example is and has been used as a key case to teach the principles of systems engineering in many colleges and universities.

The methodological aspects that should be highlighted include the following:

- *Systems thinking*: breaking down the object into comprehensible parts that can be temporarily isolated and processed without losing the overall context
- *Delimitation* of the design object and associated tasks
- A scanning for the appropriate *system levels* to begin with planning (regional, trans-regional, etc.)
- The stepwise proceeding *from the general to the detail*, in combination with
- Consequent creation of *solution variants* at all levels under consideration
- The exemplary demonstration of important procedural steps within the *preliminary study*
- The *problem-solving cycle* in micro-logic
- The method of *cooperation between the commissioning party and the design team,* especially when *formulating the mandate* and during the *evaluation phase*.

To better illustrate the various positions and individual viewpoints of the design team and the commissioning party, and to capture human interactions better, we use dialogs, which are printed in *italics* wherever possible. Remarks about methodological systems engineering aspects and explanations are added as commentaries.

---

The original version of this chapter was revised. The correction to this chapter is available at https://doi.org/10.1007/978-3-030-13431-0_17

© Springer Nature Switzerland AG 2019, Corrected Publication 2021   291
R. Haberfellner et al., *Systems Engineering*,
https://doi.org/10.1007/978-3-030-13431-0_9

This case study is often used as a role play in seminars to teach practical systems engineering aspects to students and practitioners.

The roles to play/the acting parties are as follows:

– The commissioning party is represented by *Mr. Conrad Perkins*. In this case he is just a single person, but should be rather seen as a spokesman for a group of political entities and representatives of public interests.
– Planning is supported by three external consultants from G&M Design Group:

*Andrew Albright*, who is the leader of the consulting team, represents his firm vis-à-vis the commissioning party.
*Betty Bridgewater* is a very experienced Systems Engineer.
*Charly Chase* is a junior team member whose task is information gathering and analytics.

The case study is structured into the following steps:

– Motivation for the project and the preliminary study
– Concretization of the project mandate
– Situation analysis, structuring of the object, identification of influence factors, definition of area, and scope of investigation
– Rough formulation of objectives
– Creating principal solution variants, defining design or intervention areas
– Lifting/raising problem and solution levels, critically questioning the chosen area of design
– Dividing the projects into phases
– Dividing the preliminary study into subject blocks/planning circles
– Applying the problem-solving cycle
– Relationship of situation analyses in various planning circles
– Identification of additional detail objectives
– Creation of solution variants
– Assessment and decision
– Further planning

## Mr. Perkins

(The representative of the commissioning party describes his difficulties to the representative of the design group).

*You are certainly familiar with the problems that our existing airport causes, and presumably you also know that we are under increasing pressure to build a new airport for our metropolitan area "Toorich". The main arguments being put forth from various sides are:*

– *There is too much noise.*
– *Wake vortices caused by starting planes have damaged some roofs of private homes.*
– *Media reports on psychological harm to children in the "Noisevalley" district.*
– *The airport's capacity is too small.*
– *The access roads to the airport are always clogged, even though it is located on an expressway.*

- *Users on the expressway are distracted by the flight traffic; this has already led to severe accidents.*
- *The city of Growtham is also expanding its airport capacity (possibly with new construction) and will totally eclipse our position with direct connections.*
- *Fully loaded large jumbo jets cannot take off from our airport because our one and only runway is too short.*

### Andrew
*If I understand you correctly, the problem actually revolves around four main issues:*

1. *The noise pollution is too severe for the population because the planes are flying too low over residential areas. This is even causing harm to people (children) and damage to objects (e.g., roofs).*
2. *The capacity of the existing airport is already utilized to the maximum. The single runway for takeoff and landing allows no increase in aircraft movements during peak times. As it is also too short, takeoff weights for large aircraft are limited.*
3. *The connections for individual local transportation are poor. There is no connection to the local rail network and no connections to long-distance passenger rail.*
4. *There is a certain competitive situation with the airport serving the city and metropolitan area around Growtham.*

### Mr. Perkins
*You have understood the problem correctly. We need a new airport and want to commission you to work up the initial design concepts. We have already thought about the location and we feel that the area around Moorfield or in the Darkwood forest would be good solutions.*

### Commentary
The Planner leaves the meeting with mixed feelings. On the one hand, he is happy about the new assignment, but on the other he is not entirely convinced of the appropriateness of the solutions suggested and feels that he is being forced into the role of an "accomplice" for carrying out a premade plan.

Certainly, the assignment has a pre-bias for certain solutions. Otherwise, the solution ("… a new airport in the Moorfield area or in the Darkwood forest") would not have been given/predetermined. Ideally, only the desired effects of such solutions (e.g., less noise pollution, greater capacity, etc.) are given. The planner therefore has to create an "open," unbiased formulation of objectives and use it to search for other potential solutions.

### Andrew
(in his office; has explained the situation to his team in detail)
*I suggest we use the following formulation internally to describe the goal of our assignment:*

*"We are seeking solutions to the current problems with the Toorich airport:*

- *Intolerable noise pollution*
- *Insufficient capacity*
- *Inadequate transport connections*

*For the time being, we will not deal with the competitive situation with the Growtham airport.*

### Betty

*I like this formulation! You are right: before thinking of a new site we should also consider the possibility of improving the situation at the existing location.*

### Charly

*Let's first get an overview of what we should design and about what we are really talking about. We should divide the airport into characteristic and comprehensible parts, also taking the environment into account, and work up a clearer idea of the unsatisfactory situation, the requirements, expectations, and hopes on the part of the people involved (in and around the airport). We should not limit ourselves to the present situation, but also plan ahead for possible future developments.*

### Commentary

The planners agree that it is reasonable to split the workload. Charly will attempt to structure the situation. Betty will think about possible solutions for the current location.

### Charly

(He has drawn the three charts in Figs. 9.1, 9.2, and 9.3 and uses them to explain the results of his considerations):

*First, I thought about the functions and the relationships at the airport itself. I think we can go over this chart* (Fig. 9.1) *quickly without further explanation.*

*I came to realize how multi-layered and multi-dimensional the environment is as I attempted to represent it graphically. I have not been able to fit it all into a single figure.*

*As you can see on my next chart* (Fig. 9.2) *the airport region spreads out at the level around the airport, with the relevant ground transportation and the airspace above it. The steeper conical section represents the approach sectors with landing and takeoff areas, and the flatter surface above should represent the air traffic network. Projected onto the plane the boundary between the outer air traffic network and the approach sectors and landing/takeoff areas is shown as the inner circle around the airport system. This circle marks the scope of the surrounding system "affected parties," which can be divided into physical/psychological effects (e.g., noise) and economic effects (e.g., transport restrictions).*

*The surrounding system of participants consists of those causally involved (passengers, cargo/mail, and airline companies) and those indirectly involved, such as utilities and air traffic control. The participants on the ground are represented by the outer circular disc around the airport.*

*The surrounding systems involving the economy, technology, legal system, and society span overlay everything.*

*I have represented the involved and affected parties in a chart, which is shown in Fig. 9.3.*

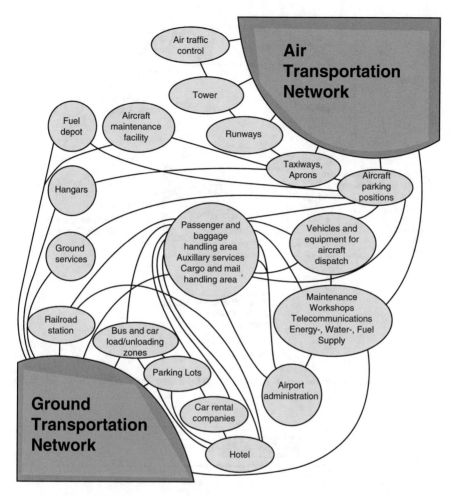

**Fig. 9.1** Airport with the essential functions and their relationships with one another

*To understand the challenges the airport project is facing and to be able to define the relevant requirements, I have tried to identify as many of the parties involved as possible and their presumed interests. Such parties can be groups of individuals, committees, and institutions.*

The result of this elaborate analysis is shown as an influence analysis in Fig. 9.4.

**Commentary**

Such representations are an excellent starting point for understanding complex systems. They visualize the parties involved, their interests and the interrelationship of components and parties involved. Furthermore, they serve as the basis for the planning process and as the basis for discussions with both the commissioning party and others involved. Sometimes, when visited later in the project, they can also serve as a source of inspiration for possible solution scenarios.

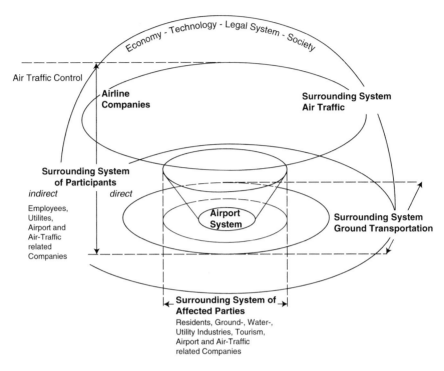

**Fig. 9.2** Representation of the area of investigation in the *system/surrounding system* view of the airport

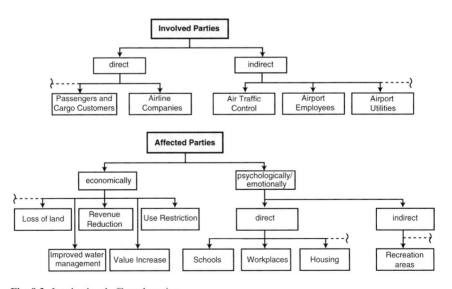

**Fig. 9.3** Involved and affected parties

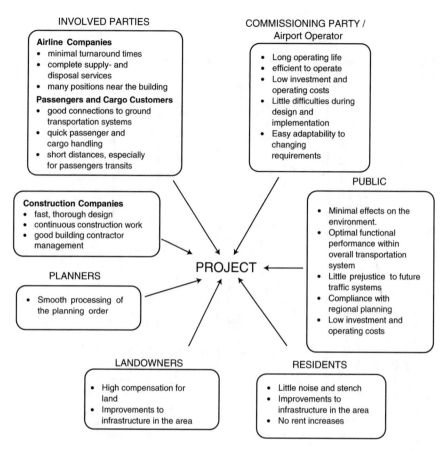

**Fig. 9.4**  Influence analysis – stakeholders and their project interests

After her presentation, Betty shares her view on possible ways to move forward.

**Betty**
*In principal, we have three conceivable possibilities to continue:*

(a) *The existing airport remains unchanged as it is.*
(b) *The airport will be remodeled.*
(c) *A new airport will be built at a new location.*

*I tried to sketch the structure of the problem with the airport and the constraints/ conditions under which it can be solved (Fig. 9.5). Fields where we have to do more research I have left blank. A rhombus indicates that we need a response from Mr. Perkins.*

A couple of days later, Mr. Perkins comes by the planning office to discuss some information he received.

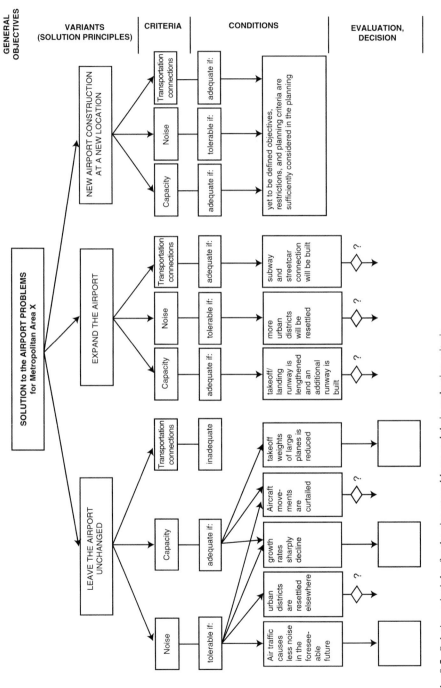

**Fig. 9.5**   Solution principles for the airport problems and their evaluation criteria

**Mr. Perkins**

(doubtingly):

*I have become aware of a professional article in which there is the following prediction about airport traffic: "For medium distances airplanes will be increasingly superseded by high-performance rail traffic." If that is correct, then in 10–15 years, we may have no use for our beautiful, huge new airport!*

**Andrew**

*You are right. If you do nothing at all, your medium- to long-term capacity problem would become less severe and could even be solved. In the short term, however, it would continue to exist, along with the noise problem and the inadequate transportation connections.*

**Charly**

*We have come to the conclusion that we have not adequately studied the medium- to long-term developments in the transportation system. I think we are too focused on developing solutions for the current or short-term situation, which is indeed precarious. Furthermore, these solutions are based on assumptions that still need further study.*

**Mr. Perkins**

*What would those be, for example?*

**Charly**

*You have already questioned the growth rates of air transportation over medium distances. Further questions that need clarification could be:*

- *Will air transportation in the future come with that degree of noise that we experience today?*
- *Will passenger and cargo rates in air traffic grow at the same rate as they are currently?*
- *Are the locations you suggested for a new airport construction (Moorfield area or the Darkwood forest) appropriate and politically feasible?*
- *Are there any feasible locations for a new airport in our metropolitan area at all?*

**Mr. Perkins**

(having regained his composure)

*You are right to bring up these questions. But consider that – no matter how the answer turns out – we have an acute problem with the airport today, and the affected parties will not be satisfied with vague future prospects of a reduction in air traffic or less noisy planes. Not to mention the expected growth in air transportation and the associated operational challenges.*

**Andrew**

*The discussion of these questions makes it possible for us to assess possibilities for improving the situation at the current location. If this analysis shows that nothing can be changed, we have an idea of the sizing of the new airport and its possible location.*

*We have already considered some basic solutions and we would like to hear your opinion on them.*

### Betty

*If we focus on the current airport problems and start from a solution-neutral task formulation, the following three fundamental variants are possible* (she explains Fig. 9.6 to Mr. Perkins and the assumptions concerning the three criteria: noise, capacity, and transportation connections).

After a lively discussion, the following assumptions are agreed:

– Even if there is a tendency toward quieter aircrafts, we cannot expect that flight operations will produce significantly less noise in the foreseeable future (see a in Fig. 9.6).
– In the longer term, it is just as unlikely to expect that the number of aircraft movements will decline drastically – (see c).
– Because of air traffic routing, the airline companies are not able to reduce the takeoff weights of wide-bodied aircraft. If this fact is not taken into account, jumbo jets will not be able to land at our airport (see e).
– The possibility of resettling urban districts was rejected by Mr. Perkins as "not subject to discussion" (see b).
– This also applies to the possibility of reducing the frequency of takeoffs and landings by regulations (see d).

### Andrew

*In the planning of a new airport these days, another problem arises: because of the great need for space, the significant noise pollution, and the high population density, airports nowadays can be located closer than 25–35 miles from the center of source and destination of the traffic volume. Therefore, a high-performance rail system is needed for ground transportation to connect the passenger and cargo flows to the airport efficiently. Such new rail transportation systems would also reduce transit times between metropolitan areas. One consequence might be that air traffic between the metropolitan areas could be largely replaced by these means of transportation.*

*A regional airport could handle air transportation on European and international routes for two or more neighboring metropolitan areas. This possibility could further influence considerations about an ideal location and the nature of the airport.*

### Mr. Perkins

(he is quite surprised about Andrew's suggestion)

*So your question is whether you should design an airport for our local needs, or for the neighboring metropolitan areas as well? I am sorry, but that sounds really crazy! Think of how complicated the discussions and the decision-making process*

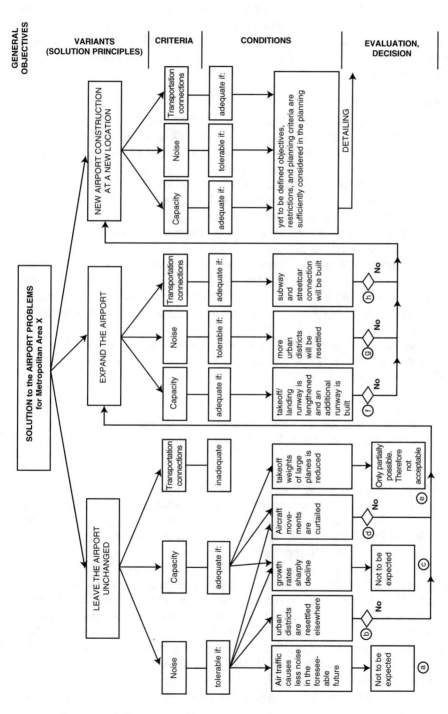

**Fig. 9.6**  Solution principles for the airport problems. Analysis and decisions

*would be if we had to reach agreements with the neighboring federal states or countries. I can't imagine that we would want to get involved in that – neither would our neighbors probably. But let me think about it and talk it over. I don't want to make the decision on the spot. Have you already prepared something? You are good about such things.*

### Charly

(is very proud that he can present the findings of his elaborate analysis)

*Yes, Sir, we certainly have. So far we have talked almost exclusively about possible solutions at the local level, in other words for our metropolitan area. The characteristics of airport operations and the requirements on the part of the surrounding systems would be the same for all variants.*

*Now, if we proceed to the next highest level, the regional level – the best variant up to this point, S3 (a new airport in a new location), is only a partial solution. A presumably better solution, at least from a regional viewpoint, would be to construct a new airport for metropolitan areas X, Y, and Z.*

*The intercity traffic between the metropolitan areas would then be shifted to the planned intercity express system on the rails so that only continental and intercontinental connections would be handled by air transport. Domestic air transportation currently accounts for approximately 50% of the passenger transportation capacity of our national airline company.*

*Figure 9.7 shows the possible consequences for handling traffic with reference to traffic types and means of transportation, in addition to the requirements for the use of these means in design, construction, and political contexts.*

### Commentary

Mr. Perkins left to discuss these issues with the committees and commissions responsible. After a few weeks, he reported back the results of the consultation.

### Mr. Perkins

*After extensive discussions, we have reached the conclusion that we should construct an international airport, but one that is designed exclusively for the needs of our metropolitan area. At least 50 years should be assumed as the planning horizon. As this surely will be the last airport built in our metropolitan area, you should foresee options for possible future extensions.*

*Our decision regarding an international airport is based on the following considerations: the implementation of new types of ground transportation, in addition to their operation in a network that allows travel times over medium distances that are comparable to air travel times, will hardly be possible before the end of the next decade. An airport among metropolitan areas X, Y, and Z would therefore be dependent for approximately 20 years on service with conventional ground transportation – primarily on the existing train routes. This type of integration would be insufficient and unsatisfactory in many respects.*

*Also, the commissioning of a regional airport would not be conceivable for at least another 20 years – because of our federal state and administrative structure*

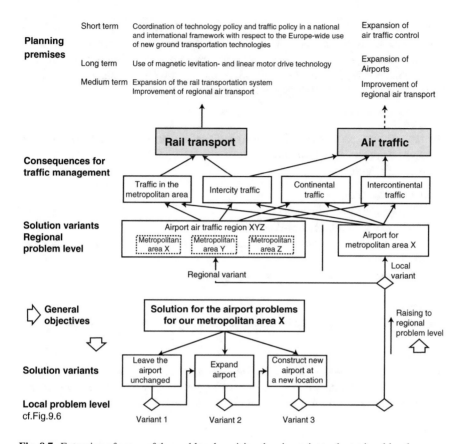

**Fig. 9.7** Extension of scope of the problem by raising the viewpoint to the regional level

*and the international agreements required. But we cannot wait that long for a solution to the pressing noise and capacity problems of the present airport.*

*Therefore, we must construct an airport that can handle the international connections for our metropolitan area. But in designing the final size extension stages, expansion of the transfer of medium-distance transportation to other means of transportation and a regional implementation of certain types of transportation should be considered.*

## Commentary

This decision left the planners feeling uneasy. They felt that a regional airport that went beyond the requirements of the metropolitan area with respect to environmental problems and future traffic conditions would be a much better solution. At the same time, they had to admit that solving the problem was urgent, and that building a regional solution would take much longer. In addition, they feared losing the consulting contract if they insisted on their point of view. Thus, they accepted Mr. Perkin's decision for better or for worse as the basis for continuing.

Here, it must be emphasized that this problem is not caused by systems engineering or its methods. Rather, it has to do with the fact that our planners are also part of the system themselves and are currently *planning under pressure*. Finding the right balance between *compromises in work ethics* and personal/firm values is a topic by itself, and depends on individuals and on company values.

**Andrew**
*For the sake of completeness let me repeat the project formulation that shall be effective from this date (formulation 3):*
    *We are designing:*

– *In a new location*
– *A major international airport*
– *For the needs of our metropolitan area*
– *For the air transportation needs of the next 50 years with some spare capacity*
– *With less noise emission*
– *With better connections to ground transportation*

**Mr. Perkins**
That's right. Now, I would be interested in knowing how you will continue.

**Andrew**
*We will proceed in accordance with the systems engineering concept, which provides a stepwise planning and decision-making process. We have tried to assign the most important project tasks to the individual phases* (Fig. 9.8).

**Betty**
*We have devoted detailed thought to the preliminary study. It should produce the following results:*

• *The problem-solving principle, which in this case we already know (a major international airport in a new location for our metropolitan region).*

  – *A decision with respect to the location of the new airport. This issue is closely connected to the best possible runway configuration (number, length, and arrangement of the takeoff and landing runways).*
  – *Defining rough layout variants for the most important areas (passenger, cargo, and service area, aircraft maintenance facilities, access roads, exits) and deciding on these.*

• *Initial rough cost estimates.*

**Charly**
*We have identified four distinct planning circles (areas) that can be considered almost independently. It makes sense to deal with them in the following sequence* (Fig. 9.9):

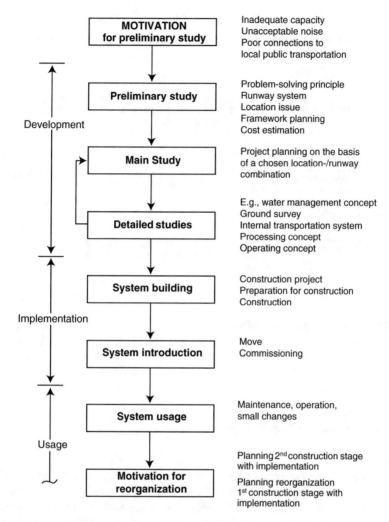

| | |
|---|---|
| **MOTIVATION** for preliminary study | Inadequate capacity Unacceptable noise Poor connections to local public transportation |
| **Preliminary study** | Problem-solving principle Runway system Location issue Framework planning Cost estimation |
| **Main Study** | Project planning on the basis of a chosen location-/runway combination |
| **Detailed studies** | E.g., water management concept Ground survey Internal transportation system Processing concept Operating concept |
| **System building** | Construction project Preparation for construction Construction |
| **System introduction** | Move Commissioning |
| **System usage** | Maintenance, operation, small changes |
| **Motivation for reorganization** | Planning 2nd construction stage with implementation Planning reorganization 1st construction stage with implementation |

**Fig. 9.8** Project phases in accordance with the systems engineering approach

### Andrew

*We will treat every planning circle in accordance with the logic of the problem-solving cycle.* Figure 9.10 *shows the procedure within each planning circle of the preliminary study.*

*The first problem-solving cycle (planning area A) is already finalized, and the decision has already been made (major international airport in a new location for our metropolitan area).*

*In planning circle B variants of runway systems are to be developed – still "location neutral" (independent of their possible location). The underlying solution variant is number 3, which has been agreed upon (see Fig. 9.5).*

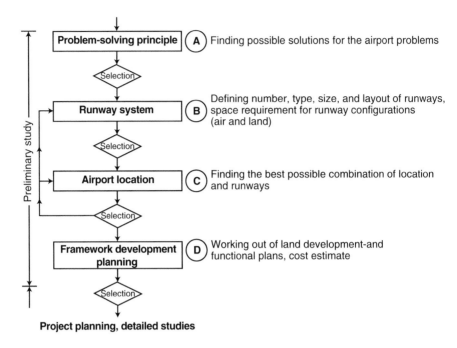

**Fig. 9.9**   Planning circles and their sequence within the preliminary study

**Mr. Perkins**
*What do you mean by "location-neutral"?*

**Andrew**
*At first, we want to work out abstract variants of runway systems independently of specific locations, considering only objectives relevant to air transport. We then want to transfer these variants, which are certainly not equivalent in many respects, to the individual locations to check them for their suitability. Of course, we want to do this together with you, Mr. Perkins, so that we can ultimately find the best combination of location and runway system.*

**Betty**
(gets up and walks to a large whiteboard where she has prepared an analysis plan)
*If you follow me all to the whiteboard, I can show you an overview of the necessary situation analyses that need to be carried out in the preliminary study (Fig. 9.11).*

**Charly**
*Next, the objectives for the results in each of the planning areas must be formulated. The general task and the results of the situation analyses form the basis for this. The situation analysis for developing the runway configurations showed that, for*

| Planning Circle / Stepsin the problem-solving cycle | (A) Problem-solving principle | (B) Runway system (location independent) | (C) Airport location | (D) Framework development planning |
|---|---|---|---|---|
| **Task Assignment** | • New airport<br>• New location<br>(Formulation 1) | Task formulation 3:<br>*new location, major international airport, for the needs of our metropolitan area, for the air transportation needs of the next 50 years with a built in spare capacity, with less noise exposure and better ground transportation connections.* | | |
| **Situation analysis** | • System<br>• Surrounding system<br>• Analysis of influence variables | Capacity<br>Air traffic control<br>Wind conditions | Noise<br>Local public transportation<br>Regional planning<br>Meteorology<br>Building Site | Operating procedures<br>Local restrictions<br>Determinations to date |
| **Objectives** | Solution-neutral formulation (formulation 2) | General objectives | | Detailed objectives |
| **Synthesis/ analysis** | Solution variants for airport problems | Runway system variants | Transfer of runway systems to locations | Layout variants |
| **Evaluation** | Simple, not formal | (not shown) | Stepwise elimination | |
| **Selection decision** | • new Location<br>• int. Major international airport<br>• regional<br>• up to > 2060<br>• less noise<br>• good connections to local public | 4 Variants | Variant 3<br>Location A | |

**Fig. 9.10**  Applying the "problem-solving cycle" in planning circles

*example, the following important trends and circumstances cannot be influenced or changed:*

– *Development of international air transportation with respect to route networks, development of transportation volume, and fleet mix.*
– *Wind conditions in metropolitan area X, etc.*

*The generally formulated objectives are thus to be operationalized in such a way that there will ultimately be subobjectives or criteria that allow an evaluation of solution drafts later on (see Fig. 9.12; Table 9.1).*

Mr. Perkins, who brought in his technical expertise at this point, discussed with the team many issues in great detail, gained a lot of confidence in the capabilities of the planners and essentially agreed to the objectives that Charly suggested.

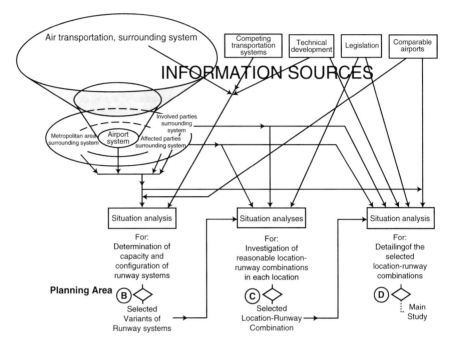

**Fig. 9.11** Situation analyses in the context of the preliminary study

## Commentary

We do not go into much more detail explaining the synthesis/analysis and evaluation of this planning area, as the principal concern there is with technical issues such as handling landings and takeoffs, taxiing management, noise generation, and similar effects. The results are the following four runway configurations, which were presented to Mr. Perkins (Fig. 9.12).

## Andrew

(arrives at Mr. Perkins' office):

*As you suggested in your phone call, we have contacted the air traffic control authorities. There are no objections of any kind to variants 1, 3, and 4, but they have some reservations regarding variant 2. Presumably, the air traffic control authorities would object to it. They raised safety concerns about that variant, but we are not totally convinced that they are valid. However, as we already have enough difficulties, we will revisit this variant only if it has clear advantages for a particular airport location over other variants. Variants 1, 3, and 4 differ mainly in capacity (three or four runways) and the shape of the noise immission curves, which can be very important for a specific location.*

## Mr. Perkins

(is very pleased with the result – especially as he has suggested contacting the air traffic authorities) *Well done! I agree with you. Now what?*

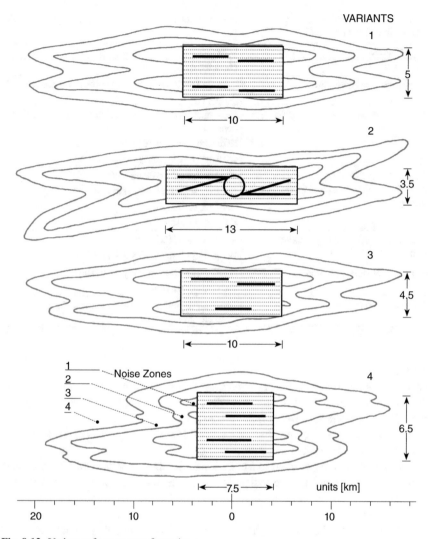

**Fig. 9.12** Variants of runway configurations

**Table 9.1** Generally formulated objectives and operational formulations for the development of ideal runway configurations in planning circle B

| Generally formulated objectives | Subobjectives, criteria |
|---|---|
| Runway system for air transportation needs (movements every peak hour) | Runway number (maximum 4) |
| –   Up to year 2060 with spare capacity | Space requirement (as little as possible) |
| –   In regional, continental, and international transportation (mix of airplanes) | Reserve capacity (as large as possible) |
| –   In our region | |

**Table 9.2** Subobjectives and criteria (operational formulations) for determining the airport location in planning circle C

| Subobjectives | Criteria |
| --- | --- |
| New location | Distance from downtown |
| Minimal noise interference | Type and number of affected parties in noise zones 2, 3, and 4 |
| Good accessibility, i.e., integration with ground transportation system | Frequency of connections and conditions for making connections in public transportation |
| | Direct connections with tourism centers |
| | Density and efficiency of the road system |
| Broadest possible compliance with overall concept of regional planning and policy | Agglomerating economic effect |
| | Interference with recreational areas |
| | Land requirements |
| | Forestry measures |
| Minimal impediments by weather conditions | Days/hours of fog |
| | Frequency of snowfall and snow depth |
| Best possible airspace situation | Interference from military air traffic |
| | Proximity to civil airways |
| Lowest possible costs | Location-dependent costs, redemption fees, compensation payments |

**Andrew**

*As we now have specific ideas about possible runway configurations, we can turn to the question where to put the airport. We will try to transfer the various runway systems to the conditions at the locations and see what fits best.*

*To decide on one configuration we use the following set of goals and criteria and ask for your comments and input at this stage.*

Andrew shows a paper chart to Mr. Perkins, who reviews it carefully. They change some minor details and add the cost perspective. The final chart is shown in Table 9.2.

**Commentary**

Now, the planners begin with the synthesis/analysis steps in planning circle C (site-specific investigations). Possible airport locations are identified in collaboration with the regional planning authorities. Then, the individual runway configurations are investigated at these locations with the help of land use and population density maps. The results are analyzed critically in view of the subobjectives and criteria in Fig. 9.14. The questionable runway system 2 shows no clear advantages over the other variants with respect to the criteria and is thus eliminated.

**Andrew**

(approaches Mr. Perkins):

*Of all the locations that we investigated, there are only three left that qualify: Swampville (Location A), Moorfield (Location B), and Waldegg (Location C): Darkwood forest is eliminated. But not all runway systems are appropriate even for*

**Table 9.3**  Evaluation criteria for location-runway combinations

| Evaluation criteria for location-runway combination | | |
|---|---|---|
| 1 Noise<br>1.1 Number of residents, school children<br>  – Noise zone 2<br>  – Noise zone 3<br>1.2 Number of jobs<br>  – Noise zone 2<br>2 Interference with recreational areas<br>  – Land need (hectares)<br>  – Forestry measures<br>  – Sports facilities | 3 Integration with ground transportation system<br>3.1 Local public transportation<br>  – Multiple connections<br>  – Conditions for making transfers<br>  – Frequency<br>3.2 Long-distance public transportation<br>3.3 Individual road traffic<br>  – Density of relevant road system<br>  – Efficiency<br>  – Airport connection | 4 In accordance with the overall concept of regional policy<br>  – Economic conglomerating effect<br>  – Water management adverse effect<br>5 Minimal impediments by weather conditions<br>  – Number of day/hours of fog<br>  – Frequency of snowfall<br>  – Snow depth<br>6 Reserve capacity<br>  – Of runway system<br>  – For further on-ground expansion<br>7 Costs<br>  – Related to location<br>  – Unrelated to location |

*the remaining locations. We have conducted appropriate analyses of the concepts and have eliminated all combinations that violate important restrictions. Thus, those remaining are:*

- *A1 and A3 (Swampville location, runway systems 1 and 3)*
- *B1 (Moorfield location, runway system 1),*
- *C4 (Waldegg location, runway system 4).*

*At this point, it is no longer possible to make a decision between these variants without further consideration. Specific criteria must now be derived from the project goals so that each variant can be evaluated. In this step, we are again dependent on your experience and your political instinct.*

**Mr. Perkins**
*I think that is important here! How should we go about this?*

**Andrew**
*Together, we should define the partial aspects that make it possible to understand the remaining variants with respect to their most important characteristics. As a basis, we can use the list of criteria that we originally created from the task formulation. During the course of the concretization and the creation of variants, we have refined and extended that list* (Table 9.3). *Supplementary documents (measurement methods, assessment aids, etc.) are contained in a separate appendix.*

**Mr. Perkins**
(is quite impressed after looking at the list)
*This looks complete, but we probably have to study it in further detail. So, how will things move forward from here?*

**Andrew**

*You and the appropriate committees you represent should carefully evaluate the individual criteria and prioritize them by assigning "importance points" – the more points, the more important the criterion is to you. In total, you can assign 1000 points.*

*The weighting should be completely neutral with respect to a solution variant! In other words: when assigning the importance points you should not have a preference for a solution in mind!*

**Mr. Perkins**

*Sounds interesting.* (he can't hide his skepticism) *And what is your role in that step?*

**Andrew**

*We, as experts in airport planning, evaluate how good a variant is with respect to the individual criteria. We work on the basis of the following scores or awarding of points: very good (10), average (5), very poor (1 or even 0).*

*The scoring takes place independently of how you have weighted the criteria. A variant's partial utility value of a specific criterion is the product of score × weight. The total utility value is then the sum of all partial values. The variant with the highest number of utility points is the best one and should be chosen.*

**Commentary**

In the final evaluation process, very often such a "value-benefit analysis" is used that involves both the planners and the commissioning party: the commissioning party is often asked to contribute mainly to allocating weights to the criteria, whereas the planners are often scoring the different variants. After the total utility value is calculated for each variant, both the criteria weights and scores are iteratively changed if needed (within reasonable limits). That way, "flaws" or misconceptions in the allocation of weights or scores can be eliminated. The value-benefit analysis is an excellent tool for decision-making not only because of the iterative process: its major advantage is that is makes subjectivity transparent. So everybody can see which criteria are used, how much importance (weight) is given to each criterion, and how the different variants are scored with regard to each criterion. However, one should keep in mind that this is a subjective tool that is very sensitive to individual assessments or group dynamics. Applying value-benefit analysis in a workshop format requires well-developed moderation/facilitation skills.

After several iterations, an agreement is reached on the following result (Fig. 9.13).

**Mr. Perkins**

*Finally! Variants A1 and C4 clearly have little to offer; I never liked them anyway, so we should eliminate them for the moment. This leaves A3 and B1, with just a minimal difference in points, but I would not want to make a decision on that basis.*

| Variants → | | A1 | | A3 | | B1 | | C4 | |
|---|---|---|---|---|---|---|---|---|---|
| Criteria | Weight | Pt. | Partial Value Benefit | Pt. | Partial Value Benefit | Pt. | Partial Value Benefit | Pt. | Partial Value Benefit |
| 1  Noise | [150] | | | | | | | | |
| 1.1  Number of residents, school children | (130) | | | | | | | | |
| – Noise zone 2 | 80 | 4 | 320 | 6 | 480 | 7 | 560 | 8 | 640 |
| – Noise zone 3 | 50 | 4 | 200 | 6 | 300 | 10 | 500 | 6 | 300 |
| 1.2  Number of jobs | (20) | | | | | | | | |
| – Noise zone 2 | 20 | 2 | 40 | 4 | 80 | 8 | 160 | 2 | 40 |
| 2  Interference with recreational areas | [50] | | | | | | | | |
| – Land need (hectares) | 20 | 6 | 120 | 8 | 160 | 6 | 120 | 1 | 20 |
| – Forestry measures | 20 | 6 | 120 | 8 | 160 | 2 | 40 | 3 | 60 |
| – Sports facilities | 10 | 6 | 60 | 6 | 60 | 4 | 40 | 7 | 70 |
| 3  Integration with ground transportation system | [350] | | | | | | | | |
| 3.1  Local public transportation | 160 | | | | | | | | |
| – Multiple connections | 40 | 10 | 400 | 10 | 400 | 8 | 320 | 4 | 160 |
| – Conditions for making transfers | 60 | 10 | 600 | 10 | 600 | 8 | 480 | 4 | 240 |
| – Frequency | 60 | 10 | 600 | 10 | 600 | 8 | 480 | 4 | 240 |
| 3.2  Long-distance public transportation | 40 | 8 | 320 | 8 | 320 | 8 | 320 | 4 | 160 |
| 3.3  Individual road traffic | (150) | | | | | | | | |
| – Density of relevant road system | 50 | 8 | 400 | 8 | 400 | 10 | 500 | 6 | 300 |
| – Efficiency | 70 | 10 | 700 | 10 | 700 | 8 | 560 | 8 | 560 |
| – Airport connection | 30 | 4 | 120 | 8 | 240 | 4 | 120 | 6 | 180 |
| 4  According with the overall concept of regional policy | [80] | | | | | | | | |
| – Economic conglomerating effect | 30 | 6 | 180 | 6 | 180 | 10 | 300 | 2 | 60 |
| – Water management adverse effect | 50 | 8 | 400 | 8 | 400 | 3 | 150 | 6 | 300 |
| 5  Minimal impediments by weather conditions | [70] | | | | | | | | |
| – Number of day/hours of fog | 50 | 4 | 200 | 4 | 200 | 6 | 300 | 10 | 500 |
| – Frequency of Snowfall | 10 | 4 | 40 | 4 | 40 | 8 | 80 | 10 | 100 |
| – Snow depth | 10 | 2 | 20 | 4 | 40 | 2 | 20 | 1 | 10 |
| 6  Reserve capacity – of runway system – for further on-ground expansion | [50] | 8 | 400 | 4 | 200 | 8 | 400 | 8 | 400 |
| 7  Costs | [250] | | | | | | | | |
| – Related to location | 100 | 2 | 200 | 6 | 600 | 8 | 800 | 6 | 600 |
| – Unrelated to location | 150 | 6 | 900 | 8 | 1200 | 6 | 900 | 6 | 900 |
| Sum of weights | 1000 | | | | | | | | |
| Sum of partial value benefits | | | 6340 | | 7360 | | 7150 | | 5840 |

**Fig. 9.13**  Value-benefit analysis of location/runway combinations

**Andrew**

*You are right; this small difference in points should not be decisive. Now, we must find important distinguishing characteristics so that we can differentiate between them more effectively.*

- *With A3, it is mainly the limited reserve capacity of runway system 3 (criterion 6)*
- *With B1, it is the adverse effect of water management at Moorfield (location B; criterion 4).*

*We must closely investigate both variants again in light of these criteria.*

**Mr. Perkins**

The matter is starting to become tricky. With respect to possible effects on water management at Moorfield, we must really examine these effects in detail, because this location has already been discussed publicly.

**Andrew**

We were also a little surprised that of everybody, YOU mentioned this location in your press conference!

**Mr. Perkins**

(is getting a little angry) *You cannot understand that because you do not need to win any elections. And you cannot blame ME for coming up with the idea of the three-runway system (variant A3), which, as we don't know enough about future traffic development, could bring us severe capacity problems!*

**Andrew**

(slightly defensive, but insisting on the matter) *I am sorry, I did not want to offend you. You know that at location 4 (Swampville) only runway variations 1 and 4 are possible. Variant A1 would have had enough capacity, but because of the high noise immission, it is clearly worse than A3* (Fig. 9.13). *And therefore we have eliminated it – at your suggestion, by the way.*

**Mr. Perkins**

(is surprised at the emotional discussion and gives in) *Alright. I think we (me and you) have reached a point where we cannot make any progress. I suggest that we elaborate somewhat more on the unknowns in water management at Moorfield and the future traffic situation. So I would like to bring in two independent and distinguished external experts to deliver expertise about both delicate questions. Please don't take that decision as a lack of confidence in your work, we simply have to make such an important decision on the basis of as much evidence as possible.*

**Commentary**

Mr. Perkins hired, as announced, for both detailed investigations, two expert-consulting companies, which were independent of the planners:

The *TraffCon* Company is supposed to provide a forecast of future developments in rail transportation technology (the most current wheel-rail transportation, magnetic cushion railroad with linear motors, and others). The idea behind this analysis is the hypothesis that a promising development of those technologies will lead to a reduction in the amount of air traffic over medium distances. This would then mitigate the disadvantage of the rather limited expansion possibilities of variant A3. After a thorough analysis *TraffCon* delivers their study, which is more optimistic with respect to the development of competing ground transportation than Andrew's team (G&M planning group) has worked out in their preliminary and conservative estimate.

The Hydrological Institute from the country's most famous university was asked to evaluate the extent of the water management effects at Moorfield. Here, the results of the investigation are even more negative than those that the G&M planners used initially as the basis of their drafts.

**Mr. Perkins**
(comes back to Andrew's team after the specialists have turned in their studies a couple of weeks later):
*Based on the assessment of the future development of rail transportation technology and the critical hydrological situation at Moorwood, we have finally decided on variant A3 (Swampville, runway configuration 3).*

**Commentary**
So far, the course of the planning and decision-making process has led to the choice of building the airport in Swampville with runway configuration 3 (Fig. 9.14). This decision largely determines the overall architecture of the solution.

**Mr. Perkins**
*What is next on our planning agenda?*

**Andrew**
*To complete the preliminary study we have yet to do the framework planning and based on that we have to work out more concrete cost estimates.*

**Mr. Perkins**
*What do you exactly mean by "framework planning"?*

**Andrew**
*First, we will work out various layout variants for the main areas (Fig. 9.15).*
*For the passenger and cargo area, we will then work out alternative handling concepts for passengers, luggage, and cargo (Fig. 9.16). We will combine these elements with one another and choose an optimal combination with you or your representatives.*

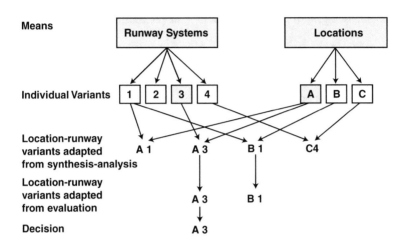

**Objective**                          **NEW AIRPORT in a NEW LOCATION**

*Capacity-relevant objectives for runway system*     *Location-relevant objectives*
For International traffic                             Minimal noise pollution
For metropolitan area X                               Good connections to local public transportation
For traffic needs > 2060                              Agreement from regional planning

**Fig. 9.14** Selecting the best possible combination of location and runways (overview)

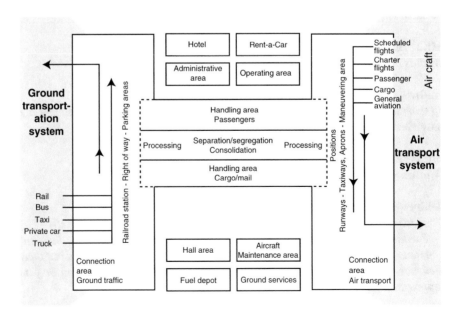

**Fig. 9.15** Airport. Structure – schematic view

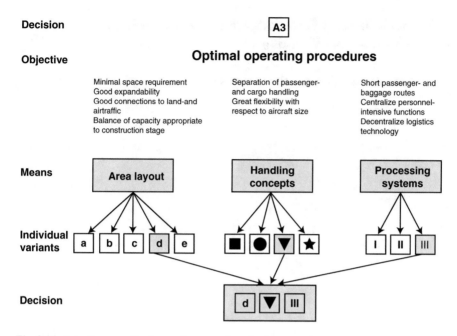

**Fig. 9.16**   Selecting an optimal operating procedure (basic concept)

## Commentary

Here is where we leave our planners and where our case study ends. We have shown and elaborated on the most relevant methodical aspects. Continuing the case study would require a significant amount of addition domain knowledge in transport, logistics, and civil engineering, which is not our focus here.

A summary of the key learnings and expected insights could look like this:

- Objectives/goals should primarily be used to direct the search for a solution and must not be invented later on to justify solutions.
- The formulation of objectives should be solution-neutral.
- An efficient system and problem structuring:
- Makes it easier to get an overview of a situation
- Supports the search for solutions, as it allows subproblems to be dealt with separately without losing the overall context.
- A stepwise planning and decision-making process helps to focus on the essential, always maintaining an overview.
- Holistic, comprehensive thinking and the encouragement of variant creation at all stages help to prevent hasty solutions.
- Close cooperation between planners and the commissioning party represented by Mr. Perkins speeds up the planning process and facilitates decision-making.

## 9.1   Self-Check of Knowledge and Understanding: Case Airport Planning

1. Can you find the abovementioned basics in the case study? see (Chap. 7).

# Chapter 10
# Case Study 3: Smart City and Science Tower, Graz

In this case study, we describe a very interesting urban development project (Smart City Graz) which includes an attractive construction project, in terms of aesthetics and in terms of technology (Science Tower).

Here, we demonstrate important systems engineering principles, such as systems thinking, top–down approach, phase concept, PSC, etc.

## 10.1 What Is a Smart City?

The term Smart City is used in several different ways.

- Some authors use this term to explain that a city has a *modern and efficient IT infrastructure* and can be connected with the whole world, which is now the space for activities of its citizens.
- Others highlight the *urban development*: they want to express that the town or an urban district are well mixed and are hosting pleasant residential areas, with professional jobs nearby, utility services, shopping for daily needs, and entertainment areas and facilities, within walking distance or easily reached by means of public transport.
- Others want to express properties such as mobility, infrastructure, energy efficiency, environmental compatibility and resource efficiency, economic attractiveness, citizen-centered administration, and a *high quality of life* for residents.
- Others still attach importance to the chance of *further development of their living and working environment*. To be able to establish their technological focus points, and the commercial exploitation.

Edited by Dr. Mario Müller, CTO at SFL technologies GmbH www.sfl-technologies.com, www.mj.mueller@fibag.at and DI Markus Pernthaler, Architect ZT GmbH, www.pernthaler.at, www.architekt@pernthaler.at

In the case of Graz, one can take almost *any mentioned aspect* for characterizing the term "Smart City": in a former industrial area, well-mixed areas (for living, working, education, daily shopping, leisure time, relaxation, etc.) should emerge through the smart cooperation of property owners, investors, public administration, planners, etc.

This should create a *high quality of life* for residents through:

– A *clever architectural programmatic design* of the urban districts
– Transformation of *buildings and their covers into producers of energy*
– Increasing the use of *nonfossil fuels*
– Promoting the *sharing idea* (vehicles, gardens, public spaces, etc.)
– Targeted *exchange of energy* between commercial areas and living places (weekends)
– Further *promotion of recycling*, etc.

Further research and development in the field of "*green technologies*" should be intensified, supporting and giving an additional push. By this are meant the development, trying and testing of technologies and organizational forms for production, storage, distribution, proper management of new energy systems, their autonomous and continuous development, commercialization, etc.

An indispensable prerequisite for all of that is a proper data and service infrastructure for big data solutions.

## 10.2   Initial Situation in Graz

Graz is the capital city of the province of Styria, situated about 200 km south of Vienna, in a climatically privileged location, in the south east corner of the Alpine region.

**Some Facts About Graz (2017)**
– Number of inhabitants: town 286,000, surrounding area included: 405,000
– Four universities, two universities of applied sciences, two university colleges for education. A total of approximately 53,000 students. Fourteen research competence centers
– UNESCO world heritage (Graz historic old town, Eggenberg castle), European Capital of Culture (2003), City of Design (2011)
– Graz-based companies with international significance are among others: *Andritz AG* (hydropower, pumps, tidal turbines, pulp and paper, metals, separation, etc.); *AVL* (powertrain engineering, simulation technologies, testing solutions); *MAGNA-Steyr* (automotive, 4WDrive technology, customer-oriented development, and production of vehicles); *Siemens* (world competence center for rail vehicle bogies), *Anton Paar* (scientific instrumentation); *SFL technologies* (multifunctional facades and glass technologies), etc.

Of course, there are also *problems* in Graz, as in many cities worldwide:

- Graz has the *highest population growth* of all Austrian cities: 226,000 (2001), 286,000 (2017), 330,000 (forecast 2036). The reasons are well known: the existing economic dynamism, one of the highest research ratios of all European regions, associated with a high quality of life.
- Unchanged continuing *growth in the surrounding areas* would result in more urban sprawl, increasing distances between leisure, schools, work places, etc.
- Today's *transport problems* would increase: traffic jams at peak times, exhaust emissions, noise pollution, intensified by the location of Graz in a basin with *little exchange of air*
- Modes of transport (modal split 2015): 47% motorized private transport (drivers 37.5%, passengers 9.3%); 20% public transport; 14.5% bicycles; 19% pedestrians[1]

From this starting point, a pleasant change of people's awareness has occurred: public administration and politics, owners of properties, planners, investors, future users, universities, energy suppliers, municipal transport services, were approaching one another, pulling in the same direction and thus facilitating an integrated development of an urban region.[2]

To understand this "smart" and positively evolving situation, one should envision urban development in the past (according to architect Pernthaler): affluent suburbs (Speckgürtel) developed without any land use planning, thereby causing problems such as the destruction of pasture land and complications in traffic, etc. Many cities nowadays are struggling with these problems.

In the whole of Graz, there are *several urban development* projects that all share the same basic approach: holistic planning and realizing. The biggest areas are "Reininghaus" and "Smart City Waagner Biro." In this case study, we limit ourselves to Waagner Biro and the flagship project, the "Science Tower."

## 10.3   Why Smart City Graz?[3]

Graz is combining world heritage, Mediterranean flair, and increasing economic power with the creative "City of Design". An interface to the south-east European region, it has a dynamic economy and is a location of research and education with international ambitions.

The aforementioned high population growth meets only limited land reserves. In the new urban development concept, together with the 2013 municipal council resolution, one finds the following course-setting properties for the *"Smart City Graz"*:

---

[1] Source: https://www.graz.at/cms/beitrag/10192604/8032890/Mobilitaetsverhalten.html

[2] Source: M. Pernthaler in "Wie baut man eine Stadt," Kleine Zeitung Graz, 7.10.2017.

[3] Source: Smart City Graz, Folder 2016 http://www.smartcitygraz.at/wordpress/wp-content/uploads/2013/09/folder_DE_EN_200x120-quer_neu.pdf

"*low emissions*, *conservation of resources* and *energy-efficiency*, for the first time to be realized in the Waagner Biro area, as a compact urban quarter for *mixed use*, *attractive public space*, and *top quality of life*."

## 10.4   Vision for Urban Development of the Smart City Graz 2050

The Directorate of Town Planning of Graz published its *long-term vision for 2050*[4,5]:

Quality criteria for the *activity areas* economy, society, ecology, mobility, energy, supply, and disposal of buildings were defined.

Note that the following explanations relate to the entire town of Graz and not only to the Waagner Biro area, which naturally has a more narrow horizon. *Common objectives* for the whole of Graz are:

– Establishment of *high-quality living space*
– Provision of *high-quality open space*
– Formation of *green walkways* and *bicycle tracks*
– Best *public transport connections*
– *Reduction of private motor traffic*
– Greater *independence from fossil energy sources*

We suggest the following partial visions in a shortened and straightforward mode: the basic idea of this vision is to "strengthen the strengths," reinforcing positive properties in a purposeful way.

For the *planning horizon 2050, the following partial visions* were formulated to explain the direction of the planned adjustments.

*Economy*
Graz is developing further on the way to a dynamic, livable, medium-sized city and this is highly esteemed by its residents. As a place for *research, qualification measures, and economy* in *the Green Tech Valley* the Smart City Graz shall become a future role model for *added value* by the use of *green technologies* (energy, mobility, resources), *health*, and *design*.

*Society*
Graz is an attractive place to live and work for each period of life. The society is a young, open, democratic and collaborative community, where the residents shape

---

[4]VISION = future scenario of an organization to describe its unique nature and to give a special identity. A vision shows the sense and benefit of their behaviors to the employees. A vision has to be accepted and lived by them and stimulate them to work toward this future picture. http://www.manager-wiki.com).

[5]Attention is drawn to the experiences of the so-called Lighthouse Cities Stockholm, Cologne, and Barcelona, and the five Follower Cities: Graz (Austria), Porto (Portugal), Suceava (Romania), Cork (Ireland), and Valetta (Malta) – http://www.grow-smarter.eu/lighthouse-cities/

their living environment. There is a *high awareness of saving resources* and a *sustainable way of life*

## Ecology

Graz shall evolve to a *low-emission, self-powered, and waste-free city.* In addition, the quality of soil and water, noise pollution, and biodiversity shall undergo a massive improvement. There is a tight network of freely accessible green spaces, and the river Mur is a central element of the urban space. The compactness and quality of the city support and demand emission-free and nonmotorized mobility.

Graz is being steadily rebuilt, following the principle of balance between building density and the quality of open spaces. The residents should have access to private open spaces, which can be used individually or in shared mode. Many spaces offer urban ecological gardening.

## Mobility

The year 2050 mobility in Graz safeguards the *activities of the citizens with minimal resource utilization* and ensures the *support of social contacts.* Local supply with goods, services, basic education, leisure offers, together with an urban structure of short walking and cycling paths, in addition to public transport utilities, offer self-motivated mobility with low consumption of resources.

Areas that today are used for motorized private transportation can be partly regained for public use. Changed offers should influence the choice of transport in urban and regional sectors (nonmotorized and even shared services).

## Energy

In 2050, Graz shall be well-balanced in terms of sustainable energy. The total amount of energy that is needed (including mobility, production, and business) shall be produced from 100% regional renewable energy sources. Citizens know the value of energy and are behaving in a consciously energy-efficient manner. Public energy providers provide cost-efficient infrastructure for the balancing and storing of energy. An important infrastructural project is the hydraulic power station under construction on the river Mur, because it not only produces sustainable energy but also offers recreational park areas near the banks of the river.

## Supply of Goods and Waste Disposal

Graz 2050 shall be largely waste-free and show a resource-saving management of water. Wastewater shall no longer be a burden for people and nature. The citizens of Graz will have taken the step from a throwaway society to a recycling society. The urban economy will meet all the requirements of a comprehensive recycling sector.

## Buildings

Throughout its lifetime, each building is seen as a chance to promote higher urban qualities. These relate to energetic, macro-economic, and societal objectives for all areas of co-existence: living, working, recreation, and mobility.

All buildings in Graz 2050 shall be characterized by quality architecture, in particular design quality and high-quality workmanship, multiple functionality, and its harmonious integration into the environment, taking into consideration the criteria of building culture.

## 10.5   Objectives and Means for Smart Urban Development Until 2030

On the basis of the *visions*, the following *objectives* on the way to 2030 were set:[6]

– Energy efficiency
– Saving resources
– Low emissions

This is to be pursued until 2030 in the three Smart City areas, Waagner Biro, Reininghaus (= Graz West), and Graz South by the following *means:*

– *Heating*, *cooling*: use of new (nonfossil) energy sources in the architecture of the buildings and active energy management (storage, distribution, exchange).
– *Building, architecture:* establishment of compact and tight structures by the densification of building zones and the development of fallow land, where public infrastructure exists.
– *Traffic*: mixed use, establishment of living space, job spaces, educational institutions, shopping and supply facilities for daily needs, development of public transport, reduction of (motorized) private transport.
– Development of *attractive public areas* (green spaces, open spaces, reduction of traffic areas), which can be used as contact and communication zones.

*Explanation of the Measures*
The energies mainly used in Graz in 2013[7] were: district heating (33%), oil (25%), electric energy (20%), gas (15%), renewable energy from biomass and other alternative energy sources (5%), coal (2%).

In the future, energy consumption shall be reduced massively by the reduction and avoidance of consumption (investment into existing and new buildings) and the use of renewable energies.

To create additional living space, urban areas with an adequate infrastructure (streets, public transport, water, sewage channels, power connection) shall be exploited and used for the creation of new "green" urban areas (Fig. 10.1).

A city with short distances and well-functioning mixed use (living, working, leisure time) brings about a rethink of traffic behavior.

---

[6] Note: a vision may sound like a handful of good wishes, its representing the "guiding star" and has to be substantiated by *objectives* and *means* that enhance its credibility.

[7] More recent data are not available.

**Fig. 10.1**  Smart City zones: Graz West = Waagner Biro (above) plus Reininghaus (below). Red lines = public transport. (Authorized copy © Stadt Graz, Stadtplanung)

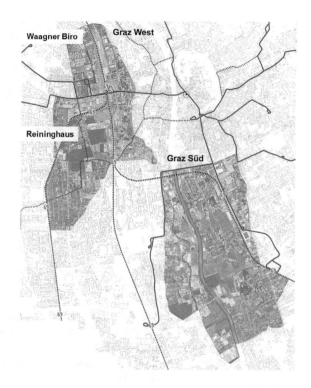

By a modest densification of the built-up spaces the size of the street area (traffic and parking) shall be reduced by 25% and used for public green areas.

Attractive public spaces are a basic prerequisite for a livable city because they are important links between spatial structures and the living space of the inhabitants.

The best possible supply of social infrastructure (schools, kindergarten, care facilities) needs a minimum building density. The number of people and the links with the public transport system are as important as social balance against the formation of ghettos.

## 10.6   The Project Area "Smart City Graz Waagner Biro"

This area is located around 2 km *northwest of the city center* of Graz. This former industrial zone is a large land reserve for future construction, which shall be developed into a living and working area with a high quality of life.

In close proximity (1000 m), the *State Hospital* and the *Emergency Hospital* are located. The *main railway station* is situated 800 m to the south. In the immediate surroundings, a *secondary school* (*Oberstufenrealgymnasium*) is to be found. The Helmut List Halle, an *attractive venue* for concerts, theatre performances, ball events, special readings, etc., has been located inside the project area since 2003.

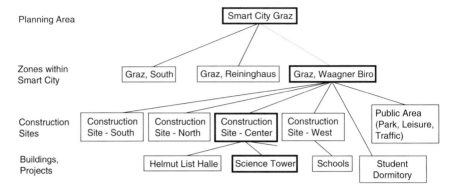

**Fig. 10.2** Activities at different hierarchical levels

East of the area lies the highly frequented train route (Südbahn: Vienna, Bruck/Mur, Maribor SLO), and in the north there is an important east–west road link (Peter-Tunner-Str.). The planning area of the zone covers 127,000 m² and can easily be reached from the south and north by public and private transport. Further public transport offers can easily be realized by the extension of the existing tramline 6, with no need for special buildings such as bridges.

The neigboring residential area is heterogeneously built up. There are several industrial enterprises – mainly along the railway line – sales areas and commercial premises, some areas for single homes, rather high residential buildings from the 1960 and 1970s, and some newer dwellings.

For the "European Capital of Culture year 2003," the cultural facility Helmut List Halle was realized through the reconstruction of an existing industrial building at the center of the construction site. The latest project is the Science Tower, which opened in September 2017.

The previously mentioned measures are located at four strategic levels of activity (Fig. 10.2).

Note: the Visions 2050 relate to the Smart City Graz as a whole (upper level in Fig. 10.2). Our further considerations are focusing on the zones (Graz Waagner Biro and others) with a time horizon until 2030 (lower levels).

In a preliminary study, the architect Markus Pernthaler and the owners of the properties established some proposals for the development of the areas. In consultation with the urban planners and some experts, a masterplan was designed as a basis for the reclassification of the properties.

Citizens were involved by an active district management, i.e., offers for information and participation in addition to an interdisciplinary platform of experts. Regular exchanges with national and international partner cities took place to support learning and reflection processes in addition to the distribution of results. The meeting point "on-site" was opened in April 2014 and is located in two containers next to the Science Tower, sponsored by SFL (= technology provider for the Science Tower).

## 10.7   Masterplan Smart City Graz Waagner Biro

The above-mentioned and jointly developed *masterplan* was the basis for the reclas-sification and all further developments. The area, therefore, was divided into clearly structured *building plots*, which differed with respect to their size, position, and favorable utilization.

Owing to the change in the *land use plan*, a *more flexible use of land plots* was enabled. For all building plots, a development plan was mandatory with the obliga-tion on quality assurance measures (such as architectural competitions). Furthermore, the diverse structure of ownership and utilization had to be respected (Fig. 10.3).

*Key aspects* to be considered in the masterplan are:

- *Green spaces and open spaces* – public spaces designed to a high quality and as places of social integration for present and future residents. Standards for future development of the building plots were established.
- *Routes across the sites* and *traffic measures* – the open space is characterized by two large public areas. To improve them, new walkways, cycle tracks, and access roads were established.
- For *public transport* a *turning place* for tramway and busses near Peter Tunner and Waagner Biro streets is required.
- The *Science Tower* is located next to the *Helmut List Halle*. The ground-breaking ceremony was held in May 2015. Bordering the railway line, a construction for noise protection was built. Up to 2015, a powerhouse for the energy supply of the Smart City Waagner Biro area – the Science Tower included – was planned. Because of its cancellation, a stand-alone solution for the Science Tower was necessary (heating, cooling, emergency power supply).
- In the area opposite the Helmut List Halle, a *school campus* with a primary school (2019) and a junior high school (2022) is planned.

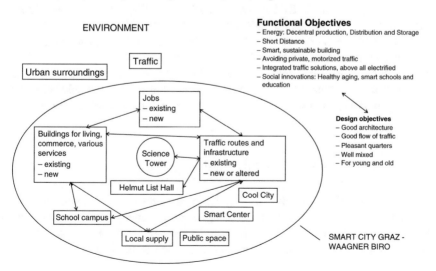

**Fig. 10.3**  Smart City Waagner Biro in the Graz system (abstract)

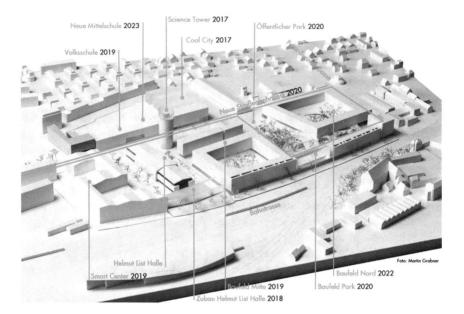

**Fig. 10.4** Rough architectural design of Smart City Waagner Biro. (Authorized photo ©Martin Graber)

## 10.8   Architectural and Design Competitions

To guarantee a *high-quality building culture*, urban development and architectural competitions are carried out for several building plots (Fig. 10.4), which we, however, are not going into here. The main characteristic is that – in spite of the different land owners – a homogeneous and consistent concept is arising.

In the field of *governance*, an accompanying district management is established to embed all parties concerned and to create urban awareness.

The public authorities and the private investors find agreements in *private law contracts* for the common implementation of the objectives in the fields of energy, mobility, buildings, public space, building culture, and district management.

## 10.9   Flagship Project Science Tower

In the Smart City Waagner Biro area an *energy-autonomous district* will arise. In an integral planning process, energy technologies for intelligent "zero emissions" shall be installed and demonstrated. The Science Tower plays a key role because it serves for practical tests and the further development of new technologies.

*International Project Consortium*
Twelve national and international partners, headed by the city of Graz, are forming a project consortium and are jointly realizing the first Austrian Smart City model project.

*Role Model Science Tower*
An international jury selected the Science Tower for Austria's first lead project in the Smart City Program. With its integrated technologies, it demonstrates new solutions for buildings of the future to be used in Austrian and in networked European Smart Cities (Stockholm, Berlin, Amsterdam, and Helsinki).

The Science Tower shall be a flagship and demonstration project: for new and partially disruptive technologies and methods of *energy production*, for *trying* out and *further development*, but also for the development of independent new technologies and concepts.

Reduction and avoidance of $CO_2$ in the production of energy is the highest premise in the entire Waagner Biro Smart City quarter. This goes right down to local $CO_2$-sinks in the form of plantings or by the chemical bonding of $CO_2$ by the decentralized production of methane by synthesis. Storage and management of energy shall be tried out by using thermal and electric storage facilities: thermo water tanks, buffering systems in the ground with underground pipes as heat exchangers (up to a depth of 200 m; second-life lithium-ion batteries from e-mobility vehicles; hydrogen-electrolysis generators; and fuel cells.

In general, the energy produced is consumed within the Science Tower; surplus energy will be given off to neighboring buildings or fed into the network of the local energy provider.

*Trying Out New Technologies and Concepts*
The following technologies and concepts were installed into the Science Tower to be tested:

– *Energy glass* (= disruptive new technology developed by Prof. Grätzel, EPFL Lausanne, which uses solar energy with the help of transparent glass – see *Grätzel cells* below)
– *Integration of depth probes* with *bidirectional thermal pumps into the entire system*
– *Facade-integrated solar tracking* (= technology for the production of electricity and simultaneously shadowing  to prevent overheating in summer).
– *Thin glass technology*, with the lowest ecological footprint
– *Free cooling* to precool buildings at night
– *Integration into the Smart City grid and in SFLenergrid* (connection of the Science Tower with renewable energy sources outside of town into an energy network)
– *Research and development institutions* in the field of "*green technologies*" will be established in the Science Tower step-by-step and will create an inter- and transdisciplinary platform.

- Moreover, concepts for *sustainable urban mobility* including electric mobility are elaborated
- *Building technology*: first-time realization of *"Grätzel cells"*[8] in the form of energy glass on a large scale. This is a new disruptive glass technology that transforms light into electrical energy. Grätzel cells work on the principle of photosynthesis, which enables metabolism by light in every plant leaf. Light is captured and converted into electricity. This effect produces growth in the plant and electricity in the Grätzel cell.

Compared with the silicon cell the *Grätzel cell* has the following *advantages*:

- It not only works under *direct sunlight*, but also in *diffuse light conditions* and shows *lower yield losses in low-light conditions*
- It does not need a *metalloid such as silicon*, which is *expensive in processing*. It uses cost-effective *base materials* and has a far *cheaper production* process
- This causes *less of an environmental impact*
- It is made of *transparent material*. Mounted in front of the window with the light passing from outside to inside, it acts as *sun protection*. In the evening, it uses the artificial light coming from inside. Furthermore, the Grätzel cell is not sensitive to partial shading situations.

After a trial installation as an "indoor"-photovoltaic system at Geneva airport, the Smart City Graz project is the first large-scale application of the Grätzel technology worldwide. It covers the fifth part of the surface of the Science Tower. Later on, it will also be integrated into the acoustic barrier at the railway track.

- For the construction of the *façade, CVG–ionic glass®*[9] is used, which is thin and lightweight and shows an extremely high flectional and breaking strength.
- For the first time, a *local energy grid* will be realized
- Furthermore, *multi-modal mobility solutions*, where *public and private traffic* will be brought together to optimize the joint offer of infrastructure and services: in particular, cars-on-demand; e-bikes; car-sharing; tramway; future autonomous traffic systems; battery charging infrastructure, etc.

The purpose, function, and characteristics of the Science Tower are shown in Fig. 10.5.

---

[8] http://www.smartcitygraz.at/moretext-was-ist-eine-gratzel-zelle/

[9] The facade of the Science Tower is realized by a new glass technology, developed by SFL, which is called "SFL ionic glass®." Oversized pieces of thin glass allow ultralight oversized façade glass elements, which nevertheless are heavy duty and highly robust. The name CVG stands for "chemically pre-stressed glass". SFL has established the largest European CVG plant. Glass up to the dimensions 6×3,21 m (which is the largest standard dimension in the glass industry) is plunged into molten salt at temperatures of up to 450 °C and stay there in a diffusion process of ions for several hours. At the surface, small components of glass (sodium) move out and large components (potassium) move from the molten salt into the glass. As the larger parts require more space, the surface tension becomes greater and this causes greater hardness and additional strength. For practical application, this means that thin and light-weight glass meets the same requirements as thick and heavy glass.

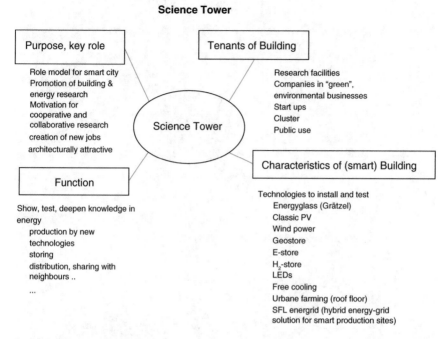

**Fig. 10.5**  Purpose, function, and characteristics of the Science Tower

Further details concerning the purpose and functionalities of the installed technology can be found at http://info.science-tower.at/. By clicking on the link you will see Fig. 10.6. After clicking on one of the dark blue buttons you will receive more detailed information on the selected topic (Smart Lift, LED lighting, etc.). Thus, you will gain a good insight into the wide range of topics and technologies that are applied and tested in a practical manner in the Science Tower.

**New Concepts and Technologies**
The establishment of the Science Tower started immediately after completion and opening in September 2017. On a total floor area of 4500 m², green innovators from science and business cooperate in a vertical network: organizational units from Technical University Graz, Joanneum Research, University of Applied Science Joanneum, a field office of the European Space Agency, and also business partners such as the "Green Tech Cluster", start-ups, and the rocket holding "Green Rocket" as a platform for financing companies in the fields of energy, environment, mobility, and health care by means of crowd-funding.

The effect on the public lies in visibility for the citizens and the possibility of showing the tower to interested groups and external delegations.

*Architecture*
The architecture of the building is of particular importance (Fig. 10.7). All new and visible technologies of the Science Tower are integral parts of the architectural concept.

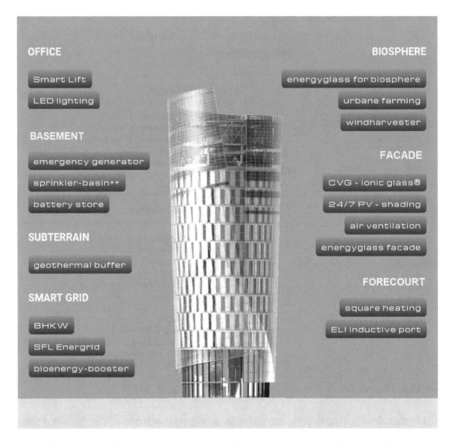

**Fig. 10.6** Overview of interesting thematic topics in the Science Tower. (See –http://info.science-tower.at/)

**Fig. 10.7** The Science Tower as an attractive high-tech role model – in addition to the Helmut List Halle. (Both buildings were designed by the architect Pernthaler. Helmut List Halle was previously a factory floor ripe for demolition, but is nowadays an elegant multipurpose hall with outstanding acoustics for up to 2400 persons). (Photo credit authorized ©paul-ott_photograph)

The Science Tower has a double-shell façade, which – like a coat – is thrown around the basic shape. Eighty percent of the coat is made of large-scale thin glass and 20% consists of energy glass, reddish and transparent with Grätzel cells. The inner façade up to the third floor is a base façade, above a module façade, which, up to the 12th floor extends cylindrically/conically from a diameter of 20 to 23 m. The façade thus serves as a test field for the latest technology in energy-producing façade elements.

Up to the 12th floor (at a height of 44 m) there are offices mainly rented by companies in the field of "green sciences," which form a "science cluster" together with universities and other companies and research institutes. A pool of know-how, technologies, and ideas regarding an ecological and sustainable future shall emerge.

On the 13th floor, the "Urban Farming Lab," is located, where methods of food production in the cities shall be developed.

In the middle of it, on the 14th floor, there is a meeting room.

## 10.10  References to the Systems Engineering Methodology

In the following chapters, we highlight some of the modules of our systems engineering-concept that are apparent by way of examples:

– *Systems thinking* as an overall thinking approach with cause and effect relations
– The *process model* with the *modules*, top–down, phases model and PSC
– *Project marketing*
– *Project organization*

*Systems Thinking*
Especially remarkable is the *holistic view* of many individual aspects:

– Graz is considered to be an *overall system* in which living, working, and traffic are supposed to interact very well.
– Within this system, subsystems are searched for, which are *particularly suitable for further urban development*: former industrial zones, fallow land, barely built-up areas, located near the center, and with a good infrastructure.
– The *interests and potentials of several stakeholders* (adjoining property owners, urban planning office, public authorities, etc.) are seen to be part of a larger framework and united in a targeted manner to find larger, well-balanced, and better overall solutions.

*From the General to the Detail*
This planning principle is easy to identify in the project "Smart City Graz":

– First, the situation of *Graz as a whole* is addressed.
– Then, the degree of detail is lowered to the *Smart City zones* in Graz (Waagner Biro, Reininghaus, etc.).
– Finally, the construction sites and building projects such as "Science Tower," "Smart Center," "Cool City," "School Campus," etc.

- Also in all phases of the "*Science Tower*"- *project* the approach "from the general to the detail" is to be seen. Each phase started with one or more research questions, whose results were the basis for the engineering work.

When treating the principle of thinking in variants we will not go into detail, although variants were considered at any level of planning: when planning the functional zones, in negotiations with regulatory authorities, in defining the lead projects and aesthetic options, in the course of architectural competitions, etc.

*Breakdown into Project Phases*
When dealing with the project phases, we are limited to the subproject "Science Tower," because it is easier to understand than the overall project with the forming of consortia, agreement on a joint proceeding, etc.

The name and the content of the several phases can be seen in Fig. 10.8:

1. *Preliminary phase 1*: preparation of submission at KLIEN (= Austrian climate and energy fund)

   - From 2010 to 2018, in several research projects, different questions concerning façade technologies, the decentralized production of energy, and the management of energy were studied. Prototypes of solutions or intermediate products were developed.
   - In 2010, Mario J. Müller first summed up the insights and findings of the first phase: to establish a competence center for urban technologies. Together

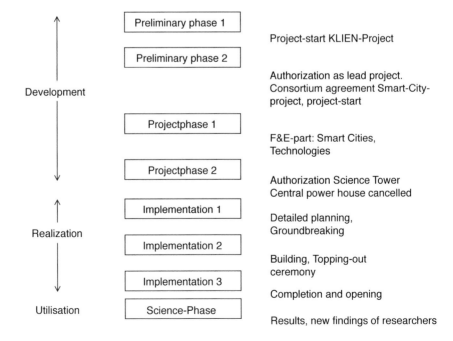

**Fig. 10.8** Phases of the subproject Science Tower

with architect Markus Pernthaler, they brought this idea into the form of a building, to which they gave the working name "Science Tower". The idea of this competence center was submitted to KLIEN within the Smart Cities Program 2011.

2. *Preliminary phase 2*: authorization and start

   – The projects "Smart City Graz" as a whole, and the "Science Tower" in particular, were presented at various conferences. In a jury session in March 2011, the project became the leading project of Austria.
   – *START of the project*
   – Based on a consortium contract between 15 partners, the team of Smart City Graz was able to start the project in April 2015.

3. *Project phase 1*: research and development questions and answers

   – The KLIEN project was started as a research project and as an investment project. The research project covered questions regarding energy technologies (80% concerning the Science Tower) and also regarding energy networks and future smart mobility. It was co-financed by the Austrian Research Promotion Agency (FFG).
   – Investments were planned for a central powerhouse, a multimodal traffic hub, and an upward wind power station.
   – The research and development project brought new questions and topics for smart cities, mathematical model calculations, and simulations, which provided new insights, scenarios of implementation, and worthwhile paths. This included the scientific evaluation of an upward wind power station. At a height of only 40 m and the planned technical structure, no technically and economically feasible configuration could be found – only at a construction height of 200 m might it have been possible. As this height was out of question, this part of the project was cancelled.

4. *Project phase 2*: planning until approval

   – Based on scientific studies, planning was started and approval for the implementation of buildings was obtained. This primarily concerned the Science Tower; the consortium could not make up its mind about the power house until the completion of the Science Tower.

5. *Implementation 1:* detailed planning and the start of construction work

   – On 15 May 2015, the ground-breaking ceremony for the Science Tower took place. After years of developing concepts and planning, the foundations were finally laid.

6. *Implementation 2*: building until the topping out ceremony

   – On 17 June 2016 the reinforced concrete construction was finished and a as the supporting construction for the unique energy glass, a steel crown was placed by a 500-tonne crane in a single stage.

7. *Implementation* 3: completion of the construction

– On 21 September 2017, the Science Tower was opened in the presence of prominent persons from politics, science, and economy. Some users had already moved into and used the Science Tower as their work place.

8. *Science phase:* exploitation of the project outcomes

– Once the Science Tower has been put into operation, it can start to perform the functions for which it was created. As a place for scientific work, the horizontally and vertically linked users shall generate new know-how in the cooperation of economy and technology.

*Application of the Problem-Solving Cycle in Each Phase*
As an example, we consider the *design of the staircase* in a detailed study.

– *Search for objectives*: what do we want?/where do we want to go?/

Considering local construction methods, the staircase would have taken up much useful space because of the cylindrical shape of the high-rise building, with emergency power, sprinkler systems, etc. The requirements were: (i) complete functionality (i.e., a safe escape from each floor) and (ii) in as little space as possible.

– *Search for solutions:* what are the possibilities?

Variant a) continuous staircase, connecting all floors
Variant b) American staircase: only the even or the odd floors are interconnected

– *Selection/decision:* which variant is the best/most appropriate?

Variant b) because it needs a smaller volume of the building and offers more floor space.

*Comment:* the so-called American staircase, in fact, consists of two staircases circling around each other. The first staircase is entered from the south and one can reach floors 0, 2, 4, 6, 8, 10, and 12, the second from the north leading to floors 1, 3, 5, 7, 9, 11, and 13. One can only change to the other staircase on the same floor when passing the main entrance to the escape routes and vice versa. The stairs have a low construction depth because of their steepness and thus offer a more usable area (Fig. 10.9).

*Project Marketing*
Affected residents and authorization instances (government representatives, political entities, funding bodies) were taken into account and regularly contacted and informed from the very beginning of the project. The investors were strongly embedded anyway.

– To integrate the affected residents into the development, the building owner SFL sponsored a double container, which had already been "on site" since 2014. It was used as a place where requests, suggestions, and wishes could be

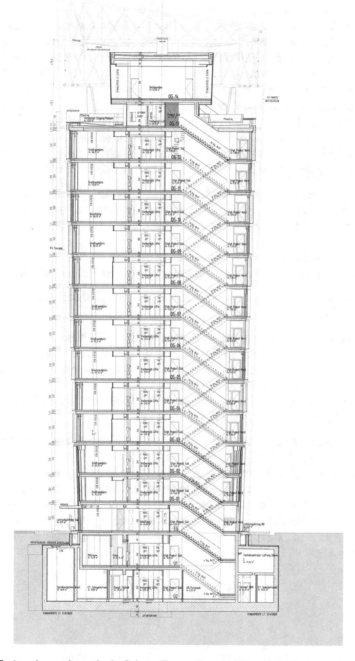

**Fig. 10.9** American staircase in the Science Tower. (Source: SFL authors' homepage)

deposited – either personally to the three employees who were financed by the City of Graz or by using a postbox.
- Two or three times per year, the consortium as a whole invited up to 200 persons to provide transparent communication to the stakeholders. These events were moderated, included presentations, poster stands, information boards, and always panel discussions with questions and answers. Also, a "Smart Christmas" invitation to the "Smart City," where Christmas punch was offered. As soon as possible, the residents were invited to enter the tower to have a look at their homes. For decision-making, active lobbying was carried out, and political decision-makers were also integrated into social communication. At an early stage the (Austrian) EU commissioner J. Hahn had been invited to a major event. In a crane drive up to 60 m, the new perspective was shown to the political stakeholders.

All decision processes were also medially escorted and the political decision-makers were medially imbedded. As a consequence, all decisions of the Graz municipal council were taken by a unanimous vote in favor of the project.

**Straightforward and Efficient Project Organization**
The project organization was very simple and therefore efficient and powerful.

- *Steering committee:* two SFL representatives (owner Höllwart and CTO Müller), plus the architect Pernthaler.

  During the preliminary phases, until project phase 2, Mr. Müller played a leading role. Every 2 weeks a *jour fixe* meeting took place to clarify open questions and to take decisions. The group consisted of research and development experts and was headed by SFL (Müller).
  From the start of implementation 1, the building owner SFL (Mr. Höllwart) was more involved and played the dominant role.

- *Project lead:* central coordination was carried out by the architect Pernthaler; the decision-maker was SFL owner Höllwart.
- *Project team:* the team composition changed depending on the particular project phase: at the beginning it was rather research and development-driven, with university people and consultants. Later on, planning and approval-driven team members (architect, public officials) were involved, and finally, the project team was implementation-driven (contracted construction companies under SFL management).

# Part V
# SE for Practice

# Chapter 11
# Seven Basic Recommendations

Our systems engineering model consists of the components indicated in Fig. 11.1.

In our systems engineering concept we recommend the consistent application of these seven simple rules:

1. ***Apply the systems approach***
   Graphically represent the object and scope of the project, in addition to the assumed relationships and influencing factors, as *systems* (*bubble approach, bubble charting*). The use of systems thinking means:

   - Understanding the *starting situation* and *solutions* as systems
   - Recognizing and representing the elements that build and influence a system, along with their relationships and demarcation – which is a good basis for collective understanding within the project group.

   Do this especially if at first it seems difficult – because this is an indication that elements, relationships, demarcation, and environment are unclear.

2. ***Adhere to the systems engineering process principles***

   - From the *general to the detail*
   - *Thinking in variants*
   - *Problem-solving cycle*

3. ***For a large project, apply the phase model*** (*forming stages*)
   If you have a large project ahead of you (one that will take 6 months or more), consider whether you should divide it into project phases to which you assign a main focus. This is to give a temporal structure to the project, which is beneficial for the client (internal or external) and for the project team as well. This is also valid, if you choose an agile approach.

---

The original version of this chapter was revised. The correction to this chapter is available at https://doi.org/10.1007/978-3-030-13431-0_17

© Springer Nature Switzerland AG 2019, Corrected Publication 2021
R. Haberfellner et al., *Systems Engineering*,
https://doi.org/10.1007/978-3-030-13431-0_11

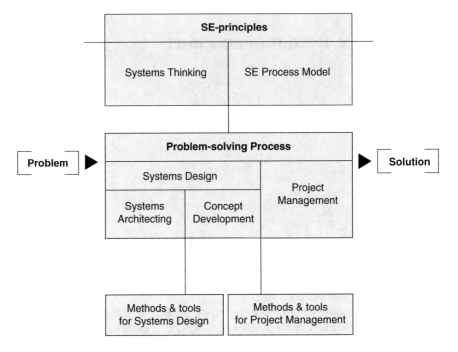

**Fig. 11.1** The systems engineering concept

4. ***Before you begin the work,*** *always* ***clarify the content to be assigned to the particular phase***
   What questions do we have to answer at the end of the phase (what is the specific purpose of this phase, about what should we have more clarity, which decision has to be made)?
   Only then consider which activities will be necessary.
5. ***It is expedient to agree on a rough project plan and a project assignment only for a preliminary study at first***
   This reduces the risk for all parties involved. In the worst case scenario (termination after the preliminary study) only the costs of this phase are expended.

   – The *following points should be agreed upon in a project order*:

     *Starting situation*: why are we dealing with this matter? What is the starting situation that decisively determines what we are to do?
     *Design objectives*: what important effects/characteristics should the system have (= solution, end result) as soon as it is available?
     What is the *specific task for the project group*: what contribution that now has to be agreed, must we provide? Which questions will we have to answer?
     Note: If you are choosing an *agile process model* for a project (such as Scrum), the assignment would *not cover the next phase but the next SPRINT* (1 week, 4 weeks, etc.).

- *Project boundary*: what are we going to deal with? What is the area of investigation? What is the area of intervention (need not be the same)? Perhaps also: what are we, by agreement, not going to explicitly deal with?
- *Boundary conditions*: what in particular do we have to keep in mind as we work on the project? (with regard to content, organization, time, personnel, e.g., related projects, capacity restrictions, taboo areas).
- *Personnel, organizational topics*:

  Always clarify which people represent your client and who competently speak for him (setting the direction, decision-making, support), and if possible identify one person (as the main sponsor of the project, who can be approached directly). This can affect questions of project content, deadlines, budgets, personnel, etc.

  Who is the *project manager* (driving force)?

  Who is working in the *project-team* (names, functions, departments)?

- *Rough outlay*: in terms of working days and/or money.
- *Deadlines*: start, finish, milestones.
- *Information and reporting systems*: at what points and/or on what occasions is a report to be delivered?

6. **Dealing with any problem in the project**, always act in accordance with the thought process (micro-logic) of the **problem-solving cycle**:

   - *Search for objectives (= situation analysis and formulation of objectives)*: what do we want and why?
   - *Search for solutions (= concept synthesis and analysis)*: what possibilities are there for accomplishing it?
   - *Selection (= evaluation and decision)*: which solution holds the best promise of success?

7. **Create a regular** (e.g. *monthly*) **progress report** that also contains expenditure to date and to be expected. Communicate it to the project group and the steering committee (whether or not they ask for it). This also benefits the project manager and the team.

## 11.1 Self-Check of Knowledge and Understanding: Seven Basic Recommendations

1. Do you agree with the recommendations?
2. Would you modify, extend, or reduce the list? The authors would like you to share your judgment with them.

# Chapter 12
# Typical Weak Areas in Projects (Stumbling Blocks)

Difficulties, weaknesses, and deficiencies in projects in accordance with various key subjects are listed in Table 12.1. It reflects the experiences of the authors and their project partners in countless client projects. It can be comforting to learn that nobody is left alone with these and similar difficulties, but is in good company, even with well-respected firms.

## 12.1 Self-Check of Knowledge and Understanding: Typical Weak Areas in Projects (Stumbling Blocks)

1. Does the list presented in Fig. 11.1 reflect your experience? Do the arguments apply to it?
2. Do you have suggestions for modification, extension, or reduction of the list that arising from your experience? The authors would like you to share your experience with them.

---

The original version of this chapter was revised. The correction to this chapter is available at https://doi.org/10.1007/978-3-030-13431-0_17

© Springer Nature Switzerland AG 2019, Corrected Publication 2021
R. Haberfellner et al., *Systems Engineering*,
https://doi.org/10.1007/978-3-030-13431-0_12

**Table 12.1** Typical weaknesses in projects and their thematic breakdown (by way of example and simplified)

| Key subject | Examples of deficiencies and shortcomings |
|---|---|
| Project objectives | Goals unclear or continually changing |
| | Disagreement on essential issues |
| | Not accepted, or only accepted in theory, but not practically supported by relevant authorities or their representatives |
| | Seen as exaggerated, unrealistic, perhaps even unnecessary |
| | Project not "sold" or unsalable |
| Procedure | No identifiable logic in the process, e.g., lack of separation of the project into phases, with clearly developed interim results and decision-making situations |
| | Excessively rigid or bureaucratic procedure (methodology overrides problems and ideas) |
| | No reasonable method of working with respect to management and organization of meetings, capturing of results and agreements and their implementation |
| | No project management |
| Instruments/ methods/tools | Inadequate, perhaps even excessive (unintelligent) use, for example, with respect to the structuring of the project, information gathering, structuring of decision-making (variants and their advantages/disadvantages), project planning (process logic, expenditure, deadlines), project monitoring, risk assessment, project information systems, etc. |
| Organization | Inefficient embedding of the project group in the hierarchy of the company |
| | Unclear, inadequate regulations and competencies (lack of organization) |
| | Excessively detailed, needless regulations perceived as unnecessary (over-organization) |
| | No (functioning) project committee |
| | Inadequate incorporation or embedding of the user in the project group or in the project committee |
| | Over-organized |
| Personnel/ human resources | No (identifiable) project manager (cannot "pull," does not want to, is not allowed to) |
| | Unsuitable, wrong project manager |
| | Unmanageable double workload of project manager or members of the project team (everyday business versus project work) |
| | Unresolved conflicts between project and departmental interests |
| | Excessive demand with respect to qualifications (expert knowledge, ability to work in a team, management capability) |
| | Inadequate communication within and outside the project |
| | Fear of innovation or shared responsibility on the part of the user |

# Chapter 13
# Activities Checklists

The conceptual approaches described in Parts I through IV are summarized and linked to one another below as checklists in an incremental process. For each of the six checklists, the particular project phase (Fig. 13.1), purpose, object (of the abstract research or the concrete work), and results are described.

The core of each of the checklists are the required *activities,* which are separated into *systems design* activities (field of activity 1) and *project management* activities (field of activity 2). Above all, the checklists should provide stimuli and illustrate activities that must be carried out independently of the project focus. They can also be viewed as guidelines for working up project- or company-specific checklists.

## 13.1 Preliminary Study: Activities Checklist (Table 13.1)

*Purpose*
The proposed project should be clarified with relatively little effort. This particularly applies to:

- Determining the analysis and design areas, and important boundary conditions
- Formulating the requirements of the solution
- Setting out promising solution approaches or system architectures and their evaluation with respect to feasibility, costs/benefits, probability of success, etc.
- Unpromising projects should be identified as such in the preliminary study and terminated in a timely fashion

*Object*
This is the problem area and the rough draft solutions in addition to their reasonable demarcation or imbedding in the environment.

© Springer Nature Switzerland AG 2019
R. Haberfellner et al., *Systems Engineering,*
https://doi.org/10.1007/978-3-030-13431-0_13

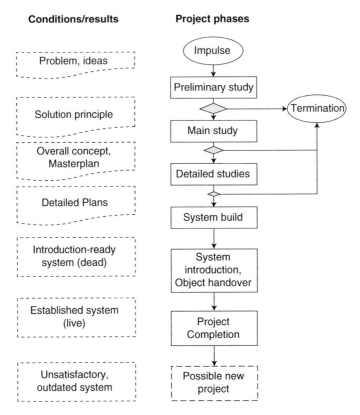

**Fig. 13.1** Project phases (according to Fig. 3.5)

**Table 13.1** Activity checklist for preliminary study

| Field of activity 1 (primarily derived from the problem-solving cycle) | Field of activity 2 (primarily derived from the list of project management activities) |
|---|---|
| *Understanding the cause and impulse*<br>– Are there problems that must be remedied?<br>– Does this involve a future concept, an opportunity that could provide competitive advantages?<br><br>*Describing the problem and problem contexts*<br>– What does the initial situation look like? | *Initialization of the project*<br>– Decide on the project leader (driving force)<br>– Set the objective for the preliminary study stage (what type of decision is to be taken at the conclusion of the preliminary study; what questions should the answer address?) |

(continued)

**Table 13.1**   (continued)

| Field of activity 1 (primarily derived from the problem-solving cycle) | Field of activity 2 (primarily derived from the list of project management activities) |
|---|---|
| – What are the characteristic symptoms or defects? Opportunities?<br>– What type?<br>– If there are defects, where do they occur?<br>– When? How often?<br>– To what degree?<br>– What is the problem connected to?<br><br>*Define and structure the area of investigation*<br>– Visualize the viewpoint (system aspect)<br>– Rough structuring into main topics (super-, subsystems)<br>– Establish the elements and their relationships<br>– Define areas of investigation (what is significant with respect to the problem?)<br>– Visualize other possible viewpoints (system aspects)<br><br>*Investigate process operations and substantiate the need for change*<br>– What characteristic processes are taking place in the problem area? (Which ones are not?)<br>– What are the characteristic behaviors of the system (positive, negative)?<br>– With which internal or external influences do they occur?<br>– Localize symptoms of problematic situations in system representations<br>– Search for causes<br><br>*Identify affected parties and participants, explore their interests*<br>– Who are the key people and opinion shapers?<br>– How do they act?<br>– Do they agree that a problematic situation is present? (If not, why not?)<br>– Do they support the need for change? (If not, why not?)<br>– Who (department, groups, people, internal/external) has which positive or negative connections to the planned undertaking? (Influence analysis)<br>– Who are the current users?<br>– What ideas do they have about the future solution?<br>– Will the user group change?<br>– If so, does this have an effect on the needs and requirements? | – Derive a list of tasks for preliminary study (partial tasks)<br>– Determine project organization, plan approach (logical dependencies of subtasks, priorities)<br>– Create a time schedule (end date, significant milestones)<br>– Determine personnel requirements (qualitative, quantitative)<br>– Determine the budget for the preliminary study (with buffer), consult with affected parties<br>– Agree on project assignment<br>– Engage the project team; configure the group<br>– Project kickoff (startup meeting)<br>– Agree on the working style and the internal organization of the project group<br>– Determine and record the organizational structure in the project (select organizational model, steering committee, decision-making bodies, responsibilities, competencies, etc.)<br>– Determine the method of project reporting<br>– Don't forget project marketing (look for project sponsors, provide information on objectives and intentions)<br><br>*Keeping the project running*<br>– External anchoring of the project: communicate progress and results to client and users (objectives, ideas, design area, evaluation criteria, etc.). What really makes sense, what is desired, supported? What is both rational and intuitively correct?<br>– Internal anchoring of the project: organize regular project meetings, determine results, develop ideas and concepts, clear up/resolve conflicts<br>– Leading, steering, moving the project forward: determine progress, have it recognized, stick to the design objectives and monitor project procedural objectives, and modify or confirm as necessary |

(continued)

**Table 13.1** (continued)

| Field of activity 1 (primarily derived from the problem-solving cycle) | Field of activity 2 (primarily derived from the list of project management activities) |
|---|---|
| *Assess future development*<br>– How will the situation develop in the future if no action is taken now?<br>– What environmental developments can be expected?<br>– Possible effects on the problem area?<br>– What development trends are identifiable in the solution area?<br>– What are their positive aspects? Why?<br>– What are their negative aspects? Why? | *Organizing information system*<br>– Make contacts<br>– If necessary, clarify the personal effects on the requirements and composition of the project group and decision-making body (changes, additions?) |
| *Identify possible interventions*<br>– What interventions are essentially conceivable?<br>– What effects might they have?<br>– What measures mitigate the problem; which ones make it worse?<br>– Where have similar problems already been solved?<br>– Comparable starting situations or the same sector?<br>– Others?<br>– Sharing of the experience or access possible?<br>– Has anyone already (unnecessarily) determined a solution direction? | |
| *Initially, define the design area (adjust later if necessary)*<br>– What should be actively influenced (changed), what should not?<br>– Observe federalist principle (solve the problem at the deepest possible level)<br>– Capture important environmental connections | |
| *Define the design object, e.g., using a work breakdown structure* | |
| *Develop requirements (objectives)*<br>– Who has influence on setting objectives?<br>– Whose interests should also be considered? (determine affected parties, participants)<br>– Determine or negotiate demands on new solution (design objectives)<br>– Desired effects with respect to functional, financial, and personnel; location of effect (system/surrounding system), time frame (short, medium, long term)<br>– Undesired effects?<br>– Divide into compulsory, recommended, and desirable goals | *Organize process for agreeing on objectives*<br>– Resolve possible conflicts of objectives and interests or enable clarification |

(continued)

**Table 13.1**  (continued)

| Field of activity 1 (primarily derived from the problem-solving cycle) | Field of activity 2 (primarily derived from the list of project management activities) |
|---|---|
| *Develop solution variants*<br>– Can design spheres be separated from one another to allow a substantially isolated search for approaches to a solution in the preliminary study? If so, the following steps apply not only to the framework concept to be developed, but also to every sphere of planning<br>– Develop variants of framework concepts / system architectures/solution principles (= synthesis)<br>– Give priority to important objectives<br>– Verify meaningfulness/justification of boundary conditions and compulsory objectives<br>– Search for alternative solution principles (thinking in variants)<br>– Use list of resources and examples as the stimulus for the solution search | |
| *Analyze solutions with regard to*<br>– Formal aspects (compulsory objectives adhered to? Solutions complete, i.e., available for evaluation? Solution variants comparable?)<br>– Procedures (internal view)<br>– Capable of being integrated with surroundings, including superordinate concepts (external view)<br>– Operational efficiency<br>– Necessary assumptions and conditions<br>– Consequences<br>– Investment costs (inflation factor), operating costs, operational benefits | *Interim orientation of client/ decision-making body*<br>– Transfer ideas<br>– Seek opinions, views |
| *Evaluate solutions (include decision-makers)*<br>– Complete list of criteria<br>– Evaluate solution variants | *If necessary refer to the client's contact person(s)*<br>*Plan and implement evaluation and decision-making procedures*<br>– Prepare presentation of variants<br>– Plan further procedure (possibly variant-specific), including budgets for further phases<br>*Bring about decisions on framework concept, system architecture, solution principle*<br>– If the decision deviates from the proposal, consider the consequences<br>– Modify the next steps if necessary<br>– Document the decision |

*Result*

An overview of the situation that does not go into detail, a list of requirements (list of objectives), conceivable and reasonable draft solutions (framework concepts, system architectures, solution principles), the evaluation of these with respect to the feasibility and preferability (functional, economic, social, human resource-related, ecological, etc.) plus a recommendation on how to proceed further.

Final check for evaluating the quality of a *preliminary study:*

- Is the problem sufficiently clear, is there agreement on it with the client and the users?

  Does one know which problem one wants to solve? What opportunity they are
      pursuing? And why?
  Is the demarcation reasonable and adequate?
  Are the relationships to the environment clear?

- Is the design area adequately defined and known? Is this agreed with the client?
- Are the objectives clear with regard to the requirements of the solution (which functions should be achieved, economic goals, human resource-related/social, time-frame, ecological, etc.)?
- Is there an adequate overview of the basic solutions (solution principles)?
- Can these variants be evaluated with respect to their appropriateness (including requirements and consequences)?
- Are the yardsticks clear and accepted by all?
- Does this facilitate deciding on a specific system architecture (solution principle)? Can it be justified logically and in a way that can be understood? Is it rationally *and* intuitively correct? Are there options open? Which ones?
- Are the critical assumptions and consequences known?
- Has the psychological component been adequately taken into account? Do the key people and groups see a reason to take action?

## 13.2   Main Study: Activities Checklist (Table 13.2)

*Purpose*

Facilitating a decision on an architecture/an overall concept that can be built on and that a decision can be made on completing the task.

*Object*

This is the system (solution) itself: with important subsystems or system aspects.

*Result*

A functional overall concept (master plan); defined functions, important subsystems or system aspects and their functions; an implementation plan; economic considerations/base for investment decisions. Functional specifications for solution components should be deducible from the overall concept.

**Table 13.2**  Activity checklist for main study

| Field of activity 1 (derived primarily from the problem-solving cycle) | Field of activity 2 (derived primarily from the list of project management activities) |
|---|---|
| *Draw conclusions from the decision on framework concept and on the basis of development already carried out (w.r.t. content)* <br> – Have new ideas or concepts come up for consideration? <br> – Objectives, boundary conditions still the same? <br> – System boundaries? <br> – Changes of affected parties, participants? <br> – Is the overall concept workable or does it need adjustment? <br><br> *Detail and concretize requirements of the overall concept (system objectives)* <br><br> *Develop overall concept* <br> – Determine object structure in greater detail (subsystem aspects) <br> – Identify relationships within the system and interfaces with the environment; describe qualitatively and quantitatively <br> – Strive for modular design <br> – Think in variants <br> – Consider interplay of hardware and software components <br><br> Hardware: function, requirements, type and number, arrangement, procurement possibilities, etc. (buildings, machines, devices or components) <br><br> Software and organization: operational procedures, organizational measures, IT programs, etc. <br><br> *Analyze overall concept with regard to* <br> – Formal aspects (compulsory objectives, etc.) <br> – Ability to be integrated <br> – Functions and processes <br> – Operating ability and efficiency <br> – Requirements and conditions <br> – Consequences, etc. | *Initiate main study* <br> – Draw conclusions from decisions based on preliminary study (organizational components) <br> – Check effects, especially with regard to <br><br> Task formulation <br> Milestones for the main study <br> Project organization (subtasks, action plan, priorities) <br> Personnel requirements or changes <br> Organizational setup of the project <br> Budget (time/costs), etc. <br><br> – Project mandate still valid? If not, reach a new agreement <br><br> Further activities in accordance with the preliminary study phase <br><br> *Keep project running (analogous to the preliminary study)* <br><br> *Subactivities to support the activities in field of activity 1 (analogous to the preliminary study)* |
| *Evaluation of the overall concepts* | *In addition* <br> – Determine priorities and next steps for detailed handling (detailed studies) <br> – Work out implementation plan for system build and introduction <br> – Regulate the organization and coordination of further execution |

After deciding on an overall concept, it is very helpful to carry out tiered, overlapping detailed studies and implementation of solution components

Final *check* for evaluating the quality of a main study:

- Is the suggested architecture and the overall concept convincing and practicable (functional, economical, organizationally, etc.)?
- Is there an overview of the conceivable alternatives?
- Are the critical components known? (Critical with respect to, for example, functionality, risk, security, manufacturability, availability, disposal)
- Is the situation ready for a decision? Is the decision reasonable and viable internally and externally?
- Are the priorities for further detailing and implementation clear?
- Are the future users or "supporters" of the new system convinced and on board?

## 13.3   Detailed Studies: Activities Checklists (Table 13.3)

*Purpose*
Detailing and specification of subsystems/function groups of the overall concept determined in accordance with the main study. Generating all technical and organizational requirements to make it possible to carry out system building. (Note: this does not involve a single detailed study in which the overall concept is detailed, but rather a number of detailed studies. Every subsystem/module of the overall concept is to be concretized in a detailed study.)

*Object*
These are the component parts of the solution (subsystems, system aspects) and their interrelationships (interfaces).

*Result*
Fully functional concepts for subsystems or system aspects that fit into the framework of the overall concept and thus are concretized so that they can ultimately be "built." If the facilities required for system building must first be created, their planning and preparation must similarly be an object and a result to be worked out in parallel.

**Table 13.3** Activity checklist for detailed study

| Field of activity 1<br>(derived primarily from the problem-solving cycle) | Field of activity 2<br>(derived primarily from the list of project management activities) |
|---|---|
| *Draw conclusions from the decision on the overall concept, or adjustment if required (content type) of*<br>– Overall concept<br>– Objectives, restrictions, interfaces, etc., for further detailing<br><br>*Determine effects on partial solutions or solution components (subsystems, system aspects)* | *Initiate detailed study phase*<br>– Draw conclusions from the decision in accordance with the main study (organizational components)<br>– Determine priorities for detailed study (start with critical or logically preferential partial solutions)<br>– Subdivision into subprojects<br>– Formulate assignments for subprojects (content, budget, deadlines, etc.) |

(continued)

**Table 13.3** (continued)

| Field of activity 1 (derived primarily from the problem-solving cycle) | Field of activity 2 (derived primarily from the list of project management activities) |
|---|---|
| *Define, determine, describe demarcation of partial solutions or solution components*<br><br>*Concretize requirements for partial solutions or solution components* | – Corporate and project organization, structure, personnel aspects for more detailed development (generally: expansion of the circle of people involved, creation of subproject groups)<br>– Determine coordination mechanisms among subproject groups<br>– Further activities analogous to preceding phases |
| *Develop detailed concepts for solution components*<br>Search for a solution and a choice in accordance with preceding phases (but at a deeper, more detailed level). Direct special attention to the integration of solution components into the overall concept | *Keeping the project running*<br>– As an analogy to the preceding phases, paying particular attention to coordination, control, and supervision (results, deadlines, expenditure, personnel capacity, etc.) |
| *Create functional specifications for system building*<br><br>*Possible requests for quotation and evaluation of outsourcing* | *Subactivities to support activities in field of activity 1: analogous to the preceding phases*<br>Additionally:<br>– Planning and preparation for system building<br>– Regulate the organization and coordination of further processing (usually a change in responsibilities for system building) |

Final *check* to evaluate the quality of each detailed study:

– Are the requirements of the overall concept satisfied by the detailed concepts?
– Can the individual detailed concepts be integrated into the framework of the overall concept? Are they capable of being integrated? Do they fulfill their intended functions? Do they exhibit any qualities that are undesirable from the point of view of the overall concept?
– Are the detailed concepts concretized in such a way that they can subsequently be built?
– If the requirements for the system are met, could the production requirements have repercussions for the design?
– Are the cost estimates made within the main study sustainable, or must they be modified?

## 13.4 System Building/implementation: Activities Checklist (Table 13.4)

*Purpose*
Concrete implementation or building of the solution or partial solutions.

*Object*
Partial solutions, solution components (subsystems, system aspects) and the necessary infrastructure for subsequent introduction/implementation.

*Result*
Partial solutions ready for introduction/implementation or solution components and their successive assembly.

*Notes*
The systems engineering process model is without doubt most effective in the development phases (preliminary, main, and detailed studies). For these phases, it is also possible to specify generally valid activities in the form of a checklist.

However, this is more difficult for the system build and introduction phases, as in these instances project-specific aspects must be very clearly expressed (product development, construction projects, IT projects, planning and information systems, etc.).

**Table 13.4**  Activity checklist for System Building/Implementation

| *Field of activity 1*<br>(derived primarily from the problem-solving cycle) | *Field of activity 2*<br>(derived primarily from the list of project management activities) |
|---|---|
| *Preparation or additional work (if necessary, make adjustments to the detailed concepts or to the concept)*<br><br>*If the system build is being outsourced*<br>–  Create specifications and functional specification document<br>–  Gather and compare bids | *Initiate system build*<br>–  Create a schedule for implementation<br>–  Make concrete tasks of release decisions<br>–  For outsourced tasks, finalize and complete contracts (specify goods and services, quality, prices, deadlines, measures if there is failure to complete, acceptance procedures, etc.)<br>–  For internal manufacture, similar consideration with regard to goods and services, quality, expenditure, deadlines |
| *Work preparation, acquisition of means of manufacture*<br><br>*Manufacture (possibly prototype)*<br><br>*Individual tests*<br><br>*Successive system integration and system tests* | *Keeping the project running, especially project control and management*<br>–  Check achievement of targets (performance, quality)<br>–  Expenditure<br>–  Deadlines<br>–  Coordination between subproject groups, with users, contracting authorities, etc.<br>–  Intervention if there are deviations, including plan adjustment |
|  | *Gradually prepare the user for takeover, and where necessary, start training* |

## 13.5    System Introduction: Activities Checklist (Table 13.5)

*Purpose*
Handover of the (partial) solution to the user. Know-how transfer. Guarantee that agreed objectives and functions will be met.

*Object*
(Partial) solution ready for implementation along with the environment in which it is to be imbedded (technically, personnel, organizational)

*Result*
Operable, functioning (partial) solution, operable infrastructure, and trained users.

*Notes*
For this phase, the division into field of activity groups 1 and 2 is dispensed with, as the systems design should be largely finalized and as such, no more real conceptual activities need to be carried out.

**Table 13.5**  Activity checklist for System Introduction

*Preparation for introduction*
– Create detailed implementation plan
– Plan possible decommissioning and replacement procedure of the old system
– Create user's manual and instructions for use
– Create organizational conditions for the operational phase (detail operational concepts, create the necessary jobs, allocate personnel and materials, etc.)
– Specify detailed maintenance concept
– Determine organization of safety/catastrophe (measures in the case of breakdown or disruption
– Create necessary infrastructure (e.g., buildings, connections to required services, etc.)
– Prepare hand-over procedures and conditions

*Introduce solution*
– Install hardware and software
– Train operators and users
– Conduct test runs
– If necessary, conduct parallel operation (old and new system at the same time)
– Correct any defects, fine tuning
– Ensure hotline support (quick ad hoc support) for service personnel and user, etc.

## 13.6   Project Completion: Activities Checklist (Table 13.6)

*Purpose*
To conduct an orderly closure, and to use debriefing to facilitate lessons learned.

*Object*
Agreements with the client and their comparison with reality.

*Result*
Completed, documented project with balanced accounts; project group disbanded.

**Table 13.6**   Activity checklist for Project Completion

| |
| --- |
| *Completion of handover procedures and implementation* |
| *Control and debriefing (lessons learned)* <br> – Objectives met or not met? Remaining deficiencies? <br> – Opportunities for improvement, maintenance requirements? <br> – Learning gains on the part of the project group, the client, the users of similar systems |
| *Complete final documentation (system- and user-oriented)* |
| *End of project party* |
| *Disbanding of the project group and (re-)incorporation of the personnel involved into another organization* |

Adapt the terminology/wording to your business/industry.

# Chapter 14
# Characteristics of Successful Project Management

In his comprehensive, empirical analysis, W. Keplinger formulated the characteristics of a successful project.[1] His results are summarized below:

**Project Success**

Before looking for success factors, we have to first make it clear at what point we want to speak of a successful project. This question is more difficult to answer than it would at first appear. We need to ask:

- What are the *criteria* by which success should be measured? Is it primarily the quality of the result with regard to content? Do costs and schedules also play a role? Is it all of these together? How much weight should be assigned to each criterion?
- *Which person or persons* should judge the success? The project manager, the client or, his representatives (e.g., the steering committee), the project group?

- *When, at what point in time,* does the question regarding success arise? Immediately upon completion of the project or later (for example, after a year)?

- *Against what should or could comparisons be made*? With the originally formulated objectives? Or with those that were aligned last?

The above questions are obviously not independent of one another. For example, it is shown during the course of the analysis that the timing of when the question regarding a project's success is posed has a significant influence on its assessment:

- A project that is significantly delayed or that has significantly exceeded its cost limits is often judged not to have been successful if the assessment comes immediately after its completion.
- If it turns out that, approximately 1 year after its completion, the project results are of good quality (performance, stability, reliability, adaptability, etc.), then it

---

[1] Keplinger, W. (1991): Merkmale erfolgreichen Projektmanagements.

© Springer Nature Switzerland AG 2019
R. Haberfellner et al., *Systems Engineering*,
https://doi.org/10.1007/978-3-030-13431-0_14

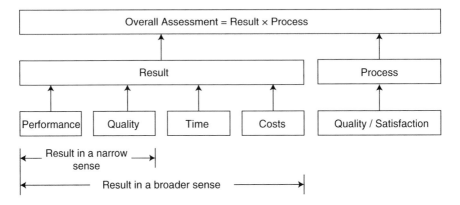

**Fig. 14.1** Criteria for determining success. (According to W. Keplinger)

is these results that count, whereas the failure to adhere to budgets and schedules is forgotten and considered "water under the bridge."
– Or: project managers tend to judge their own projects more critically than the users or the client.

Keplinger's research approach is not elaborated on here; suffice to say that he ultimately describes it as the product of results and processes. According to Keplinger, success is measured not only by the result but also by the processes, strategies, and collaborative efforts that have led to this result (Fig. 14.1).

**Characteristics of Success**
Figure 14.2 shows an overview of Keplinger's 14 success criteria. The sequence bears no significance regarding the importance of the criteria, which may vary from project to project.

1. **The *(top) management* supports the project:**

   – Objectives tend to be supported.
   – Allocation of resources is easier.

     → If this support is not (initially) in evidence or available, the project manager **should** look for a patron (promoter). If one cannot be found, the point/intention of the project may have to be called into question.

2. **Good *external relationships:***

   – With users
   – With implementers

     → If these relationships are present, there is a greater probability that the project will be supported by users and that the subsequent results will be accepted.

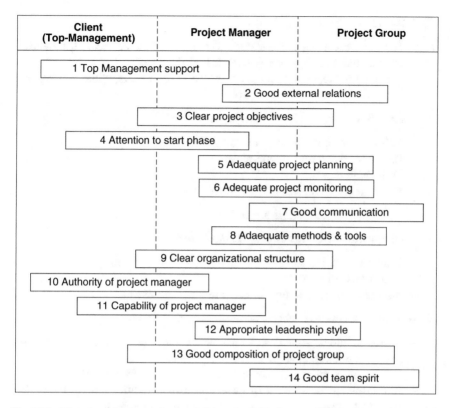

**Fig. 14.2** Allocation of success characteristics to the field of responsibility of the project participants. (According W. Keplinger)

3 **Clearly agreed** *project objectives:*

- This does not mean that objectives may not be adapted during the course of the project, however.
- It is possible to mutually agree upon adaptations to the objectives, but new agreements should be made quickly and clearly.

4. **Pay attention to the** *start phase***, in which:**

- A functioning team should be created.
- With a common understanding of the problem and
- A common planning of procedures and activities takes place

   → A kick-off meeting is a visible sign of the project start

5. **Adequate** *project planning* **(organization, approach):**

- Conducted together, short, intensive
- At the beginning, avoid detailed planning over a broad scope
- Observe phase objectives: focus initially on the results to be achieved in the next phase (which questions should it be possible to answer at the end of the phase, what decision should it be possible to make?); keep it approximate and not too detailed.

6. **Adequate** *project monitoring:*

   – A regular reporting of the status and progress of the project helps the team, even though this is primarily intended for the client
   – Whereby the reporting should be limited to key factors, without recourse to bureaucratic and detailed controls

7. **Open, direct** *communication* **and** *information:*

   – Volunteer information, do not wait to be asked
   – Internal and external communication
   – Written memos
   – But: oral communication is also important

      → Do not forget project marketing

8. **Appropriate** *implementation of methods and tools:*

   – Network planning and IT-supported project management systems are not the (only) success factors
   – Simple aids, consistently applied, are often successful

9. **Appropriate, unbureaucratic** *organizational structure:*

   – Teams should be kept small (often what comes out of a meeting with many people is nothing more than … many people!)
   – Involve other groups of people via open channels of information
   – A transparent assignment of tasks and responsibilities among the project team and between the team and the decision-making body

      → The core team should be allowed to engage intensively in the project

10. **Sufficient** *authority* **vested in the project manager:**

    – Successful project managers have authority – or simply assume it
    – The corporate culture should be of a type that will accept this, that is, it will not display an adverse reaction to it
    – Successful project managers exercise their formal authority only in exceptional cases. They manage by persuasion and not by decree

11. *Capability***, authority, and experience of the project manager:**

    – The person who is the project manager is, quite simply, *the* key success factor.
    – The ability to manage is the most important factor; technical competence is secondary.

      → The more important the project, the more important the person who is the project manager.
      → The most capable project managers belong in important projects (their efforts should not be wasted on projects of a lower priority).

→ Projects should also be viewed as learning opportunities and vehicles for personal development: Younger and inexperienced employees should be specifically assigned to lead small projects (experienced persons contribute professionally to the project and provide support – if required).

12. **Appropriate *leadership style* of the project manager:**

   - Cooperative in routine cases
   - Authoritarian in exceptional circumstances
   - Primarily task-oriented
   - Conflicts are quickly addressed and resolved; there are no simmering problems
   - The best way to resolve conflicts is with open discussion
   - Those continually seeking conflict may be removed from the team (in an equitable/a fair manner)

13. **Formation of an *adequate project group:***

   - Qualified employees who are able to work in a team are required
   - Technical skills are of primary importance
   - Followed by the ability to work as part of a team
   - Pay attention to group stability (if possible, no unforeseen personnel changes)
   - Create enough time for project work
   - Avoid, as far as possible, the havoc that can be created through deputization (if a particular person never has time and continually sends deputies, a formal replacement should be effected without delay).

   → The project manager should have a say in team formation (and should not be given team members who are simply expendable elsewhere).

14. ***Motivated project teams:***

   - The commonest motivators are:

   The work itself and the associated challenge
   Progress, successful results, recognition
   Independent design opportunities

   - The commonest demotivators are:

   Setbacks, failures
   Lack of support from above
   Subsequent, unforeseen changes
   Slow or sluggish progress, lack of intensity in the work

   → Collaboration in the project team should, as far as possible, be voluntary (no forcing team members into something; the project concept may also have to be "sold" to the team)
   → The project also has to be marketed internally
   → Arrange for the project kick-off along with time for socializing (sharing common experiences)

With regard to Fig. 14.2 it is interesting to note that:

– None of the characteristics of success belongs exclusively to the area of responsibility of only one party; the blocks always extend beyond the responsibility boundaries of the various parties
– There is no characteristic of success in which the project manager does not appear. He or she is *the* central success factor

Even if these statements cannot – scientifically speaking – be proved, they may help to inspire a project manager to find his or her role.

## 14.1   Self-Check of Knowledge and Understanding: Characteristics of Successful Project Management

1. Are these considerations in line with your experience?
2. Are you missing something?

The authors would like you to share your experience with them.

# Part VI
# Methods and Tools (M&T)

# Chapter 15
# Survey of Methods and Tools

Below we show, in an encyclopedic format, common methods and tools (M&Ts) that, on the one hand, support the work of *systems design* (*architectural design* and *conceptual design*) and, on the other, the work of *project management*.

We deliberately limit ourselves to a short characterization of each, which should allow the reader to judge if certain methods/techniques are suitable in the context of a present project or working step, and if so, which ones exactly. More detailed descriptions or user instructions can be found quickly and easily on the internet, from current information sources such as Wikipedia and internet search engines.

In contrast to other systems engineering promoters we do not regard M&Ts such as risk management, evaluation and decision-making techniques, validation, and verification, as core components of the systems engineering concept. They are important, but in our opinion, the application of certain M&Ts depends heavily on the kind of project or sector and the local practices, norms, binding regulations, customary partners, supplier commitments, and much more.

We have assembled M&Ts from different types of offers and have located them as the "feet" of the systems engineering manakin (Fig. 15.1). Without specifying how they are to be applied, we leave it to the users to apply them and appraise their suitability. Thus, we follow the principle, mentioned several times, that methodology and methods *do not replace* but rather *support* independent thinking.

The survey in Sect. 15.1 lists selected M&Ts in columns 1–8 according to the relevant steps of the problem-solving cycle in which they have their main application. This allocation is explained in the following text. However, a definite allocation is not always possible because several techniques are applicable to different steps. Techniques assigned to project management are found in group (9), general techniques in group (0).

### M&Ts for Situation Analysis (1) (2) (3)
Here, the focus is particularly on techniques of information acquisition (1), information editing (2), and information illustration (3). Techniques used for the analysis

---

The original version of this chapter was revised. The correction to this chapter is available at
https://doi.org/10.1007/978-3-030-13431-0_17

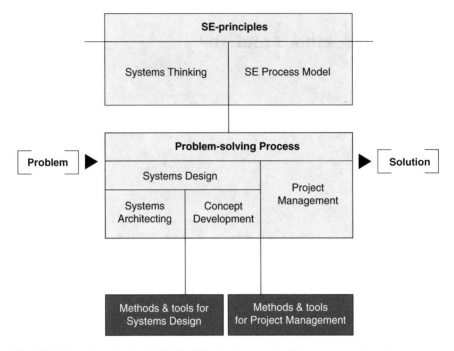

**Fig. 15.1** Methods and tools (M&Ts) within the framework of the systems engineering concept

and evaluation of solutions can usually also be used for the analysis and evaluation of the actual condition, which after all represents a special solution.

Creative techniques (5) can be applied universally. They are not only applicable in the solution search but are also useful in situation analysis (for example, brainstorming in the search for problem causes) and in project management (brainstorming with regard to activities to be conducted in a single step). Mathematical methods for optimizing solutions, such as simulation, may also be used to analyze system behavior in the actual condition.

### M&Ts for the Formulation of Objectives (4)
Besides the techniques listed in column 4, one can use the techniques of information acquisition (1) and editing (2), for example, interviews and survey techniques, plus, because the formulation of objectives substantiates the value system underlying the decision, the techniques of evaluation and decision-making (8).

### M&Ts for the Solution Synthesis (5) (6)
The focus is on creativity techniques (5) and, possibly, optimization techniques (6). In addition to these, techniques of information procurement (1) and processing (2) can also be applied.

### M&Ts for Solution Analysis (7)
Here, our concern is of course primarily with the techniques listed in column 7, plus with those of information procurement (1) and processing (2) and possible simulation or other optimization techniques (6).

**M&Ts for Evaluation and Decision (8)**

The focus is naturally on evaluation techniques (8). Once again, techniques for information procurement (1) and processing (2) are feasible, optionally also techniques for optimization and analysis (6).

The individual M&Ts will be described in alphabetical order in the glossary.

**Table 15.1**  Methods and tools (M&Ts) assigned to the thematic areas of systems engineering

| Information procurement (1) | Information editing (2) | Information illustration (3) | Formulation of objectives (4) |
|---|---|---|---|
| – Information acquisition techniques<br>– Information acquisition plan<br>– Questioning<br>– Observation techniques<br>– Checklists<br>– Forecasting techniques<br>– Delphi method<br>– Questionnaire<br>– Interview, interviewing techniques<br>– Activity scanning, (multi-moment observations)<br>– Panel polling, survey<br>– Forecast and predictions<br>– Sample tests<br>– Scenario analysis | – ABC analysis<br>– Flow chart<br>– Analysis techniques<br>– Benchmarking<br>– Capability maturity model, CMMI<br>– Illustration techniques<br>– DSM<br>– EFQM<br>– Flow diagram<br>– Histogram<br>– Information editing techniques<br>– Ishikawa diagram<br>– Kaizen<br>– Key indicator system<br>– Correlation analysis<br>– Mathematical statistics<br>– Sampling<br>– Process analysis<br>– Regression analysis<br>– SysML<br>– UML<br>– Value analysis | – Information preparation/processing methods<br>– Modeling and representation techniques<br>– Failure-causes-measures analysis<br>– Mind mapping<br>– MTBF<br>– MTTR<br>– MTTF<br>– Input–output model<br>– Polarity profile<br>– Network thinking, cause-effect networks<br>– Allocation problems<br>– Visualization techniques (information graphics) | – Operationalization<br>– Target costing<br>– Use case |

*CMMI* capability maturity model integration, *DSM* design structure matrix, *EFQM* European Foundation for Quality Management, *SysML* systems modeling language, *UML* universal modeling language, *MTBF* mean time between failures, *MTTR* mean time to repair, *MTTF* mean time to failure

| Project management (9) | | | |
|---|---|---|---|
| – Bar chart, Gantt chart<br>– Work breakdown structures | – Project audit<br>– Project management body of knowledge<br>– Configuration management | – Network planning<br>– Critical path method<br>– PERT<br>– Project management software<br>– Milestone monitoring | – Timing trend diagram<br>– To-do list<br>– Lessons learned |

*PERT* program evaluation and review technique

| General (0) | | | |
| --- | --- | --- | --- |
| – Eisenhower method<br>– Heuristic methods<br>– Design structure matrix<br>– System dynamics | – Re-engineering<br>– Agile systems engineering<br>– Analysis techniques<br>– Business process model and notation | – Quality management<br>– Safety management<br>– (Total) quality control<br>– Total quality management<br>– Continuous improvement process | – Virtual product development<br>– Digital factory<br>– ISO 9001<br>– Just in time<br>– Kaizen |

| Synthesis of solutions – creativity (5) | Synthesis of solutions – optimization (6) | Analysis of solutions (7) | Evaluation + decision (8) |
| --- | --- | --- | --- |
| – Analogy method<br>– Bionics<br>– Brainstorming<br>– Just in time<br>– Card technique<br>– Creativity technique<br>– Method 635<br>– Morphological analysis<br>– Synectics<br>– TRIZ, TIPS | – Operations research<br>– Dynamic optimization<br>– Decision tree<br>– FMEA<br>– Linear programming<br>– Monte Carlo method<br>– Real options<br>– Sequencing problems<br>– Simplex algorithm<br>– Simulation technique<br>– Numeric simulation<br>– Game theory<br>– System dynamics<br>– Queuing models<br>– Assignment or allocation of problems<br>– Use case<br>– Just in time, Just in sequence | – FMEA<br>– Systems theoretic process analysis<br>– EFQM<br>– Fault tree analysis<br>– Risk analysis<br>– MTTF<br>– MTBF<br>– MTTR<br>– Ishikawa diagram<br>– ISO 9001<br>– Reverse engineering<br>– Risk management<br>– Weak point analysis<br>– Safety analysis, management<br>– Total quality control<br>– Total quality management<br>– Value analysis<br>– Reliability analysis<br>– Six sigma | – Valuation techniques<br>– Analytic hierarchy process<br>– Evaluation techniques<br>– Decision theory<br>– Decision tree method<br>– Value benefit analysis<br>– Cost-benefit analysis<br>– Cost-effectiveness analysis<br>– Economic value analysis<br>– Real options<br>– Sensitivity analyses<br>– Economic efficiency calculation<br>– Economic feasibility calculation<br>– Scoring methods<br>– Criteria plan |

*TRIZ* Teorija Rezbenija Izobretatelskib Zadach, *TIPS* theory of inventive problem-solving, *FMEA* failure mode and effects analysis

## 15.1   Self-Check of Knowledge and Understanding: Survey of Methods and Tools

1. Name at least two methods for each step of the problem solving cycle
2. Which M&Ts have you already applied in your projects?
3. What are you missing and why?

# Chapter 16
# Encyclopedia/Glossary

An → arrow in the text refers to a corresponding key word in the encyclopedia.

**ABC Analysis (Synonyms: Pareto Analysis, 80–20 Rule)**
An important categorization technique in information processing that makes it possible to identify groups of items that contribute in certain intensities to a chosen output measure. For example: 80% of the sales are made with 20% of the articles (or clients), or 10% of the population has control over 90% of the assets.

*Reference:*
Koch, R. (2004): *Living the 80/20 Way.*

**Activity Sampling (Synonyms: Work Sampling, Multi-Moment Observations)**
Activity sampling is a statistical method for determining the proportion of time spent by workers in various defined categories of activity (e.g., setting up a machine, assembling parts, waiting, etc.). Its great advantage over other statistical techniques is the efficiency with which it measures and analyzes the nature and performance figures of complex processes and interactions.

In an activity sampling study, a sufficiently large and representative number of random observations is made during a specified amount of time. The nature and frequency of observed activities are recorded and later analyzed.

Activity sampling is frequently used when calculating standard times for manual manufacturing tasks or for analyzing extremely complex process interactions in socio-technical systems.

*Reference:*
Groover, M. P. Work Systems and Methods, measurement, and Management of Work.

---

The original version of this chapter was revised. The correction to this chapter is available at https://doi.org/10.1007/978-3-030-13431-0_17

© Springer Nature Switzerland AG 2019, Corrected Publication 2021
R. Haberfellner et al., *Systems Engineering*,
https://doi.org/10.1007/978-3-030-13431-0_16

**Agile Systems Engineering**
Refers either to an AGILE Systems ENGINEERING Approach or to Engineering of AGILE SYSTEMS. See Sect. 1.3 on the "Agility of Systems."

**Analogy Method**
→ Creativity technique for finding solutions. The main idea is to look for possible solutions to a problem outside the problem scope by searching for analogies. These can be similarities of form, characteristic, and function. → Bionics, for example, is such a method that systematically screens nature for principles that are transferable to mechanical design improvements or which can be used to develop products with completely new characteristics.

*References:*
Gerardin, L. (1968): Bionics.
Rossmann T.; a.o. (2007): Bionics - Natural Technologies and Biomimetics.

**Analysis Techniques**
Techniques that are used for a systematic investigation of all aspects/components/ elements of an object (or subject) based on defined criteria. These are then sorted, structured, and evaluated – very often also with respect to their interaction. Analysis techniques are very important to systems engineering and are used to investigate the past, present, and, when designing, the future (requirements) of systems.

Mathematical methods, simulation runs, plausibility tests, and destruction analyses support the various techniques, which are often named after their purpose (→ reliability analysis, → security analysis, disaster analysis, compatibility analysis, consequence analysis, → risk analysis, cause-effect analyses, → cost-effectiveness analyses, etc.).

In designing systems analysis, techniques can either be used to predict a model's behaviors ahead of time (ex-ante) or to investigate the behavior of the realized model (object) in the real environment. In spite of all the progress in mathematical/ scientific modeling and computer performance, and despite sophisticated simulations of system behaviors, often only the construction and analysis of a real, functioning system can clarify the functionality of principles and solution ideas. When designing chemical engineering processes, for example, miniature plants are often constructed for testing first or pilot plants. When designing machines, we often see functional models, pilot samples, or prototypes. Finally, pilot runs are used to test production processes and means of serial and mass production.

With workflow planning in particular, test runs of various sizes right up to parallel operation of the old and new systems are carried out.

**Analytic Hierarchy Process, AHP**
A method for supporting decision-making processes, similar to the → value-benefit analysis (VBA). The essential difference compared with the VBA is the method of weighting criteria. These are determined along the lines of paired comparisons. The method chosen is largely a matter of taste. Besides, there is no such thing as a truly objective method for the evaluation of variants.

*Reference:*
Saaty, Th. L. (2001): *Decision Making for Leaders*

## Assignment or Allocation Problems

Special cases of → operations research, for which special solution methods were developed:

*Transportation problems* can generally be described as follows: at certain starting points physically separated from one another (points of departure), specific resources (e.g., vehicles, goods for shipment) are available in certain quantities. At certain endpoints (recipients) there is a specific need for the same resources. The connection paths including transport times and costs of the transfer from each starting point to each endpoint are given. The available resources are to be sent from the starting points to the endpoints so that the shipping expense (costs, times, vehicle use) is minimal.

An *allocation problem* can also be seen as a special instance of a transport problem. At each starting point, there is just one unit of the required resource available, and at each endpoint just one unit is requested. (Example: the best possible allocation of $n$ persons to $n$ workplaces or the transfer of $n$ vehicles from $n$ starting points to $n$ endpoints so that the overall mileage is minimized).

*References:* See → linear programming, → operations research.

## Bar Chart (Synonym: Gantt Chart)

An aid for the graphic representation of the *duration* (by the length of the bar) and the *temporal arrangement* of activities, e.g., in a project (by the length of the time axis). In their original form, bar graphs contain no logical dependencies of the represented activities. However, these can be added. IT-supported projects management systems permit the automatic generation of bar graphs from a → network plan.

## Benchmarking

Benchmarking is a continual process of comparing entities such as one's own products, performance, practices, processes, with others to set goals to perform as well as or better than the best and thereby achieve a competitive advantage.

The origin of the term (according to Wikipedia): a cabinet-maker's workbench with a mark for making, for example, all the legs of a table the same length.

*References:*
Boxwell, R.G. (1994): Benchmarking for Competitive Advantage.
Walleck, A.S., a.o. (1991): Benchmarking world-class performance

## Bionics

A special → creativity technique belonging to the group of → analogy methods. The essential idea: looking for analogies in biology and using them for the solution of technical problems. Familiar examples are material surfaces that exhibit the *lotus effect*, that is, minimal wettability and a high degree of self-cleaning. Also: wing shapes for airplanes (so-called winglet shape), which are based on bird wings (eagle, buzzard, condor). Also: *Velcro tape* based on the adhesive properties of burdock.

*References:*
Rossmann T.; Tropea C.; Vincent, J. (2007): Bionics.
Gerardin, L. (1968): Bionics.
Marteka, V. (1965): Bionics.

## Brainstorming

→ A creativity technique with which a group's problem-solving skills are to be used so that the flow of ideas is encouraged by a type of game rule.

Brainstorming rules: (1) clear questioning; (2) public protocol for the remarks made (e.g., on a flipchart) to show that there is no censorship of ideas; (3) temporal division between *idea gathering* (unordered, uncritical – no judgements) and the subsequent *evaluation* (critical assessment of applicability only at this point).

*Reference:*
Osborn, A. F. (1957): Applied Imagination.

## Business Process Model and Notation, BPMN

The business process model and notation (BPMN) is a standard for business process modeling that provides a graphical notation for specifying business processes in a business process diagram (BPD), based on a flowcharting technique very similar to activity diagrams from → unified modeling language (UML), which is also developed and standardized by the object management group (OMG). The objective of the BPMN is to support business process management, for both technical users and business users, by providing a notation system that is intuitive to business users, yet able to represent complex process semantics. The specification also provides mapping between the graphics of the notation and the underlying constructs of the execution languages, particularly the business process execution language (BPEL).

*References:*
Silver Bruce (2011): BPMN Method and Style with BPMN Implementer's Guide
White, Stephen A.; Bock, Conrad (2011). BPMN 2.0 Handbook: Methods, Concepts, Case Studies and Standards in Business Process Management Notation.
Grosskopf, A.; Decker, G.; Weske, M. (2009): The Process: Business Process Modeling Using BPMN.

## Business Re-engineering, See → Re-engineering

## Capability Maturity Model, CMM; Capability Maturity Model Integration, CMMI

The CMM in its original form was a model for evaluating the quality (maturity) of software processes in organizations (software development, maintenance, configuration, etc.), plus determining measures for their improvement. The CMM was replaced by the more sophisticated CMMI in 2003.

*References:*
Gallagher B. P.; Phillips, M.; Richter, K.J; Shrum. S. (2009): CMMI-ACQ.
Chrissis, M.B.; Konrad, M.; Shrum, S. (2011): CMMI.

## Card Technique (Synonym: Metaplan Method)
→ Creativity technique in which the ideas and statements are not expressed orally, as in
→ brainstorming, but rather are written on small cards by every participant and posted
on a bulletin board. Advantage: easier evaluation, e.g., in terms of clustering similar
ideas. Disadvantage: depends on the availability of aids such as a bulletin board.

## Checklists
Lists of activities that are necessary for the completion of tasks. There is a meaningful distinction between (1) checklists of a *compulsory nature*: every activity must be
carried out (for example: airplane takeoff) and (2) checklists of a *discretionary
nature* as an aid in searching for ideas – or as a stimulus for one's own thinking
(remember the important things). Checklists are usually strongly task-/context-
oriented and therefore not universally applicable.

## Configuration Management, CM
The impetus for configuration management (CM) was the continually increasing
complexity of products caused by a variety of possible combinations involving
various modules and the continual change in product configurations. The first
solution approaches to CM were developed in the aircraft and aerospace industries.
Similar complexity problems were also evident in other sectors in which – for both
suppliers and clients – it always had to be clear what parts or modules constituted
the purchased or delivered product. Methods and instruments of CM were refined
and specialized for various application fields. Today, CM transformations are part of
many disciplines, such as product data management (PDM), software configuration
management, etc.

The American National Standards Institute (ANSI), in cooperation with the
Electronic Industries Alliance (EIA) has defined CM as follows: "Configuration
management is a management process for establishing and maintaining consistency
of a product's performance, its functional and physical attributes, with its
requirements, design, and operational information, throughout its life." (Wikipedia)

*References:*
ISO 10007:2003: Quality management systems - Guidelines for configuration
    management.
Lyon, D.D. (2000): Practical Configuration Management.

## Continuous Improvement Process
*See → Kaizen*

## Correlation Analysis
Statistical procedure in which the magnitude of the mutual dependencies of various variables out of a sample (for example: people's weights and body heights) is
modeled often linearly as in the *Pearson product–moment correlation*. Correlation
is expressed by a so-called *correlation coefficient* r, which can take on values
between $-1$ and $+1$:

- A positive (+) correlation means that variables are coupled and changing their magnitude in the same direction (such as horsepower and acceleration of a car).
- A negative (−) correlation means that variables are coupled and changing their magnitude in opposite directions (such as the dependency of fuel consumption and weight or engine power of a car).
- $r = 0$ means that the variables are not coupled, i.e., totally independent from each other, such as the body height and political beliefs of a person.

However: even a correlation coefficient $r$ near 1 cannot mean a true dependency of two variables, for an outside, third variable can influence both of them, or there can be any other, unknown relationship. Therefore, it is important not to confuse correlation with causation!

A classic example of such a *spurious correlation* is the statistically significant correlation of the number of stork nests and the number of births in a given year in Copenhagen, where one should not draw the conclusion that storks deliver babies.

*References:*
Walpole, R.E. (2007, 9.ed.): Probability & Statistics For Engineers & Scientists.
Silver, N. (2013): The Signal and the Noise.
Mann, P.S. (2012): Introductory Statistics

**Cost-Benefit Analysis, CBA**
Cost-benefit analysis (CBA) is particularly used in public services to support decision-making, whereby it aids in evaluating macroeconomic proposals (measures by public authorities) or the not-for-profit effects of microeconomic proposals. A characteristic of CBA is the expression of costs and benefits in monetary units. For more information, see Sect. 6.4.2 (evaluation methods) and an example can be found in the case study in Chap. 9.

*References:*
Sassone P.G.; Schaffer, W. A. (1978): Cost-benefit Analysis - A Handbook.
Nas, T.F. (1996): Cost-benefit analysis: Theory and application.

**Cost-Effectiveness Analysis**
→ Valuation technique for the comparative assessment of product variants, investment decisions, etc. Cost criteria are expressed as monetary factors, the benefit or the effectiveness in a key figure that is specified the same way as with the value-benefit analysis (determination of the criteria, weighting, assignment of scores, multiplication with weighting points and summation of all effectiveness criteria). In a final step, the sum of the costs applied in a single period are divided by the cumulative effectiveness key figure. The product is an abstract key figure that indicates what an effectiveness point for each variant costs. A decision-making rule is of course to choose the variant with which a point costs the least. The results of an assessment can also be represented graphically by placing on one axis the cumulative costs and on the other, the cumulative effectiveness key figure. See Part III, Sect. 6.4.2.4.

*Reference*:
Levin H M (1983): Cost-Effectiveness: A Primer.

## Creativity Techniques, CTs

Creativity techniques (CTs) are intended to overcome passive waiting for ideas and increasing the probability of finding good solutions actively and quickly. Depending on their purpose, they can be divided into those that:

- Are ascribed to purely *intuitive* processes
- Promote analogous and/or contrasting *linking* of ideas (restructuring of the solution field)
- Are intended to lead to a variety of solutions by means of a *combinational* process

Thus, individual techniques can be assigned to several categories. Creativity techniques include in particular → brainstorming, → card technique, and → method 635. The characteristic of these techniques is that criticism and discussion of the reasonableness and practicability of an idea are not allowed at first. On the other hand, it is allowable, and even desirable, to seize on an expressed idea and modify it or twist its meaning around.

This glossary also includes → analogy methods → synectics, → bionics, and → morphological analysis. *Attribute listing* seeks further manifestations of the characteristics, functions, and effects (traits) of the existing or a discovered solution based on each trait. The emphasis on applying *attribute listing* is improvement. Beyond these methods, there are comprehensive systems for finding solutions such as G. Nadler's ideals concept (see Part I, Sect. 2.1.4.2), the systematic heuristics of G. Altshuller's theory of imaginative problem-solving (→ TRIZ).

In most cases, the above-mentioned techniques are applicable to both individual and group work. Requirements for good, efficient progress are experienced moderators and a little training with the group members. With many techniques, the recording of the results from group work requires special care and tact (e.g., when preference is given to one formulation from among several statements of comparable content). While dealing with a problem, phases of individual and group work are also replaced as various techniques are applied. To achieve, for example, the broad field sought by the morphological analysis, it is useful to draw up a list of parameters using intuitive techniques (e.g., brainstorming), and a broad array of solutions to which the parameters are applied to test whether they are usable in differentiating among solutions.

*References:*
De Bono, E. (1970): The Use of Lateral Thinking.
Osborn, A.F. (1957): Applied Imagination.
Csikszentmihalyi, M. (2013): Creativity: Flow and the Psychology of Discovery and Invention.
Teramata, T.; Nijstad, B.A. (Eds. 2003): Group Creativity: Innovation Through Collaboration.

**Criteria Plan**

List of variables and their measures used for a comparative evaluation of solutions (see Part III, Sect. 6.4.3.2).

**Critical Path Method, CPM**

The critical path method (CPM) is a widely used algorithm for planning and scheduling project activities → network planning techniques.

**Decision Theory**

A branch of applied probability theory for analyzing consequences of decisions and, if possible, to find optimal decisions. Decision theory uses both analytical and qualitative methods and can handle decision situations under both *certainty* and *uncertainty*. Some widely used methods are: → decision tree methods, → value-benefit analysis and the → analytic hierarchy process, in which criteria and alternatives are represented and evaluated comparatively to find the optimal solution. These methods are based on *secure* assumptions, whereby the situation in question can be described precisely (deterministic), or if it is permissible to use such a simplification. In cases of decisions under uncertainty, *uncertainty* has to be modeled using statistical methods. An advanced technique in that area is the *real options theory* (see Sect. 2.2.5), which makes it possible to include uncertainty and risk in the planning and evaluation process. In simple cases → sensitivity analyses can be conducted to get a feeling for the variety of possible decision consequences.

*References:*
Saaty, Th. L. (2001): Decision Making for Leaders.
Schuyler, J. R. (2001): Risk and Decision Analysis in Projects.
Beer, S. (1966): Decision and Control.
Keeney, R.L.; Raiffa, H. (1976): Decisions with Multiple Objectives.
Peterson, M. (2009): An Introduction to Decision Theory.
Goodwin, P.; Wright, G. (2004): Decision Analysis for Management Judgment

**Decision Tree Method**

A tree-like graphic representation of mostly multiple, consecutive decisions, where the leaves of the tree (terminal nodes) are evaluated and weighted with the probability of their occurrence and/or the risk incurred risk. The path through the tree with the highest value at the terminal node corresponds to the optimal set of the optimal decision values.

*References:*
Magee, J.F. (1964): Decision Trees for Decision Making
Foster, Provost; Fawcett, Tom (2013): Data Science for Business. What You Need to Know about Data Mining and Data-Analytic Thinking. O'Reilly Media.
de Ville, Barry; Neville, Padraic (2013): Decision Trees for Analytics Using SAS® Enterprise Miner™. SAS Institute

**Delphi Method**

A systematic, multi-stage survey technique that is helpful in assessing future events or developments from an expert point of view. Various experts are questioned individually on a particular subject, e.g.: "How long do you think it will take for renewable energy to meet 50% of the overall energy need?" The results are evaluated and shared with all respondents. In the next round of questioning, the experts can adapt/change their opinion. After two or three rounds Delphi gives a good, harmonized view on a subject with fewer extreme or contradicting opinions.

*Reference:*
Hsu, Chia-Chien and Sandford, Brian A. (2007). The Delphi Technique

## Design Structure Matrix, DSM
(also referred to as: dependency structure matrix, problem-solving matrix, design precedence matrix)
The design structure matrix (DSM) provides a simple way of both analyzing and the managing complex systems by modeling the system structure or processes and is therefore an important tool in systems engineering. The dependencies of all constituent modules, assemblies, subsystems or activities and their interactions, information exchange, and dependencies are modeled as a matrix. This can involve, for example, product architecture or a design and development process, etc. DSM is also increasingly being used as a management tool in project management, where it also facilitates process presentations of projects whose activities present feedback and cyclical dependencies.

*References:*
Eppinger, Steven D.; Browning, Tyson R. (2012): Design Structure Matrix Methods and Applications.
Lindemann, U., et al. (2009): *Structural Complexity Management*

## Digital Factory
A concept that virtualizes all processes involved in an industrial factory as a computer model/simulation.
It can be used to design, optimize or analyze individual production or factory processes and resources associated with its products (e.g., automobiles, airplanes, process plants) without the need for a physical test/mockup. A *digital factory* can be divided into four levels – database/data core, integration platform, tools, organization, and design work flow – and consists in practice of many digital submodules, methods, and tools, including simulation and 3D visualization and is used for a holistic design, implementation, management, and ongoing improvement.

*References:*
Canetta, L.; Redaelli, Cl.; Flores, M. (Eds.) 2011: Digital Factory for Human-oriented Production Systems

## Economic Feasibility Calculation
A method of evaluating the profitability of an existing or a planned system, or for a profitability comparison of several variants. Important elements of an economic feasibility calculation are the intended *useful life*, the *interest rate* or *the discount rate*, and the *performance* expressed in monetary units, which are compared with

the *use of resources* (costs). The profitability principle demands either minimizing costs with given performance or maximizing performance with given costs.

With respect to the degree of detail at which these elements are considered, we distinguish among *static*, *dynamic*, *investment chain*, *relative profitability-* (e.g., messaging application programming interface), and *simultaneous* models. Newer models include, for example, → real options.

*References:*
Farris, P W.; Bendle N T.; Pfeifer Ph E; Reibstein D J. (2010). Marketing Metrics: The Definitive Guide to Measuring Marketing Performance.
Feibel B J. (2003): Investment Performance Measurement.
Brealey R A., Myers S C. and Allen F. (20013): Principles of Corporate Finance.

**Effect Networks/Influence Matrices**
Important components of a method for representing and analyzing complex effect relationships (see → Network Thinking) to draw conclusions from them.

Recommended procedure and suggestions:

1. Possible use of an effect network diagram in which the relations among the individual elements (variables) are illustrated (Fig. 16.1).
2. Use of a matrix in which the columns and rows represent the network (Fig. 16.2)
3. Assessment of the strengths of the influences and entries in the matrix. The meaning of the numbers is as follows: 0 = no influence; 3 = strong influence, etc. (the scale can be chosen arbitrarily).

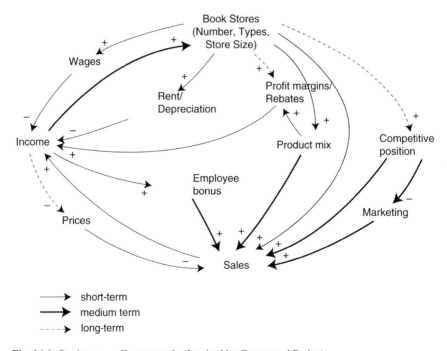

**Fig. 16.1** Bookstore – effect network. (Inspired by Gomez and Probst)

| Effect of → on | Network of bookstores | Wages | Product mix | Prices | Sales | Competitive position | Rent/Depreciation | Marketing | Profit margins/rebates | Earnings | Employee bonus | sum of active effects (SAE) | Quotient Q (SAE / SPE x 100) |
|---|---|---|---|---|---|---|---|---|---|---|---|---|---|
| Network of bookstores | – | 3 | 2 | 1 | 3 | 3 | 3 | 1 | 2 | 2 | 0 | 20 | 153 |
| Wages | 0 | – | 0 | 0 | 2 | 1 | 1 | 0 | 0 | 3 | 2 | 9 | 90 |
| Product mix | 1 | 1 | – | 0 | 3 | 2 | 1 | 1 | 2 | 2 | 0 | 13 | 144 |
| Prices | 0 | 0 | 2 | – | 3 | 1 | 1 | 1 | 3 | 3 | 0 | 14 | 467 |
| Sales | 2 | 1 | 1 | 1 | – | 2 | 3 | 2 | 1 | 3 | 2 | 18 | 95 |
| Competitive position | 3 | 1 | 0 | 0 | 3 | – | 2 | 2 | 3 | 2 | 0 | 16 | 114 |
| Rent/Depreciation | 2 | 0 | 0 | 0 | 0 | 0 | – | 0 | 1 | 3 | 0 | 6 | 46 |
| Marketing | 0 | 0 | 1 | 0 | 2 | 1 | 0 | – | 0 | 1 | 0 | 5 | 63 |
| Profit margins/rebates | 2 | 1 | 2 | 1 | 1 | 2 | 1 | 0 | –' | 3 | 0 | 13 | 100 |
| Earnings | 3 | 2 | 1 | 0 | 0 | 2 | 1 | 1 | 1 | – | 3 | 14 | 61 |
| Employee bonus | 0 | 1 | 0 | 0 | 2 | 0 | 0 | 0 | 0 | 1 | – | 4 | 57 |
| sum of passive effects (SPE) | 13 | 10 | 9 | 3 | 19 | 14 | 13 | 8 | 13 | 23 | 7 | | |
| Product P (SAE x SPE) | 260 | 90 | 117 | 42 | 342 | 224 | 78 | 40 | 169 | 322 | 28 | | |

**Fig. 16.2** Bookstore – influence matrix. (Inspired by Gomez and Probst)

4. Calculation of row sums (active effect, means "element influences"). This sum indicates the level of influence taken by the particular element in the row. Calculation of the column sums (passive effect, means "element is influenced") to show as a whole how strongly the particular element in the column is influenced by others.
5. Interpreting the results:

   (a) If the sum of active effects is high, changes in this element can have large effects on the system. As long as the element is not determined from the outside, but rather can be changed through action, this indicates a possibility for intervention. If an element has both high active effects and passive effects (the product of multiplying the total active effects by the total passive effects is large), this means that changes also produce major backlashes or retroactive effects.

   (b) If an element exhibits both low total active effects and low total passive effects (total active effects times total passive effects = small), then the element should be considered relatively neutral in comparison with other elements and has a "buffering" character.

(c) If the total active effects are high and the total passive effects are low (total active effects divided by total passive effects = large), then this result indicates that intervention is relatively ineffective in this case.

If the total passive effects are high and the total active effects are low (total active effects divided by total passive effects = small), then this element exerts a very small influence and the element is greatly influenced by other factors.

6. Application: when considering which measures to use in changing a complex causal-networked system, these considerations help in thinking through the effectiveness of measures and the desirability of the effects.

The methodology shown here constitutes an aid to quantitatively supported systems analysis in terms of system thinking. It is particularly appropriate for situation analysis and concept analysis.

*References:*
Probst Gilbert J. B. and Gomez Peter (1992): Thinking in Networks to Avoid Pitfalls of Managerial Thinking
Vester, F. (2007): The Art of Interconnected Thinking.
Colgan St. (2009): Joined-Up Thinking.

**EFQM Model**
A → Quality management system in the context of → total quality management. It was developed in 1988 by the European Foundation for Quality Management (EFQM) and is used by, according to current estimates, over 10,000 companies.

The EFQM model enables a holistic view of quality and consists of three pillars: *human resources* (management and workers) who work in *processes* and achieve *results*, which in turn benefit people (customers, society).

*References:*
Gryna, F M. (2001): Quality Planning and Analysis
Deming, W.E. (1997): *Out of the Crisis,*

**Eisenhower Method**
A matrix-based method for dividing decision-making situations according to the criteria *important/unimportant* and *urgent/not urgent*, which is attributed to the former general and US president Dwight D. Eisenhower: only *important* decisions (which also require a more thorough, methodical preparation) should be placed at the higher levels of the hierarchy. *Urgent* decisions indicate the priority with which they are to be treated. When *urgent* decisions also involve *important* issues (overall importance and risk), time pressure should be eliminated and if possible decisions broken into partial decisions: what needs to be done right away to take as much time pressure off the situation and win as much time as possible? What is important to accomplish during the time gained? Positions and bodies fairly high on the hierarchical ladder should not allow themselves to become cluttered with unimportant tasks that are not urgent.

*Reference:*
Covey Stephen R. (2004): The 7 Habits of Highly Effective People

## Failure Cause Analysis
A conceptual approach that is aimed at preventing focusing on measures too prematurely before dealing with failures and their causes. The basic idea is shown in Fig. 6.6 in Sect. 6.1.3.2.

## Fault Tree Analysis
A procedure that is used to understand how systems can fail and determine the associated probabilities. A fault tree analysis, which can be used for all kinds of systems, analyzes an undesired state of a system using Boolean logic to combine a series of lower-level events and is a major tool for *reliability engineering* and *system analysis*. It is described in DIN 25424-1.

*References:*
Roberts, N. H.;Vesely W.E. (1987): Fault Tree Handbook.
Ericson, Clifton A.: (2011): Fault Tree Analysis Primer.
DIN standard 25424-1

## Flow Charts
Show *logical* and/or *quantitative connections* between elements of a system and/or a process. The *logical* relations are usually information relations, or they express logical dependencies ("is a requirement for..."). *Quantitative connections*, e.g., in the form of energy amounts, materials of all kinds, concentrations, etc., can be represented in the form of energy flow (Sankey) diagrams, in which the amount of flow is represented by the breadth of the connecting lines.

*Reference:*
Bohl, Marilyn and Rynn, Maria (2007): Tools for Structured and Object-Oriented Design.

## FMEA (= Failure Mode and Effects Analysis)
*Synonyms: FMECA (Failure Mode and Effects and Criticality Analysis).*

Failure mode and effects analysis (FMEA) follows the basic idea of *precautionary error prevention* rather than *subsequently dealing with an error* (error detection and correction). This should be done as early as in the design phase by identification of potential error causes. Costs of controlling and error correction in the manufacturing phase or in the field (with the client) should be avoided. The goal is to lower overall costs. Using such a systematic approach and building on the knowledge and experience gained can prevent to repeat design flaws in new products and processes.

Failure mode and effects analysis is used in the development of products and processes, and in many sectors, e.g., the automobile industry, it is generally required by the supplier.

*Reference:*
SAE (2009): Potential Failure Mode and Effects Analysis in Design (Design FMEA) and Potential Failure Mode and Effects Analysis in Manufacturing and Assembly Processes (Process FMEA).

**Forecasting Techniques**
Methods and techniques with which future developments, results, or conditions can be predicted. The time frames for forecasting can be short, medium, or long term. With respect to techniques, there is a distinction between *intuitive* and *analytical.* Intuitive methods take in more subjective opinions and assessments (which may also be influenced by facts): → interview, surveys, → scenario writing, → Delphi method.

Among analytical methods there are *endogenous* and *exogenous models.* Endogenous models (in terms of a clarification of the future from the intrinsic development of the past) are extrapolations of time series, linear, progressive, digressive trend extrapolations, saturation curves, etc. Exogenous models (consideration of relevant external factors) include, for example, the consideration of multiple linear developments or changes in relevant influence factors on the issue to be predicted. Here, there is a fluid transition to → simulation methods, → system dynamics, etc.

*References:*
Armstrong, J. S. (ed.) (2001). Principles of forecasting: a handbook for researchers and practitioners.
Rescher, N. (1998): Predicting the future: An introduction to the theory of forecasting.

**Game Theory**
Subarea of → operations research. A game in terms of game theory involves a situation in which there are several players that *mutually* influence the results of their decisions. This is described in a mathematical model. Game theory in particular attempts to characterize and predict rational behaviors in games. Examples for applications include concepts for auctioning radio and mobile communications licenses and the coordination of radio frequencies for disorganized rescue efforts, etc.

*References:*
Brandenburger, A M.; Nalebuff B. J. (1996): Co-Opetition. Currency Doubleday
Nash, John (1950) "Equilibrium points in n-person games" Proceedings of the National Academy of Sciences 36(1):48–49
Harsanyi, J C. and Selten, R A. (1988): General Theory of Equilibrium Selection in Games
McCain, Roger A. (2014): Game Theory: A Nontechnical Introduction to the Analysis of Strategy

**Heuristics, Heuristic Methods**
A term for solution-seeking methods, which are practical (performant, simple. etc.), promising, but not guaranteed to find an optimal solution. Heuristics can also be

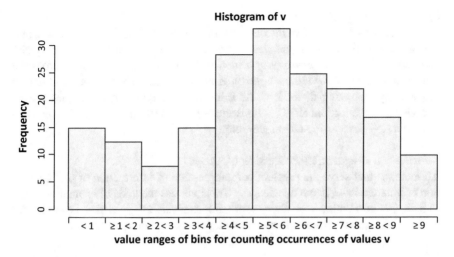

**Fig. 16.3**  Histogram

described as a way of coming up with (sufficiently) good solutions under time and resource constraints.

Example: putting together an efficient production schedule.

*References*:
Michalewicz, Z.; Fogel, D.B. (2004): How To Solve It: Modern Heuristics.
Gigerenzer, G., Todd P. M. (2000): Simple Heuristics That Make Us Smart.
Dörner, D. (1980): Heuristics and Cognition in Complex Systems.

## Histogram

A graphic representation of the frequency distribution of measurement or count values. It starts with the data arranged by size and divides the entire area of the *sample* into classes (bins) (Fig. 16.3).

The indicators of a histogram can be used for the evaluation, that is, the general curve shape, the distribution, and the centering (symmetrical, crooked distribution, etc.). In creating a histogram it makes sense to use practical instructions such as the smallest sample size, number of classes with known number of measured values, desired confidence interval, procedure in selecting measured values. Refer to the relevant literature.

*Also see References under → statistics.*

## Information Acquisition Plan

Before conducting a fairly large survey, an information acquisition plan should be prepared based on the following questions: what information is needed? Why? What information is essential? Why? What conclusions should be possible or supported? What degree of precision or detail is necessary? What timeframe will the survey cover? Subsequently, one must consider the quickest and easiest way of gathering this information.

**Information Acquisition Techniques**

There is a distinction between information oriented toward the *past, the present,* and *the future.* Additional characteristics deal with the difference between primary and secondary information: *primary information* is obtained right at the appropriate source. *Secondary information* is taken from documents already on hand and is acquired by analyzing them. Specific information acquisition techniques are → interviews, → questionnaires, → observation, → Delphi method, → scenario techniques, → forecasting techniques, etc.

**Information Preparation/Processing Methods**

All methods that serve the preparation/compression of information or the recognition of inherent laws, dependencies, etc. This includes methods of → mathematical statistics, → correlation analysis, → regression analysis, all types of → key indicator systems, etc. The results are illustrated using → visualization techniques.

**Interview**

Information gathering through oral questioning, which can be divided into procedures with various characteristics: conversation, conversation with a list of questions, interview with open answers, interview with predetermined answers ((multiple) choice).

**Interviewing Techniques**

An umbrella term for various possibilities for oral questioning (→ interview) and/or written polling (→ survey). A (regular) recurring inquiry among a steady circle of people is referred to as a → panel survey or poll. See also → Delphi method.

*References:*
Innes, J. (2009): The Interview Book.
Gordon N J., Fleisher, W L. (2011): Effective Interviewing and Interrogation Techniques.

**Ishikawa Diagram**

(Synonyms: cause-effect diagram, fishbone diagram)

  Diagram for systematic representation and analysis of possible causes of problems. The cause-effect diagram was developed by the Japanese Kaoru Ishikawa and was later was named after him. Originally used in the realm of quality management for analyzing quality problems and their causes. Today, it is also used in → security analyses, risk analyses, etc. Its fishbone-like structure inspired the synonym "fishbone diagram."

*References:*
Ishikawa, K. (1990): Introduction to Quality Control.
Tague N R. (2005): The Quality Toolbox.

**ISO 9001**

This popular standard establishes the requirements for a quality management (QM) system. Such a system is useful if an organization needs to prove its capabilities producing products that meet the demands of the customers and/or authorities, or simply seeks to increase customer satisfaction. This ISO standard describes the entire QM system in model form.

The eight basic principles of quality management are: (1) client orientation; (2) responsibility of management; (3) involvement of the relevant people; (4) process-oriented approach; (5) system-oriented management approach; (6) continuous improvement; (7) fact-based decision-making approach; (8) supplier relationships for mutual benefit.

*References:*
ISO 9001 in Plain English, (2015) by Craig Cochran
ISO 9001 - What does it mean in the supply chain? Available from: http://www.iso.
    org/iso/pub100304.pdf

## Just in Sequence, JIS
*See → just in time*

## Just in Time, JIT
Just in time (JIT) involves the delivery of *materials* (raw materials, parts, assembly groups, or products) at *precisely the right time,* with the necessary *quality* and in the desired *quantity* (including packing) to the agreed-upon location. Storage costs are largely eliminated, and the usual administrative expenses are also significantly reduced.

An extension of JIT is what is known as just in sequence (JIS). Building on the JIT principle, with JIS, the products are also delivered to the client in the right sequence. JIT and JIS are widely applied standards in the automobile industry today.

*References:*
Hirano, Hiroyuki and Makota, Furuya (2006): JIT Is Flow
Womack, James P. and Jones, Daniel T. (2003): Lean Thinking.
Takeda, Hitoshi (2006): The Synchronized Production System: Going Beyond Just-in-time Through Kaizen

## Kaizen
(means in Japanese *a change for the better*) is a philosophy of life and work with striving for continual improvement as its main idea. This concept has been further developed into a management system in Japanese industry. Another term used for this is *continuous improvement process (CIP).* The → just in time concept is a reflection of the Kaizen philosophy.

*Reference:*
Masaaki Imai (2012): Gemba Kaizen: A Commonsense Approach to a Continuous Improvement Strategy.

## (Key) Indicator System
Indicators are easily remembered, meaningful methods for presenting numerical issues or conditions. A distinction can be made among various types of indicators: *structural indicators*: ratios of sizes or characteristics, such as the portion of plastic materials in an automobile (as a percentage of weight). *Measurement indicators*:

characteristics of the same kind at various points in time (e.g., sales for 2015, 2016, etc.). *Index indicators:* the normalization of a characteristic based on 100. Conversion of other values relative to this one (cost of living in Munich, 100: Berlin, 92, Paris, 105, etc.). *Relationship indicator:* representation of relationships of different measurers to one another (e.g., parts produced per worker and day), etc. Important indicators, which are useful for controlling and steering, are called *key indicators.*

*References:*
Reichert, F., Kunz, A., Moryson, R, 2008, MAE-P3 - A System to Gain Transparency
    of Production Structure
Austin, Robert D. (1996): Measuring and Managing Performance in Organizations.

**Lessons Learned**
Analysis and determination of positive and/or negative experiences in projects to learn from them. Lessons learned, which may be part of the final project documentation, are fairly structured compilations of information and are very important for the completed project. Carefully archived in accessible form, they can be used to prepare similar projects and to improve project management.

**Linear Programming (Synonym: Linear Optimization)**
A widely used and very powerful method in → operations research, for finding optimal solutions by adjusting or combining factors. Examples: optimal *production program; waste optimizing* with parts made of sheet metal, wood, etc.; optimal *routing* in transportation and communication networks; *mixing problems,* such as steel production (in what proportions must the ingredients such as different ores be mixed so that the alloy contains the whole required combination of elements at the lowest cost?); *refining* fuels, etc.; → *game theory; nonlinear* and *(mixed) integer optimization.*

*References:*
Dantzig, G. B. (1963): Linear Programming and Extensions
Schrijver, A. (1998): Theory of Linear and Integer Programming.
Also see → operations research.

**Mathematical Statistics**
Methods for analyzing mass phenomena in terms of recognizing patterns and internal laws. There is a distinction between *descriptive statistics* (determining the characteristic of a distribution of numerical values, mean, variance, standard deviation, etc.), and *evaluating or inferential statistics* (based on descriptive statistics and → probability calculation).

On the one hand, *estimation procedures* (inference from a sampling on the parameters of the basic population), and on the other, *testing procedures* (checking whether deviations that occur and are noticed between, for example, empirical and theoretical values are random or significant are used).

*References:*
Pitman, Jim (1999): Probability.
Walpole, R.E. (2007, 9.ed.): Probability & Statistics For Engineers & Scientists

Wheelan, Ch. (2013) Naked Statistics: Stripping the Dread from the data
Field, A. (2013): Discovering Statistics using IBM SPSS Statistics.

## Mean Time Between Failure, MTBF
Measurement of the reliability of units (assemblies, devices, or installations) that
are repaired. The MTBF is the inverse of the failure rate.

*Reference:*
Jones, James V. (2006): Integrated Logistics Support Handbook

## Mean Time to Failure, MTTF
Mean time interval between malfunctions of objects that cannot or should not be
repaired.

*Reference:* also see → mean time between failure

## Mean Time to Repair, MTTR
Average time between the failure of a unit and the repair performed.

*Reference:* also see → mean time between failure.

## Method 6-3-5
→ Creativity technique in which six people are confronted with a problem described
as precisely as possible and write down three ideas each on a sheet of paper within
5 min. The paper is passed to the next person in the circle every 5 min and the same
process is repeated. In 30 min, this theoretically produces $6 \times 3 \times 6 = 108$ solutions.
In practice, one would expect fewer solutions, as there are often duplicate entries or
the papers are passed on without new ideas being added.

*References:*
Schroer, B.; Kain A. and Lindemann U. (2010): Supporting Creativity In Conceptual
    Design: Method 635-Extended
Linsey J S. and Becker B. (2011): Effectiveness of Brainwriting Techniques
See also → creativity techniques.

## Milestone Monitoring (Progress/Slip Charts)
A project management tool for tracking changes of milestones during the course
(review/report points) of a project. Project managers and stakeholders (e.g., steering
committee) can be informed graphically about impacts on the final deadline of a
project and, if there is a deviation from the original plan, about corrective measures
required or already implemented (Fig. 16.4).

## Mind Mapping
The creation of graphic memory maps, thought maps, or mind maps, coined and
first published by Tony Buzan. In the center of the mind map is the topic (or problem) to be dealt with. Extending from it are main and secondary branches that structure and subdivide the topic. On each branch there is just one term. Mind maps are
well suited for developing a thought or topics, and for documenting and sorting
results from creative sessions such as brainstorming.

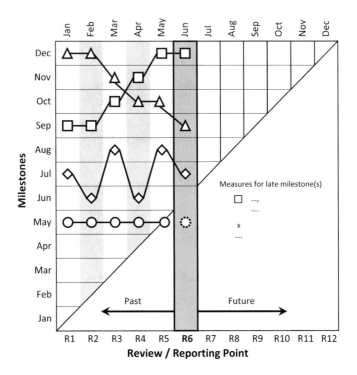

**Fig. 16.4** Milestone monitoring diagram

*Reference:*
Buzan, T. and B. (2010): The Mind Map Book

**Modeling and Representation Techniques**
Modeling and representing the results of the synthesis, i.e., of solutions, theoretically involves the techniques that have already been described in connection with systems thinking and situation analysis. These include in particular system representations of all types such as bubble charts, system hierarchical representation, effect networks, special views of systems, black box representations, and many more.

See also → effect networks/influence matrices, correlation matrices, → process charts, → flow charts, all kinds of plans, drawings, sketches, tables, diagrams, and physical modeling (reduced-scale physical models) or abstract, mathematical, and usually IT-supported modeling. Numerical methods facilitate through dynamic simulation of procedures and processes (e.g., changing parameters) the possibility of quickly evaluating and visualizing changes of system performance and behavior.

**Monte Carlo Method**
Is a powerful → simulation technique useful for stochastic processes that are dependent on random variables following often unknown probability distributions and was originally developed in the 1930s and 1940s. The Monte Carlo method assigns

in multiple runs values to these variables, which are created by random stochastic sampling from historic data (in the case of known/assumed probability distributions the values are generated in such a way that they follow their respective distribution). With these values the process is evaluated (e.g., customer waiting times in a queue) and the result recorded for each run. These results, as they are a based on the stochastic variation of (input) variables, follow a distribution function (with expected value and variance) themselves. The number of runs (iterations) depends on the desired quality/reliability of the result (confidence interval).

The term *Monte Carlo method* was chosen as a project name in reference to the Casino in Monaco.

*References:*
Rubinstein, R. Y.; Kroese, D. P. (2007). Simulation and the Monte Carlo Method
Asmussen, S. and Glynn, Peter W.(2007): Stochastic Simulation: Algorithms and Analysis
Lemieux, Ch. (2009): Monte Carlo and Quasi-Monte Carlo Sampling.
See also → operations research

**Morphological Analysis**
Analytical, systematic → creativity technique used to search systematically for the greatest possible number of solutions. The concept was developed by F. Zwicky and consists of the following steps (Fig. 16.5):

1. Defining the problem: what do we want to find? In the example: the greatest possible variety (or all conceivable) types of vehicle.
2. Setting the parameters that can be used to describe the numerous variants. In the example: *space layout* – arrangement of drive unit and cargo space (SL); type of *steering* (ST); and *drive type* (DT).
3. Generating possible parameter values: in the example, SL with SL1 to SL3, ST with ST1 and ST2, and DT with DT1 to DT4.
4. Coming up with the different configurations: all possible combination parameter values are created (at first without assessment or critical evaluation). In the example, there is a possible total of $3 \times 2 \times 4 = 24$ different combinations (configurations).
5. Creating a shortlist of technically feasible configurations – without judgment and prejudice. In our example the obvious (and known) configurations are:

   (a) SL1 + ST1 + DT2 = automobile or bus
   (b) SL2 + ST2 + DT3 = streetcar (tram)
   (c) SL3 + ST2 + DT3 = train (locomotive + rail wagon)
   (d) SL1 + ST1 + DT3 = trolley-bus

6. Evaluation and final decision about best configuration (Fig. 16.5).

*Reference:*
Zwicky, F. (1969): Discovery, Invention, Research – Through the Morphological Approach.

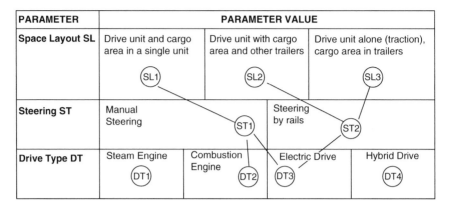

| PARAMETER | PARAMETER VALUE | | |
|---|---|---|---|
| **Space Layout SL** | Drive unit and cargo area in a single unit | Drive unit with cargo area and other trailers | Drive unit alone (traction), cargo area in trailers |
| | (SL1) | (SL2) | (SL3) |
| **Steering ST** | Manual Steering | (ST1) | Steering by rails (ST2) |
| **Drive Type DT** | Steam Engine (DT1) | Combustion Engine (DT2) | Electric Drive (DT3)   Hybrid Drive (DT4) |

**Fig. 16.5**  Morphological box

## Multi-Moment Observations, See → Activity Scanning
Text

## Network Planning Techniques
A set of methods for planning logical and temporal sequence of projects using a mathematical graph representation. Network planning techniques can be used in many ways:

1. *Structure analysis:* in a first step, the activities to be carried out are determined (questioning experts, brainstorming, etc.) and put into a logical sequence according to established rules (what activities must be completed so that a certain activity can begin, which activities can run entirely or partially in parallel, which activity subsequently starts, etc.) The result is represented graphically to create what is known as a network plan.
2. *Time analysis:* if the respective temporal duration of the individual activities is provided – or can be estimated, it is possible – based on the dependencies established in the structure – to calculate the completion times of the network plan. The earliest possible or latest allowable times for results (condition) and activities are established. The most important result is the *critical path,* the path that is crucial to the final deadline. Each delay along this path leads to a delay in the overall project completion (some methods also allow for dealing with uncertain activity times, such as → PERT). Activities on the other paths may have time buffers (*buffer times*) within which they can be postponed or preponed without having any effect on the final deadline.
3. *Cost analysis:* when costs can be assigned to the individual activities (personnel, materials, machine hours, and investment costs), it is possible to calculate and predict the project costs at each project stage. It is also possible to calculate the optimal allocation of investments or to speed up the project by optimally investing in additional resources (crash costs) using → linear optimization.
4. *Capacity analysis:* with the assignment of resources such as machines, equipment, work groups, personnel, etc., to the individual activities, useful load

overviews can be created. Their analysis allows load peaks to be smoothed out by postponing or bringing forward certain activities using their buffer times.

There are two ways of representing the network of activities as a mathematical graph. One way is called *event node network*, where activities are represented as arcs (edges) and their start and end events as nodes (CPM, PERT). The other way is called *activity node network*, where the assignment is just the other way around: the activities are on the nodes and their connections drawn as arcs (edges). Metra-Potential Method (MPM) and many others use this notation.

With respect to the time estimate for the duration of the activities, there are also various possibilities. With *deterministic time estimates* a single, constant value for the duration of a process is assumed; with *stochastic activity times* a stochastic distribution function is given – or in the simplified case of → PERT – approximates the general distribution with the skewed beta distribution (often called PERT distribution) using a three-value estimate of the duration of the activities: an optimistic one, a pessimistic one, and a most likely one. This allows probabilities to be calculated for individual and overall project completion times.

Most IT-supported → project management software systems support using network planning techniques, especially for time, cost, and capacity analysis. Graphic charts and tables, visualizing the project's network plans, capacity, and load overviews, in addition to task lists, are also standard features of these software systems.

*References:*
Milosevic, D. Z. (2003): Project Management ToolBox.
O'Brien, J.J.; Plotnick, F.L. (2010): CPM in Construction Management
Lewis, James P. (2004): Project Planning, Scheduling & Control.

**Network Thinking**
*See* → effect network thinking.

**Numerical/Computer Simulation**
Numerical systems models that are evaluated by computers (or powerful computer clusters) to simulate the system's behavior. Well-known examples are a computational fluid dynamics, finite element method (FEM) strength calculation, weather and climate forecasts, thermal, energetic or functional building simulation, to name but a few.

In virtual product development numerical simulation is used to visualize and test features or characteristics of products before a physical prototype is built. This approach can lead to a significant shortening of development or time-to-market times. Often, developers use the time gained rather to evaluate more variants of solutions to improve the product.

*References:*
Angermann, L. ed. (2011): Numerical Simulations - Applications, Examples and Theory

Awrejcewicz, J. ed. (2011): Numerical Simulations of Physical and Engineering Processes.

## Observation Techniques

A method of information gathering that is suitable for observable phenomena (states, conditions, procedures, processes, behavior, etc.). Here, the information does not have to be gathered by questioning experts or people involved, but can be collected by observation, e.g., by doing a complete survey (video surveillance) or observation at random, discrete, selected times and extrapolating the results with statistical methods (→ multi-moment observations; → activity scanning).

## Operationalization

Operationalization defines the measurement of phenomena that are not directly measurable or traceable by supplementing fuzzy, abstract terms with indicators or proxies that are easy to capture and/or measure. For example, the term *work atmosphere can* be operationalized by indicators such as: *personnel fluctuation, rate of sick leaves*, etc. *Corporate success* could be measured with indicators such as: *profit, cash flow, growth, market share, average product age, sales volume with new* (e.g.: <5 years old) *products, employer attractiveness indicator*, etc.

## Operations Research, OR

The application of mathematical methods to systems to give them an optimal shape (structure) or an optimal functional performance. The term *optimal* is always understood with respect to a target (objective) function and given constraints that must be defined. OR in systems engineering is used primarily in the solution search (synthesis/analysis).

Well-known *OR application areas* are blending problems, inventory optimization, → assignment or allocation problems, → queuing theory, → sequencing problems, job shop scheduling, maintenance planning, or transportation. Important OR *methods* are → linear programming, (mixed) integer programming, nonlinear programming, dynamic programming, simulation, etc.

See also: → game theory, → simulation technique, → Monte Carlo method, → heuristic methods, etc.

*References*:
Hillier, F.S.; Lieberman, G.J. (2010): Introduction to Operations Research.
Winston, W.L. (2008): Operations Research: Applications and Algorithms.
Baldick R. (2006): Applied Optimization. Formulation and Algorithms for Engineering Systems.
Taha, H.A. (1992): Operations research: An introduction.
Eiselt, H.A. (2012): Operations Research. A Model Based Approach.

## Panel Polling (Polls)

→ Information acquisition technique with which certain matters such as assessment of the situation, opinions, desires, estimates of future developments, etc., are reviewed and collected, repeatedly, often at regular intervals, from a particular group of people (clients, suppliers, residents, etc.). This facilitates a dynamic observation of changes and trends. Example: → Delphi method.

## Polarity Profile (Spider Web Diagram)

→ Visualization technique that can be used to illustrate characteristics of object systems, people, etc., which can be put on a scale. These diagrams visualize a wide variety of object properties in one view. This greatly simplifies comparisons.

In a *situation analysis*, this representation is helpful in characterizing various phenomena or systems. Different solution variants and the degree to which the goal is met can be presented for evaluation (see Part III, Sect. 6.4, Evaluation and Decision). It is also possible to represent areas that are suited to using methods and resources (see Part II, Chap. 4, Project Management, Figure 4.9). For operationalizing abstract or fuzzy terms see → operationalization. The example below should illustrate the idea and application (Fig. 16.6).

The coordinate axes represent the degree to which the goal is met. Possible constraints could also be plotted that way. The scales should be arranged in such a way that the best performance of each criterion moves in the same direction, good either inward or outward for all criteria. As there is often no meaningful order or correlation between the different axes or scaling factors, the area spanned by the polar diagrams does not simply correlate with the overall quality of an object and therefore cannot be used for a comparison.

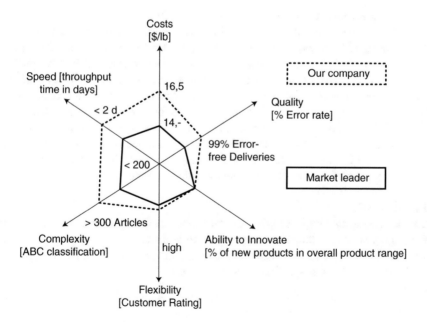

**Fig. 16.6** Polarity profile of two competitors

**Process Analysis**

Systematic investigation of business processes by decomposition into partial steps/ sub-processes, and precise analysis of the process logic and contents. The goal is often to identify opportunities for improvement such as simplifying, speeding up, reducing costs. Various → visualization techniques can be helpful. Also, methods such as → Benchmarking can help to raise awareness for problems and opportunities, and facilitate a better understanding of the process itself and/or identify weaknesses.

**Process Modelling Diagrams**

Techniques belonging to the set of → Information Preparation/Processing Methods. They show graphic representations of processes or procedures as the basis for understanding them, critical analysis, improvements, programming, automatization, etc. Examples are data flow charts, → business process model and notation, program flow charts, work flow process diagrams, technical procedures, project procedures, etc.

*References*:

Hommes, Bart-Jan; van Reijswoud Victor (2000):The evaluation of business process modeling techniques.

Mendling, J.; Reijers, H. A.; van der Aalst, W. M. P. (2010). Seven process modeling guidelines (7PMG).

**Program Evaluation and Review Technique, PERT**

A stochastic → network planning technique where activities are assumed to follow a probability distribution (PERT distribution) determined by an optimistic, a most likely, and a pessimistic duration estimate. The expected value of the total project duration is calculated similar to the → critical path method as the sum of the expected durations of the activities on the critical path (= the sequence of activities that determine the project duration – each delay of one of those activities leads to a delay of the project) – its variance as the sum of the activities' variances. The big advantage over CPM is that this technique calculates a stochastic forecast of project durations and confidence intervals for achieving them, which is essential in → risk management and far superior to the "single point estimates" of CPM.

**Project Audit**

Tool for assessing a project's progress. The objective is to determine to what extent the project is progressing according to plan in terms of time, cost, quality, etc., or whether measures need to be taken to make sure the project will be successful.

A project audit consists of the following steps:

1. Establishing a baseline based on the planned deadlines, costs, and accomplishments (target or plan)
2. Assessing the current status with respect to completion/progress, costs, and deadlines (actual situation)
3. Determining the resources still needed to complete the project (e.g., cost to complete, CTC)

4. Conducting a deviation analysis: why and to what degree is there a deviation from the planning up to this point (deadlines, costs, completion)?
5. Conducting a critical analysis of the project premise: evaluation of the assumptions in respect for plausibility, probability of occurrence, prospects of success
6. Evaluation of opportunities and risks
7. Adoption of a catalog of measures for the remainder of the project

The audit can be initiated by different management levels depending on the degree of the suspected/feared deviation and the risk associated therewith (project administration, the company's business management, executive board). The selection of the auditors takes place in accordance with the initiating management level. For reasons of objectivity, the auditors should not be part of the project team or the particular environment in which the project is situated. Entirely external auditors are often a good choice.

## Project Management Body of Knowledge Guide, PMBoK Guide
A widespread project management standard of the US Project Management Institute, by which it is published and supported. The methods described in the PMBoK Guide are usable with projects in various application fields: construction, software development, mechanical engineering, automotive industry, etc.

The PMBoK Guide is process-oriented; in other words, it uses a model by which the work is accomplished through processes. A project is carried out using the interaction of many processes. The PMBoK structures all methodological skills along processes. *Inputs* (documents, plans, designs, etc.), *tools* and *techniques* (mechanisms applied to inputs), and outputs (documents, plans, designs, etc.) are described for every process.

The PMBoK Guide is the basis for the certification examination for project management professionals (PMPs), and others.

*Reference:*
Snijders, P.; Wuttke, Th.; Zandhuis, A. (2009) PMBOK Guide

## Project Management Software, PMS
Information technology tools for supporting project management, mostly on the basis of → network planning techniques.

With the help of PMS, the structure of projects can be designed and time calculations, including determining the critical path, can be made. PMS also offers many opportunities for automatic dispositions or evaluations. Depending on the scope of their function and their purpose, various categories of PMS can be distinguished (according to Wikipedia):

– Single project management systems (usually design software for designing and tracking a single project).
– Multi-project management systems (planning software for administering and managing multiple projects).
– Enterprise project management systems (for integration into company-wide planning, e.g., enterprise resource planning software).

– Project collaboration platforms with communication solutions (e.g., groupware solutions, portal software), but also many other programs that really are not project-management-specific, but which may, for example, have interfaces with project management software, such as office solutions, creativity tools (e.g., mind mapping).
– Some software solutions are sector-specific.

**Quality Control, QC**
Quality control (QC) describes a process by which the quality of all factors involved in production are reviewed. The ISO 9000 standard defines quality control as "A part of quality management focused on fulfilling quality requirements".

Since the 1930s, different QC paradigms, which build on various concepts, have been applied. Among those are: *statistical quality control* (SQC; 1930s) using statistical methods and sampling, → *total quality control* (TQC; 1950s) favoring a holistic approach to involve all of the company's quality stakeholders, *statistical process control* (SPC; 1960s) building on process control systems, *company-wide quality control* (CWQC; 1960s) which is similar to TQC, *total quality management* (TQM; 1980s) using methods of statistical quality control for organizational improvement and (lean) *six sigma* ($6\sigma$; 1980s), which applies statistical quality control methods to business strategy.

See → total quality management, TQM, → total quality control, TQC → Six Sigma.

*Reference:*
Juran, Joseph M. (1995)): A history of managing for quality: the evolution, trends, and future directions of managing for quality

**Quality Management, QM**
Quality management (QM) ensures that an organization, product or service is consistently following defined standards or specified properties. It consists of four major domains: *quality planning*, *quality assurance*, → *quality control* and *quality improvement*. QM also includes the means and tools to achieve and sustain quality levels by carrying out quality assurance and controlling processes and product properties.

*Reference*:
Rose, Kenneth H. (July 2005). Project Quality Management: Why, What and How.

**Questionnaires**
A set of questions that can be answered orally ($\rightarrow$ interview), in writing, or electronically. The key success factor for a questionnaire is the careful design of its questions, which focus on a governing question. Before formulation of the questions it is necessary to decide on the evaluation method to be used. It is also possible to distinguish among various types of questions, e.g.:

– *Dichotomous questions*: questions that have two possible responses such as yes/no.
– *Questions based on level of measurement*: various possibilities must be evaluated according to scales or grades, etc.

- *Open-ended questions*: any kind of answer can be accepted
- *Closed questions*: the possible answers are divided into groups and must be sorted into the defined value ranges
- *Filter or contingency questions:* questions that are used to determine if respondents are qualified or experienced enough to answer a subsequent one

Generally, it is good practice to pre-test questionnaires on a small, representative group of people. The results, feedback, and experience gained in this test can be used to improve the questionnaire, but must not be used for a scientific evaluation/survey.

*Reference:*
Kreuter, F.; Presser, St. and Tourangeau, R. (2008): Social Desirability Bias in CATI, IVR, and Web Surveys

## Queuing Models
Queuing models and the corresponding theory belong to → operations research. They model situations where persons or goods arrive at service stations. Both arrival times and service times are stochastic. This can lead to the forming of waiting lines (queues) and in extreme cases to blocking or denial of service or to idling servers at the other extreme.

Examples of areas where queuing models are very useful include: the analysis of computers, telecommunication networks, traffic systems, logistics, and manufacturing systems. Depending on the field of application, the term *service station* may have different meanings.

The analytical results from waiting line theory provide the basis for the management of arrivals or for sizing of the service stations to minimize the overall costs caused by waiting and service:

The most important cost influencing factors are:

- The number of arrivals at a particular arrival rate (= average number of arrivals per time unit) and/or their stochastic distribution
- The number of service stations, with a particular service rate (= average number of service operations per time unit) and/or their stochastic distribution
- Line discipline (rules of behavior of the incoming elements)
- Service strategy (organization and sequence of processing such as: first in first out, etc.)

In classical queuing theory, inter-arrival times are often assumed to follow an exponential distribution and service times. This leads to a stochastic process that is independent of its history (Markov property: M) and allows for analytical solutions. Queues are characterized by the distribution of arrival rates, service rates, and number of servers: M/M/1 would be such a queue.

In many practical cases, such as complex networks, empirical probability distributions or complex service rules for arrivals or service times (e.g., branching waiting lines), it is not possible to develop analytical solutions. Instead → numerical/ computer simulation is used applying the → Monte Carlo method.

In systems engineering, the queueing theory is applied in the step involving synthesis/analysis of solutions.

*References:*
Asmussen, S. and Glynn, Peter W.(2007): Stochastic Simulation: Algorithms and Analysis
Giambene, Giovanni (2014): Queuing Theory and Telecommunications Networks and Applications.
See also → operations research.

## Real Options
*See Sect. 2.2.5*

## Reengineering, Business Process Reengineering, BPR
Promoted by Michael Hammer and James Champy in the mid-1990s as a concept for drastic changes in production and business processes. The result should be decisive and measurable improvements regarding costs, quality, service, and time. "Quantum leaps, not small steps to improvement." Reengineering is thus contrary to such concepts as Kaizen and CIP, which support a policy of small, continuous steps.

*References:*
Hammer, H.; Champy, J. (2003): *Reengineering the Corporation.*
Schantin, D. (2004): *Makromodellierung von Geschäftsprozessen*

## Regression Analysis
Being a method of → mathematical statistics, regression analysis is an analysis technique that is an attempt at modeling the relationship between one dependent and one or more independent variables. Related to this, a → *correlation analysis* calculates how well a given (often linear) model can represent an assumed relationship between a set of variables.

## Reliability Analyses
Used to model/describe the reliability of objects to calculate their probability of failure. These objects can be as small as electronic components or as large and complex as power plants. Reliability as a part of the quality of objects covers their behavior during or after specified time spans under given operating conditions (see DIN 55 350-11) and depends on a number of variables. These result from the type of application, usage, or operation, etc., i.e. on one hand into *external factors*, e.g., the specified user demands and effects from the environment, and on the other hand *internal factors* such as the quality resulting from the development and manufacturing of the object.

   Reliability analyses are carried out to specify reliability requirements of objects to be designed, and to investigate the reliability of existing or already developed objects. A simplified deterministic perspective of reliability leads to experience-based measures that are intended to prevent or counter failure or breakdown (e.g., *n*-fold safety coefficients of a component).

   A wide range of material characteristics or manufacturing conditions and a great variability in operating conditions; however, require a probabilistic perspective. This also applies to systems consisting of many elements that must perform a great number of functions (e.g., instrumentation and control equipment), and to systems whose malfunction could pose danger to equipment, humans or the environment (e.g.: signal equipment in a railroad operation).

There are many mathematical methods of calculation reliability. Failure probability calculations often assume an exponential growing probability distribution of failures over time – electrical components in contrast are assumed to follow a so-called "bathtub curve," which assumes high failure rates early and late in the product's life. The most commonly used measures are the mean time *between* failures (MTBF) and the mean time *to* failures (MTTF).

In systems engineering, reliability analyses are often used during *solution search* (synthesis/analysis), and if necessary in the *situation analysis* (improvement of an unsatisfactory solution).

*References:*

DIN 55350-11:2008-05, English version: Concepts for quality management - Part 11: Supplement to DIN EN ISO 9000:2005

Finkelstein, M. (2008): Failure Rate Modelling for Reliability and Risk.

Nakagawa, Toshio (2005): Maintenance Theory of Reliability.

Levitin, Gregory, G. (2005): The Universal Generating Function in Reliability Analysis and Optimization.

**Reverse Engineering**

Reverse engineering (also called back engineering) describes a process of extracting all knowledge and design information from an existing system (e.g., an industrially manufactured product) by only investigating its structures, states, properties, and behavior.

A design is reconstructed that way by trying to reverse the engineering process and moving backward, starting at the finished item. To verify the design and other gained insights, a 1:1 copy of the object can be manufactured and compared with the original. On the basis of the copy and its construction plan it is now possible to move forward with further development. This approach is used as an analysis method in a variety of fields: in *natural sciences* (genetics, biology, and physics) or as what is known as a retrosynthesis in chemistry.

In situations in which the design to be investigated is protected by intellectual property laws or trade secret laws, it is often forbidden to use reverse engineering. Legal regulations vary from country to country. Sometimes, there are exceptions for ensuring "interoperability" with other systems.

In areas where it is legal, reverse engineering can be used in many different ways:

The design of *electronic components*, such as microprocessors, could be reconstructed by mechanically taking down the silicon die layer by layer. In *software*, it is often possible, for example, to recover the source code from an executable program (using a disassembler and many efforts at analysis), or the decoding of communication protocols of network components by analyzing their actual communication.

In *mechanical engineering*, real objects/machine parts can easily be digitized using a laser scanner and are ready to be used in a digital engineering process (digital mockups, computational flow simulation, FEM analysis, etc.) or a design process to be developed further. Digitizing real components is also useful for target-actual comparison in quality management –e.g., for comparing in injection molding the shape of the mold's computer-aided design (CAD) model with the shape of the resulting finished part, which has been digitized using 3D scanners.

*References:*
Raja, V.; Fernandes, K. J. (2008): Reverse Engineering
Eilam, Eldad (2005). Reversing: secrets of reverse engineering.


**Risk Analysis**
Risk analyses are used to identify early as possible the dangers or risks – particularly to humans – that result from a product as early as possible → risk management. Another reason for thorough risk analysis is the intention to minimize product liability risk with respect to clients.

They are also part of the documentation for certifications such as the European Union compliance statement, which is mandatory for the CE test mark, or the Federal Communications Commission Declaration of Conformity, which is used on certain electronic devices sold in the USA.


**Risk Management**
Assessed potential damages or the consequences of a malfunction due to undesirable occurrences are called *risk. Risk management* is the systematic identification and evaluation of risks and the management of measures to counteract identified risks. It is a systematic process that is applicable in many areas, for example, with business risks, credit risks, financial risks, environmental risks, insurance risks, technical risks, project risks, and many more.

*Project risks* are of particular interest here: risk management in projects deals with all measures that appear necessary for preventing or dealing with unplanned results, because they may endanger the progress of the project. As these measures often require financial or time resources, it is always necessary to weigh the associated efforts against the damages in case the risks occur (weighted with the likelihood).

Risk management also deals with topics of so-called *issues management* – the handling of risks that have arisen without previously being identified. Risk analyses in systems engineering are very useful in the *concept analysis* step of the *problem-solving cycle*.

The → PMBoK Guide envisions six steps for risk management:

1. *Risk management planning:* determine processes with which the following risk processes work. These include identification methods, documentation strategies, evaluation strategies, and responsibilities.
2. *Risk identification:* risks in terms of potential hindrances are identified and documented using various methods.

3. *Qualitative risk analysis:* the risks identified are assessed qualitatively and assigned priorities based on the probability that the risks will occur, on the one hand, and on the other hand, their effect on the success of the project.
4. *Quantitative risk analysis:* then there is the quantitative evaluation of risk (in monetary terms), countermeasures, and/or necessary allowances.
5. *Risk response planning*: countermeasures are identified to minimize the occurrence of risk or to reduce the effects of risk.
6. *Risk monitoring and control*: the status of the risks (usually documented in a risk list) and the status of the countermeasures is continually monitored.

*References:*
Snijders, Paul; Wuttke, Thomas; Zandhuis Anton (2009): PMBOK Guide.
Hester, R E. and Harrison, R M., eds. (1998): Risk Assessment and Risk Management.
Hopkin, P. (2012): Fundamentals of Risk Management: Understanding, Evaluating and Implementing.

## Safety Analyses (Synonym: Security Analyses)

Safety analyses are supposed to reduce the risk of a hazard in businesses and/or projects. The objective is to recognize *threats* and assess their *likelihood of occurrence* and *damage potential* and the attendant *risk*. Important traditional methods include → fault tree analysis and → FMEA. An interesting new method is → Systems Theoretic Process Analysis (STPA), which is also designed to prevent component interaction accidents.

*Reference:*
Leveson Nancy (2012): Engineering a Safer World. Systems Thinking Applied to Safety

## Safety Management (Synonym: Security Management)

Safety management tasks in business or in projects can be categorized into *strategic* and *operative tasks*: *strategic tasks* include such things as strategic analyses (threat analyses, vulnerability analyses), establishment of safety goals, strategies and safety measures, the creating of a safety concept, assigning responsibilities for safety and the necessary competencies for the responsible management positions and employees, strategic controls, etc. *Operative tasks* include operative analyses (→ risk analyses), operative design of measures for implementation of the safety concept (i.e., organization of training and drills in which safety-related information is provided, and the proper behavior of the employees regarding dangers is explained and practiced. This improves the acceptance of safety measures significantly. Operative controls such as checking the implementation of the safety concept, implementing the measures planned, adherence to the safety guidelines, and the effectiveness of the implemented safety measures ensure the sustainability of this concept.

*References:*
McKinnon, Ronald C. (2012): Safety Management: Near Miss Identification, Recognition, and Investigation.
Ortmeier, P. J. (2001): Security Management: An Introduction.

**Sampling**

Sampling takes a representative portion of information, a material or product to test. Sampling is used frequently in statistics (e.g., market research, technology, production, quality control, natural, social, and human scientific, medical, and psychological research), because often, it is not possible to investigate all elements, such as the entire population or all manufactured examples of a product. When analyzing unknown systems, samples are often used to construct a hypothesis for the whole system following the *induction principle*, with which inference is made from the particular to the general.

*Reference:* see the reference for → mathematical statistics.

**Scenario Analysis (Scenario Planning)**

Scenario analysis (sometimes also called *scenario technique*) is a strategic planning method that has its roots in the military, but in the meantime it has also been used with economic and social questions. It is used for analyzing such things as *extreme scenarios* (best-case scenario/worst-case scenario), plus particularly *relevant* or *typical* scenarios (trend scenarios). The scenario technique is also used in psychology and psychotherapy (psychodrama, socio-drama). There, it involves both future and past scenarios. The biggest drawback in using scenario analysis is the fact that it is difficult to put realistic scenarios together: a realistic worst case scenario is not a scenario where everything that can go wrong does go wrong. Instead of going through the effort and risk of constructing realistic scenarios, often stochastic methods are applied, e.g., by assigning probabilities to input and system parameters and by evaluating system behavior using the → Monte Carlo simulation.

However, scenario analysis is still preferred in preparing strategic decisions, because of its simplicity compared with statistical methods, e.g., with reference to technological developments, business models, with market and sector developments, with orientation toward future developments, strategic development, and verification, for early recognition of possible changes through sensitization for the future. Further application fields are crisis management, project management, → risk management, and many others.

*Example:* Dennis Meadows' study for the Club of Rome for national economic scenarios: "The Limits of Growth." For strategic business planning a scenario analysis was used successfully by Shell in the 1970s to overcome the crisis in petroleum prices. The International Panel on Climate Change has worked up scenarios of what the world will look like in the future and the effects that climate change can produce.

*References:*
Kahn, H. (1967): The Year 2000
Wright, G.; Cairns, G. (2011): Scenario thinking: practical approaches to the future
Cornelius, Peter; Van de Putte, Alexander and Romani, Mattia (2005): Three Decades of Scenario Planning in Shell. California Management Review, Nov. 2005

**Scoring Method**
*Synonym for* → *value-benefit analysis.*

## Sensitivity Analyses

The purpose of a sensitivity analysis is to check the stability of an achieved result. This occurs, for example, by varying the value of the parameters that influence the result within an area considered reliable and looking at the new results thus created. Example: determining a series of variants by means of a → value-benefit analysis. Now, it is possible to change the weights of individual criteria that are considered crucial, or to vary the assignment of scores within a reasonable framework. If nothing changes, in other words, if the winner is still out front, we may have a higher degree of certainty that the results are correct. However, if the results change, we have to wonder if the cluster of parameters just chosen would also work.

In systems engineering, *sensitivity analysis* is mainly used during solution search (after an optimization process) or after an evaluation of solutions.

## Sequencing Problems

A category of → operations research problems, which involve putting activities into an optimal and feasible sequence. The objective can be, for example, to minimize the order throughput time throughout manufacturing. A well-known example is the traveling salesperson (salesman) problem, in which the shortest route for a traveling salesperson is sought for visiting $n$ specified cities exactly once.

*References:* see → operations research

## Simplex Algorithm

The most important algorithm for solving → linear optimizing tasks in practice. Developed by George Dantzig in 1947 and subsequently improved and extended.

*References:* see → operations research

## Simulation Techniques

The term *simulation* can be used in many ways. Here, we mainly examine the aspects that are significant in connection with development processes.

On the one hand, the term *simulation technique* is used in connection with tasks that are not (or not yet) solvable analytically (using an optimization algorithm). In this case, the → Monte Carlo method is a good choice. On the other hand, *modeling* and subsequently *simulation techniques* are used when the system to be developed is not yet real and only exists as a model. Experiments are conducted on a *simulated model* to learn about the real system. A simulation model thus presents an abstraction of the system to be engineered (structure, function, behavior) – usually in electronic form.

There can be many reasons for using simulations:

(a) Improving or speeding up the *development process*

- *Virtual product development*: design and construction, calculation, and testing are primarily digital, i.e., on the computer. That way, time and money are saved in constructing prototypes, and the "time to market" is reduced and/or the product quality is improved through (multiple) reworkings of a design.

- *Virtual testing:* e.g.: simulation of product tests (e.g., crash tests in the automobile industry) in a computer model, perhaps by modifying the design in several steps. Subsequently, a physical prototype is constructed and tested in a testing station. That way, the conformity of the computer model and the testing station results can be compared (and the testing station trial is also a type of reality simulation).
- Simulation of *production facilities* because repeatedly remodeling the physical facility would be too complex and expensive. Key word → digital factory.
- *City planning, traffic planning* (public and company internal).
- Simulation of *logistics systems* (warehousing, storage, retrieval, supply chain, etc.).

(b) Risk-free and economical *training in a training* simulator:

- Pilot training in flight simulators (practice with critical situations such as engine failure, emergency landing, etc.).
- Similar: driving simulator.
- Training of doctors and surgeons in operating techniques.
- Experimental business games (training for analysis and decision-making skills).

(c) The real system *cannot be observed directly*:

- System-related: simulation of a singular molecule in a fluid, astrophysical processes.
- The real system works too fast: simulation of circuitry.
- The real system works too slowly: simulation of geological processes.

Of course, there are also limits to using simulation models:

- Limited resources (time, money). Therefore, a model must be as simple as possible.
- Each simulation model simplifies reality. In particular, models of complex situations or systems are often grossly simplified. This naturally reduces the precision and sometimes the usefulness of the simulation results to be transferred to reality. With other parameters, the results can simply be wrong. This is why simulation models must be carefully tested and validated.
- Imprecision of the input data (measuring errors, randomness, etc.).

*References:*
Banks, J., Ed. (1998): Handbook of Simulation: Principles, Methodology, Advances, Applications, and Practice.
Banks, J., J.S. Carson, B.L. Nelson, and D.M. Nicol (2005), Discrete-Event System Simulation.
Cellier, F.E. and Kofman, E. (2006): Continuous System Simulation,
Sterman, J.D. (2006), Business Dynamics: Systems Thinking and Modeling for a Complex World.

Terano, T., H. Kita, T. Kaneda, K. Arai, and H. Deguchi, Eds. (2005), Agent-Based
    Simulation: From Modeling Methodologies to Real-World Applications.
Robinson, St. (2004): Simulation: The Practice of Model Development and Use.

## Six Sigma, 6σ

Six sigma is a collection of methods and tools for process improvement and belongs
to → *quality control*. It was introduced at Motorola in 1986 and later adopted by
many companies in different sectors – such as General Electric, where it became a
central element of the enterprise's business strategy.

The underlying assumption is that it is possible to improve the quality of the
process output by systematically identifying and removing the causes of defects and
by minimizing the (unintended) variability in manufacturing and business processes.
Six sigma uses a large variety of → quality management methods (data-driven,
statistical) methods.

The term "six sigma" originates from the goal to control 6σ of a processes
variability – which means 99.99966% of all possible outcomes.

*Reference:*
Tennant, Geoff (2001). SIX SIGMA: SPC and TQM in Manufacturing and Services.

## Statistics
*See → mathematical statistics*

## Systems Theoretic Process Analysis, STPA
A new hazard analysis technique with the same goals as any other hazard analysis
technique: to identify scenarios leading to identified hazards and thus to losses so
that they can be eliminated or controlled. STPA, however, has a different theoretical
basis or accident causality model: STPA is based on systems theory whereas
traditional hazard analysis techniques have reliability theory at their foundation.
However, many of the causes do not involve failures or unreliability.

Although traditional techniques were designed to prevent component failure accidents (accidents caused by one or more components that fail), STPA also was designed
to address increasingly common component interaction accidents, which can result
from design flaws or unsafe interactions among nonfailing (operational) components.

*References:*
Leveson, Nancy (2012): Engineering a Safer World. Systems Thinking Applied to
    Safety. MIT Press
Leveson, Nancy (2013): An STPA Primer    http://psas.scripts.mit.edu/home/
    wp-content/uploads/2015/06/STPA-Primer-v1.pdf

## Survey
*See → interview*

## Synectics
A →creativity technique developed by W.J. Gordon to activate the solution search.
A combination of systematic and intuitive elements based on brainstorming that is

embedded in a process of stepwise analogy creation (→ analogy method). The process consists of four phases:

1. *Preparation* phase: after intervening with spontaneous solution ideas, disassociation is used to encourage the implementation of structures extraneous to the problem and the unaccustomed combination of elements.
2. *Incubation phase:* in addition to direct analogies from technology or nature (→ bionics), personal analogies are formulated (How do I feel as…?), along with symbolic, contrary, and fantastic analogies (What would a fairy do?).
3. *Illumination* phase: the analogies are checked for their appropriateness for transfer to the problem; also called *force-fit.*
4. *Verification* phase: working up solution concepts.

Synectics places high demands on the participants, especially on the moderator. In addition to overcoming the restraints of personal analogies, the transfer of unfamiliar structures and the unaccustomed combination of elements requires some practice. This, plus the greater demands on the team and the preparation of the moderator, may be reasons why synectics is not used so frequently in practice.

*Reference:*
Gordon, William J.J. (1961): Synectics: The Development of Creative Capacity. New York. Harper & Row Publ.

**System Dynamics, SD**
Methodology developed by Jay W. Forrester at MIT for a holistic analysis and simulation of complex, dynamic systems. It is especially applied in socio-economic fields. The effects on management decisions, on structure and system behavior (e.g., business success) can thus be simulated and recommendations for action can be extrapolated (also see the brief presentation of system dynamics in Sect. 1.4).

The *qualitative method* mainly involves the identification and investigation of self-contained feedback loops. There is a distinction between loops with positive, strengthening effects (reinforcing loops) and negative, stabilizing ones (balancing loops). Causal loop diagrams are often used here as a graphical representation technique.

With *quantitative models*, the representation in *stock and flow diagrams* and their simulation facilitates a deeper understanding of the system. Stocks, flows, and auxiliary quantities are useful in describing the systems relationships, and they show how the feedback loops lead to system behavior, which in part is often not linear and counterintuitive. This is the main advantage of this method.

Special software packages such as CONSIDEO, iThink/STELLA, DYNAMO, Vensim, Powersim or AnyLogic allow a numerical simulation of the system dynamics models. The simulation runs of various scenarios foster understanding of the system behavior over time.

*Peter Senge* has identified, investigated, and classified typical structures in feedback system types, which he has called *archetypes*. Knowledge of these basic structures allows a better understanding and a good forecast of the behavior of various social systems and/or management situations and thus provides a basis for more effective interventions.

*Application areas:* system dynamics was the underlying methodology for simulating the World3 world model, which was created under the direction of Dennis L. Meadows under contract from the Club of Rome for studies on "Limits to Growth" (1972, 2004). It is also useful for simulating and explaining the complex behavior of humans in social systems. Typical examples are the investigation of the phenomenon of over-fishing and the occurrence of disasters such as the nuclear accident in Chernobyl.

Building on the methods of system dynamics, information dynamics investigates the information processing of systems and determines that information is the essential determinant for the behavior and the efficiency of systems.

*References:*
Forrester, J.W. (1977): Industrial Dynamics.
Meadows, D.; Meadows, D.; Randers, J. (1972): The Limits to Growth.
Meadows, D.; Meadows D. L.; Randers J. (2004): Limits to Growth: The 30-Year Update.
Senge, P.M. (1990): The Fifth Discipline: The Art & Practice of The Learning Organization.
Senge, P.M. (1994): The Fifth Discipline Fieldbook: Strategies and Tools for Building a Learning Organization.
Sterman, J.B. (2006): Business Dynamics. Systems Thinking and Modeling for a Complex World.

**Systems Modeling Language, SysML**
SysML is a standardized language based on → UML for modeling complex systems. Originally designed mainly for software development, the application field noticeably extends to overall product development (hardware and software). SysML supports analysis, design, and the testing of complex systems, in addition to the following steps:

– *Modeling system requirements* and making them available
– *Analyzing and evaluating systems* to solve requirement and design issues and testing alternatives
– Communicating system information unambiguously among various stakeholders
– *References:*
– Dori, Dov (2016): Model-Based Systems Engineering with OPM and SysML.
– Weilkiens, T. (2008): Systems Engineering with SysML/UML: Modeling, Analysis, Design.

**Target Costing**
Approach to guiding the product development process by previously set cost targets. The process starts with *market-driven costing* where a realistic sales price (including a desired profit margin) is estimated. This "allowable cost" is the basis

for the next step, the *product level costing*, where the internally achievable cost structure for producing a competitive product is evaluated. Here, the → value analysis technique can be of great help. If necessary, functional changes in the product specifications are made to achieve the cost target. In the third process step of target costing, this internal step is repeated at the *component level*.

During the development process, it is important to track these costs as well.

*References:*
Clifton, B.C.; Bird, H.M.B.; Albano, R.E.; Townsend, W.P. (2004) Target Costing: Market-Driven Product Design.
IMA: Implementing Target Costing (1994)

## Theory of Inventive Problem Solving, TIPS
English acronym for → TRIZ (Teoriya Resheniya Izobretatelskikh Zadatch).

## To-Do List
Also known as a *task list* or *list of open points (LOP list)*. A simple aid in project meetings for assigning the planned tasks (= WHAT) to individuals (sometimes also groups or organizational units (= WHO) and tying these *tasks* or *work packages* to a committed deadline (=WHEN). It is important to make sure that tasks are not delegated to a group or organization, but rather have exactly one person responsible.

The to-do list often has the form of a three-column list that is to be communicated to the corresponding persons and involved groups immediately after the project meeting. This creates the basis for the task supervision (*task tracking*): the next meeting starts with a report on the pending and completed (or uncompleted) tasks. Project team members whose tasks are lagging behind schedule have to explain and justify the situation and the reason to the team – which will also create (peer) pressure for improving work discipline as an intended side-effect.

## Total Quality Control, TQC
Ongoing holistic → quality control (or assurance) of all products and processes involving, in addition to classical production-oriented company functions, other departments such as design and purchasing.

*Reference:*
Feigenbaum, Armand V. (1991). Total Quality Control.

## Total Quality Management, TQM
Total quality management (TQM) is a traditional → quality control method that demands companywide efforts to establish an environment where a company's core business processes are continuously improving, which leads to high-quality services or products for customers. In different industries and regions there are variants of TQM. In recent years, some of the TQM ideas have been adopted by modern and more popular concepts such as the ISO 9000 standard, lean manufacturing, or → six sigma.

Quality management of such a kind is nowadays mandatory in sectors such as the airline and aerospace industries, medical technology, health care, pharmaceuticals, and food manufacturing.

Total quality management originated in the USA (W.E. Deming). It, became very popular in Japan after WW2 and experienced a renaissance in the USA (Baldridge Award). In Europe, TQM is known as the → EFQM Model for Business Excellence.

To give an example of some basic TQM core beliefs: not only people, but also poorly planned and poorly managed processes cause errors. It is senseless and futile to search for a single culprit. The goal has to be zero errors. Sourcing closer, more reliable relationships with fewer suppliers is important. As the entire focus is on the customers, processes must be directed primarily toward them and their needs.

*References:*

Juran, Joseph M. (1995)): A history of managing for quality: the evolution, trends, and future directions of managing for quality

Evans, J.R.; Lindsay, W.M., 1995. The management and control of quality.

Deming, E. (1997): Out of the Crisis.

Omachonu, Vincent K.; Ross, Joel E. (2004): Principles of Total Quality,

Tague, Nancy R. (2005): The Quality Toolbox.

Gitlow, Howard Seth; Levine, David M. (2005): Six Sigma for Green Belts and Champions

## TRIZ/TIPS

The TRIZ method is one of the → creativity techniques. The original term TRIZ is a Russian acronym for *Theory of Inventive Problem Solving* (TIPS), which is used in the English-speaking world. The methodology was developed by a group in Russia around G. Altshuller. It was developed during the examination of a large number of patent specifications, from which those that appeared to describe technical breakthroughs should be selected. Those were subsequently analyzed in greater detail and *three essential, common phenomena* were identified:

1. Many inventions are based on a comparatively small number of general solution principles.
2. Only the overcoming of inconsistencies makes innovative developments possible.
3. The evolution of technical systems follows certain patterns and rules.

The TRIZ method contains a series of methodical tools that facilitate more effective analysis of a technical problem and make it possible to find creative solutions.

Using this method, inventors attempt to systematize their activities to get to new problem solutions more quickly and more efficiently. The TRIZ method has become widespread.

The *classical TRIZ methods* are:

1. Principles of innovation and contradiction matrices
2. Separation principles for solving physical contradictions

3. Algorithms or at least stepwise procedures for solving invention problems
4. System of 76 standard solutions and substance field analysis
5. S-curves and laws for systems development (evolutionary laws of technical development, laws of technical evolution)
6. Principle (law) of ideality
7. Modeling technical systems using "little men" (dwarf model)

The following methods were introduced by Altshuller's followers and now also belong to TRIZ:

1. Innovation checklist (innovation situation questionnaire)
2. Function structure according to TRIZ (a type of cause and effect diagram, which is different than Ishikawa's version)
3. Subject action object function model (an expanded functional model based on work by *L. Miles* on → value analysis)
4. Process analysis
5. Materials cost time operator
6. Anticipatory error detection
7. Feature transfer (part of "alternative system design")
8. Resources

*References:*
Altshuller, G. S. (1984): Creativity as an Exact Science – The Theory of the Solution of Inventive Problems. New York: Gorden and Breach
Altshuller, G. (1999): The Innovation Algorithm: TRIZ, systematic innovation, and technical creativity.
Fey, V.; Rivin, E. (2005): Innovation on Demand: New Product Development Using TRIZ.
Rantanen, K.; Domb, E. (2007): Simplified TRIZ: New Problem Solving Applications for Engineers and Manufacturing Professionals.

**Unified Modeling Language, UML**
A *language for modeling software* and other systems developed and standardized by the OMG. UML is standardized by ISO standards (ISO/IEC 19501) and is today one of the dominant languages for modeling software systems. Owing to its software focus and despite the name its missing an unifying concept, UML is therefore not very useful for business process modeling, where → *business process model and notation (BPMN)* is used instead.
    Selected UML components are used in software project management:

– Project commissioning parties and business professionals test and confirm, for example, the requirements that the business analysts have determined using BPMN and create the so-called UML *use case diagrams.*
– Software developers create a program logic that the business analysts have described in collaboration with professionals in *activity diagrams.*
– Systems engineers install and operate the software system based on an installation plan that exists as a *deployment diagram.*

Besides its usefulness as a graphic notation, UML primarily specifies the data objects and program entities along with their attributes and relationships. Modern UML-based software development tools can automatically generate the corresponding source code.

*References:*
Weilkiens, T. (2008): Systems Engineering with SysML/UML: Modeling, Analysis, Design.
Coad, Peter; Lefebvre, Eric; De Luca, Jeff (1999): Java Modeling in Color with UML: Enterprise Components and Process.

## Use Case

Use Cases are used in both systems- and software engineering to describe all steps of a task where, for example, a person ("actor") interacts with a system. Such a use case can be seen as a scenario. The basic idea is to extract requirements by analyzing a "usage" or interaction process, which is often more efficient than a classical specification list.

The result of a use case can be success or failure/termination. Use cases traditionally are named for the goals from the agents' perspective: *enrolling a member, withdrawing money, returning a car.*

The *granularity* of use cases can be vastly different: at a very high level a use case describes what happens only very roughly and in an abstract manner. However, the technique of writing a use case can be refined, even at the level of IT processes, so that the behavior of an application is described in detail. This contradicts the original intention of use cases, but it is often very useful.

*Use cases* and *business processes* show a different view of the system modeled and have to be clearly distinguished from one another. A *use case* describes what the actor/environment expects from the system. Business processes, on the other hand, model how the system operates internally to meet the requirements of the environment.

## Valuation Techniques

Formalized procedures for evaluating and comparing, for example, solution variants with respect to meeting their objectives (see Part III, Sect. 6.4). Examples: → value-benefit analysis, → cost-effectiveness analysis, → analytic hierarchy process (AHP).

## Value Analysis (Synonyms: Value Management, Value Engineering)

Value analysis was originally an aid in *product design.* With respect to simplifying products or parts and reducing their costs, it primarily serves *value improvement* and *cost reduction.* An extension of this method added the ability to expand this basic idea to the *value creation* of new products. In a further step, the methodology of *value analysis* was also applied to researching and designing *intangible services* (e.g., organizational processes).

The core of value analysis is thinking in functions, at first at the level of the entire product and use by clients. This means "letting go" of existing or obvious possible solution patterns and opening up to innovative solutions and considering their value to a potential client. There is thus a distinction between *main* and *secondary functions*, and there may also be *undesirable functions* that the client does not want and for which he or she will not pay. In addition, there are *utilitarian functions* (utilities) and *prestige utility* (e.g., esthetics, image, prestige, without any useful advantage). A client's appreciation of a product and the desired and valued functional features should also determine the price that a client is prepared to pay in a competitive situation. Often, this is specified in the form of a target cost as a component of the value analysis goal (with functional, quality, performance, and deadline goals, etc.) for the design process. See → target costing.

In contrast to the → value benefit analysis, which is an evaluation process, value analysis is a special approach to the solution search that also seeks an evaluation of individual solution elements with respect to their optimization.

Over the course of time (and in its promoters' self-perception), value analysis has grown from an approach to improving the product or the product design process into a comprehensive problem-solving methodology (VDI Guideline 2800 or formerly DIN 69910 standard) – see, for example, the value analysis work plan illustrated in Part I, Sect. 2.2.1.4.

*References:*
Miles, Lawrence D. (1972): Techniques of Value Analysis and Engineering.
Sato, Yoshihiko; Kaufman, J. Jerry (2005): Value Analysis Tear-Down: A New
    Process for Product Development and Innovation.

**Value-Benefit Analysis, VBA (Synonym: Scoring Method)**
Evaluation method used to determine the preferability of variants.
    *See Part III, Sect. 6.4.2.3 or the airport Case Study in Chap. 9.*

**Virtual Product Development**
Development of a product within the virtual world of computers. With the help of CAD systems and PDM Systems, products are designed graphically and assigned relevant technical characteristics (material characteristics: thickness, surface finish, tensile and yield strength; mechanical and kinematic relationships; production information, etc.). Afterward, further processing, calculations, and simulations of all types can be carried out, such as strength calculations, finite element calculations, installation tests (with the help of digital mockups), crash simulations, etc. The results can have repercussions on the dimensioning or shaping of the construction, i.e., they may lead to iterations for improving the quality of the construction.

New products can be tested virtually in this way before they are available in physical form. Virtual product development is one of the most important trends in construction and development activities.

*References:*
Weisberg, D.: The Engineering Design Revolution, E-Book, www.cadhistory.net,
    May 2010
Nambisan, Satish ed. (2010): Information Technology and Product Development.
Kenneth B. Kahn, ed. (2013): The PDMA Handbook of New Product Development.

Bordegoni, M.; Rizzi, C., eds. (2011): Innovation in Product Design: From CAD to Virtual Prototyping.
Virtual Vehicle Research Center: http://vif.tugraz.at

## Visualization Techniques (Information Graphics)

Collective term for techniques for representing and visualizing all types of information: *layout plans* show the size of rooms and their arrangement, etc. *Organizational charts* show the organizational structure, the manner of structuring, hierarchical relationships, job holders, etc. *Flow charts* show the sequence of mutual dependencies, the logical order of activities. *Communication diagrams* show between which positions, how often, what type of communication takes place, etc. *Bubble charts* are used for draft visualizations of a problem or for visualizing system relationships. Bubbles = elements, and arrows indicate relationships among them. Questions: which system aspect is of interest to us? What perspective are we using? Which elements of significance, which relationships, how to demarcate them, what is not significant? Also see Part I, Sect. 1.2.5. *Tabular representations* provide an overview, structure, order. → *Mind maps* likewise order and stimulate the continuation of trains of thought.

*References:*
Tufte, Edward R. (2001): The Visual Display of Quantitative Information
Zelazny, Gene (2001): Say It With Charts: The Executive's Guide to Visual Communication
Rendgen, Sandra; et al. (2012): Information Graphics

## Work Breakdown Structure, WBS

A graphic overview that represents objects of a system or tasks (usually as a tree), that are connected with respect to their development and implementation. The structuring can be carried out in various ways.

- *Object-oriented WBSs* (object structure designs) provide an overview of the system components to be designed, which can be further subdivided arbitrarily. Example: a car divided into chassis, body, motor, interior, etc.
- *Task-oriented WBSs* describe the tasks to be completed, such as development, construction, manufacturing, assembly, marketing, distribution, etc.
- *Phase-oriented WBSs* describe the phases, which are divided according to temporal and decision-oriented aspects, e.g., *preliminary study* (feasibility study), *main study* (master plan), *detailed study*, etc.
- *References:*
- Project Management Institute (2006): Practice Standard for Work Breakdown Structures. Project Management Institute
- Haugan, Gregory T. (2003): The Work Breakdown Structure in Government Contracting

## 16.1   Self-Check of Knowledge and Understanding: Encyclopedia/Glossary

1. Do you think this glossary useful even nowadays, in the age of Wikipedia, Google, etc.? Why or why not?
2. How many percent of the described methods and tools were familiar for you?

# Correction to: Systems Engineering

**Correction to:**
**R. Haberfellner et al.,** *Systems Engineering*,
**https://doi.org/10.1007/978-3-030-13431-0**

The original version of this book has been revised since this book was inadvertently published with few errors.

Chapter 1

Page 25 questions appear as

9. Explain the difference between a complex and a complicated system.
10. What are the three main points of view that can be used to create a system model?
11. Choose a system based on an arbitrary example for further investigation:
    – Structure the system hierarchically!
    – Which system aspects would be interesting?

it should re-order to

9. Choose a system based on an arbitrary example for further investigation:
    – Structure the system hierarchically!
    – Which system aspects would be interesting?
10. Explain the difference between a complex and a complicated system.
11. What are the three main points of view that can be used to create a system model?

---

The updated online version of the book can be found at
https://doi.org/10.1007/978-3-030-13431-0

© Springer Nature Switzerland AG 2021
R. Haberfellner et al., *Systems Engineering*,
https://doi.org/10.1007/978-3-030-13431-0_17

Chapter 2

Page 97 question 20 should be deleted

Chapter 9 and 16

Red text in the figures and tables should appear in black text
Self-Check of Knowledge and Understanding

Chapter 1

Page 417 Question 1 says:

What is meant by systems engineering systems thinking?

it should say

What does systems engineering systems thinking consist of?

Page 417 Question 3 says:

In which cases is it not appropriate to use the term "System"?

it should say

In which cases is it not appropriate to use the term "system"?

Page 418 Question 4 says:

Sketch an example of a system with its components and their relationships that is imbedded in an environment and interacting with other systems.

it should say

Sketch a system with its components and their relationships that is embedded in an environment and interacts with other systems.

Page 418 Question 5 says:

What is a "system of systems" (SoS)? What is the main difference between a subsystem and an SoS?

it should say

What is the main difference between a system consisting of subsystems and one consisting of systems (SoS)?

Page 418 Question 7 says:

What is meant by "black, grey and white box" models for describing real-world phenomena?

it should say

What is meant by "black-, gray-, and white-box" models for describing real-world phenomena?

Chapter 2

Page 420 Question 2 says:

What are the components (basic principles) of the systems engineering process model?

it should say

What are the four components (basic principles) of the systems engineering process model?

Page 420 Question 3 says:

What do the individual components of the systems engineering process model focus on?

it should say

What are the individual components focusing on?

Page 421 Question 6 says:

Is it necessary to pass through all project phases?

it should say

Is it necessary to pass all project phases?

Page 421 Question 7 says:

The term "analysis" is embodied twice in the problem-solving cycle. What is meant in one case and in the other case?

it should say

The term "analysis" is embodied twice in the PSC. What is meant in each case?

Page 421 Question 8 says:

What is the difference between the analysis and evaluation of solutions steps?

it should say

What is the difference between the steps analysis and evaluation of solutions?

Page 421 Question 9 says:

What is the difference between actual-state-oriented and desired-state-oriented process models? How or where would you position the systems engineering process model?

it should say

What is the difference between actual-state-oriented and desired-state-oriented process models? How or where would you position the Hall/BWI process model?

Page 421 Question 10 says:

What role does the client have in the problem-solving cycle?

it should say

What role does the client play in the PSC?

Page 422 Question 12 says:

What is meant by the term "Vee model"? What are its characteristics? To which component of the systems engineering process model does it relate?

it should say

What is meant by the term "V-Model"? What are its characteristics?

Page 422 Question 14 says:

What is denoted by the term "simultaneous (concurrent) engineering"?

it should say

What does the term "simultaneous (concurrent) engineering" denote?

Page 422 Question 15 says:

What are the characteristics of the so-called "agile process models"? Give some examples of that category.

it should say

What are the characteristics of the so=called "Agile Process Models"? Give some important examples of that category.

Page 422 Question 17 says:

Can you find some agile properties in the systems engineering model in spite of its primarily plan-driven character?

it should say

Can you find some agile properties in our systems engineering model in spite of its primarily plan-driven character?
Chapter 3

Page 423 Question 3 says:

What do we mean by saying that systems thinking may be applied to a problem in addition to the solutions?

it should say

What do we mean by saying that systems thinking may be applied to a problem and to the solutions?

Page 424 Question 7 says:

Do you think it is possible that the admissible ignorance tends to zero over the course of the project?

it should say

Do you think it is possible that the admissible ignorance tends toward zero over the course of the project?

Page 424 Question 9 says:

Are immediate measures in contrast to a systematic systems engineering-approach?

it should say

Are immediate measures in contrast to a systematic systems engineering approach?

Page 424 Question 10 says:

Explain the different thinking levels in problem-solving.

it should say

Explain the different thinking levels of problem-solving with the help of Fig. 3.10.

Page 424 Question 11 says:

What kind of information is acquired in the single steps of the problem-solving cycle and is passed over to the next steps?

it should say

What kind of information is acquired in the single steps of the PSC and is passed over to the next steps?

Chapter 4

Page 426 Question 4 says:

Explain the term project management.

it should say

Explain the term "project management."

Page 426 Question 7 says:

Which tasks are summarized under the term project management?

it should say

Which tasks are summarized under the term "project management"?

Page 426 Question 9 says:

Which ideal typical forms of project organization (PO) do you know?

it should say

Which ideal typical forms of project organization do you know?

Page 426 Question 10 says:

What are the characteristics, the advantages, and disadvantages of these PO forms?

it should say

What are the characteristics, the advantages, and disadvantages of these forms of project organization?

Page 426 Question 11 says:

What kind of project organization seems to suitable for which kind of projects?

it should say

Which type of project organization is suitable for which type of project?

Page 426 Question 12 says:

Which methods, techniques, tools for planning and control of projects do you know?

it should say

Which methods, techniques, tools for planning, and control of projects do you know?

Page 427 Question 13 says:

What are typical demands on a project manager? From the management perspective, from the perspective of the team members, etc.?

it should say

What are typical demands of a project manager? From the management perspective, from the perspective of the team members, etc.?

Page 427 Question 14 says:

What are typical attributes of high performing teams?

it should say

What are typical attributes of a high performing team?

Page 427 Question 15 says:

What are the consequences of the increasing use of agile methods for the systems engineering concept? Which modules will continue to apply and which one will be most affected?

it should say

In which direction may project management evolve in response to the emergence of agile methods?

Chapter 5

Page 427 Question 1 says:

How would you define the term "structure of a system"?

it should say

How would you define the term "structure" of a system?

Page 427 Question 2 says:

How would you define the term architecture?

it should say

How would you define the term "architecture"?

Page 427 Question 3 says:

Give examples for architectural variants

it should say

Give your own examples of architectural variants

Page 427 Question 7 says:

Sketch the model of Henderson/Clark for architectural innovations in the context with other innovations.

it should say

Sketch the Henderson/Clark model for architectural innovations in the context of other innovations.

Chapter 6.1

Page 428 Question 3 says:

How are detected weaknesses or failures, conceivable causes and measures to resolve the problems mentally and logically interconnected?

it should say

How are detected weaknesses or failures, conceivable causes, and measures to resolve the problems mentally and logically interconnected?

Page 429 Question 6 says:

What techniques and tools do you know for situation analysis?

it should say

Which techniques and tools do you know about for situation analysis?

Page 429 Question 7 says:

Does it make sense to use a working hypothesis in the situation analysis? Does this not rather mean that one works with prejudices?

it should say

Does it make sense to use a working hypothesis in the situation analysis? Does this not mean that one works with prejudices?

Chapter 6.2

Page 429 Question 2 says:

How are the steps "formulation of objectives," "situation analysis," "synthesis/analysis," "evaluation connected" linked together?

it should say

How are the steps "formulation of objectives," "situation analysis," "synthesis/analysis," and "evaluation" connected, linked together?

Page 429 Question 3 says:

Why don't compulsory objectives play a role in the evaluation step?

it should say

Why do compulsory objectives not play a role in the "evaluation" step?

Page 430 Question 6 says:

Can certain demands be objectives and means?

it should say

Can certain demands be objectives in addition to means?

Page 430 Question 7 says:

Does it make sense to use the term "objective" goals or "objective" targets or aims or objectives? (Here: objective as the opposite of subjective).

it should say

Does it make sense to use the term "objective" (here: objective as the opposite of subjective) goal/target/aim/?

Page 430 Question 9 says:

How can we attribute different significance to objectives?

it should say

How can we give different importance to objectives?

Page 430 Question 10 says:

What can one do, if goals oppose or contradict each other? Give examples.

it should say

What can one do if goals oppose or contradict each other? Give examples.

Chapter 6.3

Page 431 Question 1 says:

What does synthesis mean; what does analysis mean?

it should say

What does synthesis mean? What does analysis mean?

Page 431 Question 3 says:

Explain Fig. 6.17 relating to the idea of abstraction and concretization

it should say

Explain Fig. 6.15 relating to the idea of abstraction and concretization

Page 431 Question 7 says:

What is the logic to differentiate between an intuitive and a formal analysis?

it should say

What is the logic in differentiating between an intuitive and a formal analysis?

Chapter 6.4

Page 432 Question 1 says:

Which steps in the problem-solving cycle are the main information suppliers for the evaluation?

it should say

Which steps in the PLC are the main information suppliers for the evaluation

Page 432 Question 4 says:

Which methods for comparative assessment do you know?

it should say

Which methods of comparative assessment do you know?

Page 432 Question 6 says:

Why does it make sense to draw the course of a utility function in a graphic representation?

it should say

Why does it make sense to represent the course of a utility function on a graph?

Page 433 Question 10 says:

Is it possible that the decision-making body deviates from the proposal of the project group?

it should say

Is it possible for the decision-making body to deviate from the proposal of the project group?

Chapter 6.5

Page 433 Question 1 says:

Do you know any other special cases you would find worth discussing? The authors would like to hear your views.

it should say

Do you know any other special cases you would find worth discussing? The authors would like to know your views.

Chapter 7

Page 434 Question 1 says:

Name some basic principles concerning (i) the application of the systems approach, (ii) the application of an accepted action model, and (iii) the application of methods, techniques, tools.

it should say

– Do you agree on the basics summarized in this chapter? Or are there too few, too many or the wrong ones?

– The authors would like to receive feedback.

Chapter 8

Page 434 Question 1 says:

Can you find the abovementioned basics in the case study? see (Chap. 7).

it should say

Can you find the aforementioned basics in the case study?

Chapter 9

Page 434 Question 1 says:

Can you find the abovementioned basics in the case study? see (Chap. 7).

it should say

Can you find the above-mentioned basics in the case study?

Chapter 11

Page 434 Question 1 says:

Do you agree with the recommendations?

it should say

Do you agree to the recommendations

Page 434 Question 2 says:

Would you modify, extend, or reduce the list? The authors would like you to share your judgment with them.

it should say

Would you modify, extend or reduce the list? The authors would like to hear about your judgment.

Chapter 12

Page 434 Question 1 says:

Does the list presented in Fig. 11.1 reflect your experience? Do the arguments apply to it?

it should say

Does the list presented in Table 12.1 reflect your experience? Do the arguments apply to it?

Chapter 15

Page 435 question 3 should be included

3. What are you missing and why?

# Self-Check of Knowledge and Understanding

## The Systems Engineering Concept in General

1. Describe the basic ideas of the systems engineering concept in your own words.

   Answer: verbal description of Figs. 1 and 3

2. What significance does the methodology have in the problem-solving process? Is methodology the only component? If not, what are the others?

   A: Methodology is only one, each of the other components may be the bottleneck. Describe Fig. 2 in your own words.

3. What is the role of project management within the systems engineering-concept?

   A: Beside systems design (characterizing the content of a solution), project management covers organizational matters (team, leadership, decisions, scheduling, budgeting etc.). Both are in mutual relationship and cannot be separated in reality, Fig. 3

## Chapter 1. Systems Thinking

1. What does systems engineering systems thinking consist of?

   A: Terminology, definitions, and most importantly, model-based approaches to illustrate real-world phenomena and tools to support understanding.

© Springer Nature Switzerland AG 2019, corrected publication 2021
R. Haberfellner et al., *Systems Engineering*,
https://doi.org/10.1007/978-3-030-13431-0

2. What defines a "system"?

   A: A "system" is a collection of elements that are inter-related/connected, constituting an entity. This entity is more than a simple summing up of elements, it has a certain purpose or function that determines the elements, their arrangement, and their relations.

3. In which cases is it not appropriate to use the term "system"?

   A: If there is only one element or if there are no identifiable relations between the elements – "chaos" may be taken as the opposite of a "system."

4. Sketch a system with its components and their relationships that is embedded in an environment and interacts with other systems.

   A: See Fig. 1.2

5. What is the main difference between a system consisting of subsystems and one consisting of systems (SoS)?

   A: An SoS is a system that can exist, function, and be developed outside the host. A subsystem has to be designed in view of the purpose and function of the system it belongs to.
   Example: a mobile phone installed into a car is an SoS. The car body parts are the subsystems.

6. When using a hierarchical, level-based approach to describe a system (Fig. 1.4), what defines the lowest level of detail?

   A: The so-called "element level." No further decomposition seems to be useful currently (at this point of time).

7. What is meant by "black-, gray-, and white-box" models for describing real-world phenomena?

   A: A "black box model" reduces a phenomenon to a most simplified input/output relationship, the content of the box is seen as black, which means that it is ignored at this point in time. This is (when appropriate) a powerful tool to simplify a situation, to reduce complexity. In "white box" models the exact input/output relationship and internal system functions are known. "Grey box" models are in between black and white box models, where some coarse structure is known or some aspects/parts are known in more detail than others.

8. Give an example where different system aspects (looking through different spectacles) can help to model the whole system with its components, properties, and relationships.

   A: An industrial company, an organism, an eco-system, etc.

9. Choose a system based on an arbitrary example for further investigation:

   A1: For example, the human body, hierarchically structured into head, torso, arms, legs, etc.

   A2: System aspects: circulation of blood, nervous system, skeleton, etc.

10. Explain the difference between a complex and a complicated system.

    A: Complex systems consist of many (often different) components with dynamically changing and highly interconnected components. Complicated systems have either few components that are dynamically connected or many, statically linked components.

11. What are the three main points of view that can be used to create a system model?

    A: (i) "Environment-oriented," (ii) "input/output-oriented," and (iii) "structure oriented" point of view. The first two points of view treat the system as a black box by focusing on the relationship between a system and its surrounding environment – in terms of interacting entities and input to output mapping. The third aspect focuses on a system's internal structure – its components and their relationship with each other.

12. What is the tradeoff when using structure-based models for subsystems and components?
    Use this trade-off to explain why hierarchical system models often use black or grey box subsystem/component models at higher levels and the aforementioned structure-based models mainly at lower levels.

    A: Structure-based models provide the most detail at the cost of simplicity of the model. Therefore, these models are used only where black box-based models are not sufficient – which is mostly the case at lower levels.

13. Give an example of "agile systems" and discuss the challenges for systems engineering in that context.

    A: The main challenges here are to design a system that can adapt or can be adapted to changing requirements (and environments) that are not entirely known at the time of design. Examples: very flexible assembly line, general purpose machine.

14. In what way can system dynamics help with modeling systems?

    A: System dynamics is a valuable tool for describing the dynamics of system behavior by using (feedback) loops to model the relationship of system components. Therefore, it can be used very well in structure-based models.

## Chapter 2. The Systems Engineering Process Model and Others

1. Describe the role of the process model in the systems engineering concept.

   A: Recommendations that have proven their suitability in practice. Thought patterns, references that help to develop and agree upon a proper model for the work of the project team.

2. What are the four components (basic principles) of the systems engineering process model?

   A: From the general to the detail, principle of variant creation, principle of structuring into project phases, problem-solving cycle.

3. What are the individual components focusing on?

   A1: From the general to the detail. Focus: first define, assess, and select rather rough goals and rough solutions. Then detailing, the overall concept is a framework and a guide for detailed development steps.

   A2: Principle of variant creation. Focus: opening the view. Do not be satisfied with the first or nearest solution that comes to mind. Search for alternatives and try to find out their advantages and disadvantages. Make an informed choice.

   A3: Principle of structuring into project phases. Focus: introduce phases into long-term operations; define milestones, decision points for the project team and the management. This reduces risk, projects with little or no promise may be terminated during the early phases. At the beginning, no decision on the realization has to be taken, only on the budget for the first phase (preliminary study).

   A4: Problem-solving cycle. Focus: repeating logic of thinking and acting, which may be used at each stage of the development.

   The elements are: search for objectives – where are we now? What do we want, intend?

   Search for solutions: what ways, solution can be found?

   Selection: which one is preferable, best, for further realization steps?

4. How are these components interconnected, related to each other? Are there any logical relationships between the components that show the modularity of the concept?

   A: Verbal description of Fig. 2.19.

5. What is the difference between project phases and life cycle phases? Does it make sense to see a difference?

A: The project phases focus on the process of development, the realization of a solution, and the subdivision of this process into appropriate procedural steps (preliminary study, main study, detailed studies, system building, implementation, project closure). The systems life cycle focus on different states of the system (in development, in realization, usage, reconfiguration, redesign, decommissioning, removal, disposal, etc.

6. Is it necessary to pass all project phases?

A: No, it depends on the size and duration of a project. In small projects, individual phases may be skipped or combined. In large projects, the number of phases may even be increased to decrease the risk (a preliminary study may be divided into a pre-feasibility and a feasibility study

7. The term "analysis" is embodied twice in the problem-solving cycle. What is meant in one case and in the other case?

A: *Situation analysis* focuses on the initial situation, the problem, the expectations, etc. The *analysis of solutions* deals with solutions (developed in the synthesis step) and their critical analysis.

8. What is the difference between the steps analysis and evaluation of solutions?

A: In the analysis of solutions, the focus is placed on each single variant, which is critically analyzed to assess the fulfillment of (mandatory) objectives, the suitability of a solution, weaknesses and potential for improvements, etc.). Unsuitable solutions are eliminated. The evaluation focuses on the comparison of variants, which are considered as suitable. The aim is to find the best of several suitable variants.

9. What is the difference between actual-state-oriented and desired-state-oriented process models? How or where would you position the Hall/BWI process model?

A1: Actual-state-oriented process models start at the actual state and try to find a better future state. Desired-state-oriented models do not take much into consideration about the actual state at the beginning. One first tries to find an ideal solution and only then is the actual situation examined to find out to what extent the ideal solution can be applied to the actual state. If necessary, the ideal solution must be adapted to the real conditions.

A2: This question relates to the problem-solving cycle. It is to position in between: without knowledge of the actual state one is not considered to be able to formulate sensible objectives. In the search for solutions, it may be appropriate to question the limitations and the boundary conditions of the actual state if that promises new solutions with tangible advantages.

10. What role does the client play in the PSC?

A: Participation in the search for objectives and responsible for the decision – see Fig. 3.13

11. What is meant by the "waterfall model"? To which component of the systems engineering process model does it relate?

   A1: Subdivision of the development and realization of a solution into phases that are limited in time. In the original version, recourses or iterations are not provided.

   A2: Relating to the project phase module

12. What is meant by the term "V-Model"? What are its characteristics?

   A: Also a phase model. The descending branch shows the development "from the general to the detail" (top–down). The ascending branch points in the opposite direction (from the components to the overall system) relating to activities such as production, test, integration, etc. Important are the additionally introduced elements validation, verification – see Fig. 2.9.

13. What is the "prototyping approach"?

   A: Two different perceptions: (i) Simplified and concrete procedure in the sense of a "quick and dirty-solution" that is ready for operation. This approach is rather problematic. (ii) A design aid in the sense of checking whether an approach is working, by using simple and low-cost means. Afterward, make it in a systematic way.

14. What does the term "simultaneous (concurrent) engineering" denote?

   A: The concept seeks an acceleration of the development and implementation process through a parallelization of processes that is as extensive as possible – see Fig. 2.16. Good communication between project teams is essential.

15. What are the characteristics of the so=called "Agile Process Models"? Give some important examples of that category.

   A1: Basic idea is to make development processes (mainly in the IT area) which were regarded as cumbersome and inflexible more agile and adaptable.

   A2: A key trigger and reinforcing factor was the so-called "Agile Manifesto". Important typical representatives are: the spiral model, extreme programming, feature-driven development, Scrum, Crystal, etc.

16. What do you think is better: agile models or plan-driven models?

   A: No general answer is possible. Small projects may be well manageable and controllable with agile methods. With the growing size of projects, an increasing number of formally agreed requirements, safety crucial features, etc., there is a tendency toward plan-driven models.

17. Can you find some agile properties in our systems engineering model in spite of its primarily plan-driven character?

   A: Yes, some examples are: (i) segmentation into four components/modules, which allow the adaption to a specific situation; (ii) principle of thinking in variants; (iii) iterations in the problems-solving cycle; (iv) dynamics of the overall concept (Fig. 2.20), etc.

18. What is the basic idea of the real options approach?

A: Classical approach – method for the evaluation of decisions based on financial mathematical models of the option price theory. Additional considerations: developers and planners should be encouraged to consider additional opportunities of a (partial) reduction or extension of concepts in combination with risk considerations.

19. What kind of risk considerations would you cite as examples?

A1: Avoiding risk – no investment in risky businesses. Diversification strategy – distribution of activities to different businesses (in contradiction with the concept of core competences). Anticipation – try to understand future changes soon and plan measures. Create or increase the scope for action – for instance, by the option to terminate a project if it does not seem to be successful any more (even if the liquidation proceeds do not reach the cumulated costs). Or: extension of a project if it seems to become more successful than expected.

A2: Another consideration is the distinction of:

- Technical risks, financial, deadline risks, personal, economic, ecological risks, etc.
- Or concerning the stage: development risk, production, sales, usage, maintenance, disposal risk, etc.

## Chapter 3. Systems Design

1. What do we mean by saying that systems thinking may be applied to a problem and to the solutions?

A: The term systems design embodies two different activities (see Fig. 3.1)

- *Systems architecting* in terms of developing a fundamental architecture, i.e., configuration of a solution
- *Concept development* as a concrete embodiment of a selected architecture. It deals with the development of a selected architectural design in a more concrete and detailed fashion

2. By which categories can we make elements and relations within a system?

A: see Fig. 3.2

3. What do we mean by saying that systems thinking may be applied to a problem in addition to the solutions?

A: Both allow the basic principles of systems thinking to be applied as: definition of elements, relations, delineation of the system against its environment, relations with environmental elements, etc. (see Fig. 3.3)

4. How does the curve of the development expenses ideally run over time? What are the conclusions to be made?

   A: see Fig. 3.6. A preliminary study is less cost-intensive than the following phases and may reduce the risk of solving the wrong problem or going the wrong way.

5. How does the curve of knowledge of a solution ideally run? What about the curve of admissible ignorance?

   A: Fig. 3.7. Describe in your own words.

6. Do you think it is possible that the knowledge at the beginning of a project may be zero?

   A: Rather unlikely: users know the current situation; specialists know a spectrum of solutions. But: knowledge is located in different persons or groups. Therefore, there is a steeper rise in the curve at the beginning.

7. Do you think it is possible that the admissible ignorance tends toward zero over the course of the project?

   A: No. This would mean that one has 100% knowledge of the solution under all conditions (present and future). But not everything is completely certain: not the assumed situation, not the future development with respect to additional requirements, system behavior, performance of operating staff, reactions of the environment, etc.

8. Is it possible to change/modify the overall concept after the decision at the end of the main study?

   A: Yes. Detailed insight into solutions (during the detailed studies) in addition to external influences may cause such changes/modifications (Fig. 3.8).

9. Are immediate measures in contrast to a systematic systems engineering approach?

   A: Not fundamentally. It may even make sense to reach appreciable effects (quick wins) with rather little effort. But this measure should not prejudge future solutions (much).

10. Explain the different thinking levels of problem-solving with the help of Fig. 3.10.

    A: Verbal description of Fig. 3.10.

11. What kind of information is acquired in the single steps of the PSC and is passed over to the next steps?

    A: See Fig. 3.11.

12. Does the flow of information in the PSC run linear or are there feedback and repetition cycles? If yes, which?

    A: See Figs. 3.12 and 3.14. Describe in your own words.

13. What is meant by the term MBSE?

    A: MBSE means model-based systems engineering. With the help of its methods one tries to accelerate development processes. This means in practice: away from a document-based development (with many Word or Excel files) where the (interim) results can hardly be integrated and kept consistent toward a digitized, model-based approach, where the results are consistent and can be integrated.

14. Which benefits are expected from MBSE? What are the expectations?

    A: Applying MBSE tools and methods, the systems in development are easier to understand, to analyze, to detail, to document, to communicate in the teams, and transfer for further processing. The models help to:

    - Support and accelerate the development process
    - Give all stakeholders access to information on products and their development and keep data consistent within complex applications
    - Facilitate change management, risk and impact analyses
    - Allow for early verification, validation or tests
    - Allow for frontloading approaches with usage contexts, use cases, etc.
    - Allow for the bidirectional traceability of requirements in the process chain
    - Facilitate collaboration and information exchange
    - Increase speed and efficiency
    - Support in mastering complexity
    - Facilitate the reuse of know-how

15. What is meant by the term SysML? Are there any relations to MBSE?

    A: SysML is a standardized graphical modeling language. It supports the requirements for modeling, structure modeling, behavior modeling, and parametrics modeling. Nowadays, SysML is the most widely used modeling language for modeling complex systems in systems engineering.

# Chapter 4. Project Management

1. Explain the term "project". What are the typical characteristics of a project?

    A: See Sect. 4.1.1.

2. Give typical examples of projects.

    A: See Sect. 4.1.1.

3. What kinds of activities would you not call a project?

   A: Repeating and ongoing tasks. Minor, short activities. Tasks that may be done by one person, etc.

4. Explain the term "project management."

   A: See Sect. 4.1.2

5. What is the difference between "systems design" and "project management"? Does it mean different jobs for different persons?

   A: (i) See Fig. 4.2 and (ii) no, just a mental differentiation.

6. What kind of message should be transported by the "iron triangle"?

   A: See Fig. 4.3. Describe in your own words.

7. Which tasks are summarized under the term "project management"?

   A: See Sects. 4.1.4 and 4.2.

8. What do we mean by institutional organizational aspects of project management?

   A: Steering committee, project manager, project team (see Fig. 4.4).

9. Which ideal typical forms of project organization do you know?

   A: Staff PO, dedicated (task force) PO, matrix PO.

10. What are the characteristics, the advantages, and disadvantages of these forms of project organization?

    A: See Sect. 4.3.2.

11. Which type of project organization is suitable for which type of project?

    A: See Fig. 4.9.

12. Which methods, techniques, tools for planning, and control of projects do you know?

    A: Work breakdown structures for structuring tasks, linear responsibility charts for dividing and assigning tasks among persons, network plans (CPM, PERT, DSM) and bar charts (Gantt charts) for planning and monitoring (intermediate) deadlines for projects, time calculations, ascertaining the critical path, etc., can also be effected with IT support, resource planning, for example, on the basis of project ratios from concluded projects (e.g., ratio scope/expenditure), time–costs–progress diagrams, progress reports, and much more. See Chap. 14 and Fig. 14.2.

13. What are typical demands of a project manager? From the management perspective, from the perspective of the team members, etc.?

    A: See Sect. 4.5.1.

14. What are typical attributes of a high performing team?

    A: See Sect. 4.5.2

15. In which direction may project management evolve in response to the emergence of agile methods?

    A: Systems thinking and three of the four submodels of the systems engineering process model will *remain important:* (i) from the general to the detail, (ii) the principle of variant creation, and (iii) the problem-solving cycle.

Only the interpretation of the phase model may change. It may be replaced by sprints, which are shorter than the phases, but have to follow the standard modules (i) and (ii).

# Chapter 5. Systems Architecting

1. How would you define the term "structure" of a system?

   A: Structure of a system is the arrangement of elements and their relationships with one another (see Sect. 1.1.2.3).

2. How would you define the term "architecture"?

   A: The term architecture denotes something that goes beyond the term structure; namely, the allocation of functions to the elements of a structure.

3. Give your own examples of architectural variants

   A: Give your own examples analogous to Sect. 5.1 (such as direct ascent versus earth orbit rendezvous versus lunar orbit rendezvous as variants for the manned moon mission. Or front-wheel drive versus rear-wheel drive)

4. Differentiate between the terms "form" and "function".

   A: See Fig. 5.2

5. At which level of a product hierarchy should an architecture be developed?

   A: At each level: final product (car), drive technology (engines), powertrain, etc.

6. Give a few examples of principles of good design characteristics for good architectures.

   A: Changeability, integrability, scalability, decentralization, autonomy, modular design, ideality, minimal prejudice, etc.

7. Sketch the Henderson/Clark model for architectural innovations in the context of other innovations.

   A: See Fig. 5.3

8.  What is the objective of architectural design?

    A: To develop an architecture that meets a defined value and purpose.

## Chapter 6.1. Situation Analysis

1.  What is the purpose of a Situation Analysis? Is it equivalent to a review of the current situation?

    A1: The purpose of a situation analysis is:

    *   To make the situation you are faced with clear, descriptive, and plausible, which means that you understand the problem and its manifestations, the underlying causes, relationships, that you see ideas, possible solutions, chances, opportunities, etc. Therefore, it gets easier to understand your task, the need for changes, the direction of thought, etc.
    *   To be able to structure and to define the problem area
    *   To define the area of intervention and/or the field of design
    *   To develop the information base for the following steps "formulation of objectives" and "search for solutions" (Fig. 6.2)

    A2: No, a situation analysis is much more than a review of the current situation (see above).

2.  Which views in the situation analysis do you know and consider characteristic and important?

    A: Systems-oriented, cause-oriented, solution-oriented, and time-oriented view.

3.  How are detected weaknesses or failures, conceivable causes, and measures to resolve the problems mentally and logically interconnected?

    A: See Fig. 6.6.

4.  Is it accepted to discuss solutions as early as in the situation analysis?

    A: Yes, if you are not talking about details of the design, but rather try to support the right understanding of the problem within the team (a problem as the difference between the actual state and a possible future state). In this case, the discussion of a solution may help to understand what one sees as the main difference compared with the present situation.

5.  Is it possible that there are different boundaries one has to observe within a special task?

    A: Yes, for instance, the borders of the problem area, of the intervention area, of the solution area, of the area of effects (Fig. 6.7).

6. Which techniques and tools do you know about for situation analysis?

   A: Techniques for acquiring information, methods and tools for processing, and representing information, etc. For details, see Sect. 6.1.7.

7. Does it make sense to use a working hypothesis in the situation analysis? Does this not mean that one works with prejudices?

   A: Working hypotheses are often based on experience, which helps to give structure to a situation, to take short cuts, to facilitate and support analyses, etc. But one should not forget to look critically at the hypothesis, whether it may be justified in this situation, too, by facts, by reasonable developments in the past, etc. One should not have a dogmatic attitude by meaning that he/she is already the holder of the truth. As soon as information and arguments are found that do not support the current working hypothesis but rather contradict it, one should move away from the hypothesis and look for a more accurate picture.

8. Does the situation analysis have the same significance in all phases of a project?

   A: No, they vary in different phases in terms of the scope of consideration and the degree of detail. The scope is larger at the beginning of a project and first at a low level of detail. This focus reverses as time progresses. In a detailed study, the focus is rather narrow and the degree of detail becomes greater (Fig. 6.9).

## Chapter 6.2. Formulation of Objectives

1. What is a goal, an objective?

   A: A statement about what is to be achieved or avoided by a solution.

2. How are the steps "formulation of objectives," "situation analysis," "synthesis/ analysis," and "evaluation" connected, linked together?

   A: Situation analysis provides situational awareness for the step formulation of objectives in an effects-based approach. The step formulation of objectives provides compulsory objectives (C = killer requirements) and recommended (R) objectives for the step search for solutions and recommended (R) objectives, and desired (D) objectives for the evaluation (Fig. 6.9).

3. Why do compulsory objectives not play a role in the "evaluation" step?

   A: Variants that do not meet the compulsory objectives have to be eliminated in the step analysis (or the conflicting compulsory objectives need to be reformulated or eliminated).

4. What is meant when one demands that objectives have to be formulated in an operational way?

   A: Objectives should be readily comprehensible (what is meant?) and its achievement should be ascertainable. This is usually the case with quantifiable objectives. But there also are examples of nonquantified objectives, which are fully operational (e.g., "reservations by phone should also be possible").

5. Why does it generally make sense to distinguish between system objectives and project course (procedural) objectives?

   A: System objectives describe the positive effects, properties, etc., of solutions as soon as they are ready and available and project course (procedural) objectives describe important characteristics of the way (milestones, project budgets, etc.).

6. Can certain demands be objectives in addition to means?

   A: Yes, see the hierarchy of objectives and means (Fig. 6.11).

7. Does it make sense to use the term "objective" (here: *objective* as the opposite of *subjective*) goal/target/aim/?

   A: No. Each formulation of goals or targets comprises verifiable facts (which ought to be "objective") in addition to "subjective" values (interests, personal priorities, etc.).
   And: one should not equate "quantified" with "objective" goals.

8. What is meant by demanding that objectives have to be solution-neutral?

   A: Objectives should describe the WHAT, the desired effects, and not the HOW, the solution itself.

9. How can we give different importance to objectives?

   A: For example, by distinguishing among compulsory, recommended, and desired objectives, or by assigning different weights to objectives.

10. What can one do if goals oppose or contradict each other? Give examples.

    A1: Example "opposing": the more goal A is achieved (low cost), the less goal B (quality) can be achieved.
    Way out: fix an upper cost limit and demand a maximum quality for that. Or: fix a minimum quality level at the lowest possible cost.
    A2: Example "contradictory": push market X and give up product Z (which has its sales focus in market X).
    Way out: eliminate one of the goals.

## Chapter 6.3. Search for Solutions Synthesis/Analysis

1. What does synthesis mean? What does analysis mean?

   A: Synthesis = development of solutions; analysis = critical examination of solutions in terms of goal fulfilment, suitability, etc.

2. What is creativity and what is its role in the search for solutions?

   A: Creativity is the human capability to create ideas, to generate concepts or products of any kind, which are new regarding essential characteristics, and are new to the creator.

3. Explain Fig. 6.15 relating to the idea of abstraction and concretization

   A: Zooming out from the real world = creating a figure by abstracting. Zooming a model into reality = creating a solution by concretization

4. Does the innovative character increase or decrease during the course of a project?

   A: It rather decreases. New concepts should increasingly be realized by using known, manageable, controllable, and mastered methods or components.

5. Which typical search strategies for solutions do you know?

   A: Linear strategies, optimizing, non-optimizing, single-step optimizing search strategies, multi-step optimizing, cyclical search strategies (see Sect. 6.3.3.3).

6. Give some examples of methods, tools, and techniques for the search of solutions.

   A: Brainstorming, morphology, analogy method, bionics, synectics, TRIZ/TIPS, etc.

7. What is the logic in differentiating between an intuitive and a formal analysis?

   Intuitive A: the very moment an idea appears for a solution it is usually viewed critically. All creative methods try to mitigate this effect by claiming "postponed judgements."
   Formal A: important planning results are systematically and critically analyzed by means of concrete queries to improve or to eliminate a solution.

8. What are the contents of a systematic analysis of solutions?

   A: We subdivide into six groups: (i) formal aspects (assessability, fulfillment of compulsory objectives), (ii) integrability (into the environment, into higher-level concepts, looking outward), (iii) functions and processes (look inward), (iv) operational efficiency (usability, ease of operation, maintainability, safety, reliability), (v) prerequisites, conditions (infrastructure, financial, personnel, ability to implement, practicability), (vi) consequences (cost, risk, impact on other solutions).

## Chapter 6.4. Evaluation and Decision

1. Which steps in the PLC are the main information suppliers for the evaluation

   A: (i) Search for objectives (recommended and desirable objectives as evaluation criteria); (ii) search for solutions (practicable, suitable solutions, maybe
   additional evaluation criteria that were figured out during the search for
   solutions).

2. What is the essential difference between analysis and evaluation?

   A: analysis = critical assessment of every single solution with respect to appropriateness or suitability; evaluation = comparative assessment of suitable
   solutions with respect to the preference for decision-making.

3. Are improvised (intuitive) decisions acceptable or should decisions always be
   methodically supported?

   A: Decisions may be improvised if there is not much risk. That is the case, if:
   (i) consequences are minor, (ii) possible solutions differ greatly and the
   champion can easily be seen, (iii) if the differences are minor so that it does
   not matter which solution one is choosing, (iv) if the following procedure
   may easily be affected subsequently.

4. Which methods of comparative assessment do you know?

   A: Balance of arguments, evaluation matrix, value-benefit analysis, cost-
   effectiveness analysis.

5. Why does it make sense to limit the sum of weights of the criteria (100 or
   1000)?

   A: It forces us to look for the overall context (where can we get weights from,
   if we want to introduce an additional criterion? Where do we give points if
   we eliminate a criterion?). The method itself of course works with every sum
   of weights. But: (i) usually we makes more careful use of limited resources
   and (ii) if we add a weight of 20 to a newly inserted criterion – in a total
   weight sum of 100 – this is equivalent to a devaluation of all other criteria by
   20% (inflation, deflation). Perhaps that is not what we really want to do.

6. Why does it make sense to represent the course of a utility function on a graph?

   A: By inserting the scores (from 1 to 5 or 0 to 10) into the graphics, one
   expresses one's values in the evaluation process to the team and offers a
   genuine discussion.

7. Why is it recommended to conduct reasonability analyses or sensitivity analyses in the evaluation process?

   A1: Reasonability analyses – compare the results calculated by way of calculation (head) with an intuitive appraisal (gut feeling). It is open with regard to

how it will turn out: maybe the gut feeling turns out to be a prejudice, maybe it includes important criteria that were not part of your evaluation matrix.

A2: Sensitivity analyses – helps to find out what the result of the evaluation looks like if you change the weights given to the criteria (within reasonable scope) or the scores to several variants.

8. Are the methods for comparative assessment "objective"?

A: No, this is shown by taking the evaluation matrix as an example – all variants you are going to evaluate are guided by properties that have been induced by your goals (partly subjective). The selection of criteria is subjective in addition to the weights (importance) you give. The scores you give are subjective too. Objectives are only the arithmetical operations (multiplication, addition).

9. Why use evaluation methods that are not "objective"?

A: It is a stopgap solution that is necessary because there is no objective value of any solution and therefore no objective method. There is only a value in a certain context. The function of these methods is to make this context apparent, transparent, and arguable. When using a method, you not only deliver a result of the evaluation, but also a brief justification for your recommendation for a decision: (i) which variants compared, (ii) which criteria used, (iii) the weights given to each criterion, (iv) the scores given to each variant in respect of a certain criterion.

If one does not agree with the result, one is free to contradict. But by doing, this one should comply with the rules of the game (= method), which, for example, means that if one wants to give more weight to a special criterion, one should tell where to get it from, where to reduce the weight.

10. Is it possible for the decision-making body to deviate from the proposal of the project group?

A: Of course, but it is less likely if the distance between the project team and the decision body is not too big. That means that at least one representative of the decision body meets with the project team regularly or is asked to participate (as a kind of godfather) in an important meeting, if his or her opinion is asked.

# Chapter 6.5. Special Cases and Situational Interpretation

Do you know any other special cases you would find worth discussing? The authors would like to know your views.

## Chapter 7. Systems Engineering Basics in Our Systems Engineering Concept

– Do you agree on the basics summarized in this chapter? Or are there too few, too many or the wrong ones?
– The authors would like to receive feedback.

> A: See text in Chap. 7.

## Chapter 8. Case Study 1: Private House Building: Additional Domicile

1. Can you find the aforementioned basics in the case study? see (Chap. 7).

## Chapter 9. Case Study 2: Airport Planning

1. Can you find the above-mentioned basics in the case study? see (Chap. 7).

## Chapter 11. Seven Basic Recommendations

1. Do you agree to the recommendations?
2. Would you modify, extend, or reduce the list? The authors would like to hear about your judgment.

## Chapter 12. Typical Weak Areas in Projects (Stumbling Blocks)

1. Does the list presented in Fig. 12.1 reflect your experience? Do the arguments apply to it?
2. Do you have suggestions for the modification, extension or reduction of the list that arise from your experience? The authors would like you to share your experience with them.

## Chapter 14. Characteristics of Successful Project Management

1. Are these considerations in line with your experience?
2. Are you missing something?

## Chapter 15. Survey of methods and tools

1. Name at least two methods for each step of the problem solving cycle
2. Which M&Ts have you already applied in your projects?

## Chapter 16. Encyclopedia/Glossary

1. Do you think this glossary is useful, even nowadays in the age of Wikipedia, Google, etc.? Why or why not?
2. How many percent of the described methods and tools were familiar for you?

# Bibliography

Ackoff, R. L. (1971): Towards a System of Systems Concepts. Management Science Vol.17, No.11, July 1971, p. 661–671. Philadelphia: University Press

Agile Project Management: Best Practices and Methodologies (n.d.) https://www.altexsoft.com/whitepapers/agile-project-management-best-practices-and-methodologies/

Alexander, C. (1964): Notes on the Synthesis on form. Cambridge MA: Harvard University Press

Altexsoft: Agile Project Management: Best Practices and Methodologies (n.d.) https://www.altexsoft.com/whitepapers/agile-project-management-best-practices-and-methodologies/

Altshuller, G. S. (1984): Creativity as an Exact Science – The Theory of the Solution of Inventive Problems. New York: Gorden and Breach

Altshuller, G. (1999): The Innovation Algorithm: TRIZ, systematic innovation, and technical creativity. Worcester, MA. Technical Innovation Center.

Amram, M.; Kulatilaka, N. (1999): Real Options: Managing strategic investment in an uncertain world. Boston, MA: Harvard Business School Press

Angermann, L. ed. (2011): Numerical Simulations – Applications, Examples and Theory. Publisher: InTech

Anthony, S. D., Johnson, M. W., Sinfield, J. V., Altman, E. J. (2008). The Innovator's Guide to Growth: Putting Disruptive Innovation to Work. Boston. Harvard Business Press.

Armstrong, J. S. (ed.) (2001). Principles of forecasting: a handbook for researchers and practitioners. Norwell, Massachusetts: Kluwer Academic Publishers

Asmussen, S. and Glynn, Peter W.(2007): Stochastic Simulation: Algorithms and Analysis, Springer.

Austin, Robert D. (1996): Measuring and Managing Performance in Organizations. Dorset House Publishing

Awrejcewicz, J. ed. (2011): Numerical Simulations of Physical and Engineering Processes. Publisher: InTech

Bahill, A. T.; Gissing, B. (1998): The Systems Engineering Process. In: Re-evaluating systems engineering concepts using systems thinking, IEEE Transaction on Systems, Man and Cybernetics, Part C: Applications and Reviews, 28(4), 516–527, 1998 (see also http://g2sebok.incose.org/)

Bahill, T.; Botta, R. (2008): Fundamental Principles of Good System Design. Engineering Management Journal, 20/4, pp. 9–17

Bahill, T. (2009): Slides for Bahill's Lectures at the Department of Systems and Industrial Engineering, University of Arizona: http://www.sie.arizona.edu/sysengr/slides/. 9. Okt. 2009

Bahill, A. T. and Madni, A. M. (2017): Tradeoff Decisions in System Design. Springer Nature

Baldick R. (2006): Applied Optimization. Formulation and Algorithms for Engineering Systems. Cambridge University Press.

Baldwin, C.; Clark, K. (2000): Design Rules, Volume I. Cambridge, MA: The MIT Press

Bandte, H. (2007): Komplexität in Organisationen: Organisationstheoretische Betrachtungen und agentenbasierte Simulation. Wiesbaden. Deutscher Universitätsverlag, GWV Fachverlage GmbH.

Banks, J., Ed. (1998): Handbook of Simulation: Principles, Methodology, Advances, Applications, and Practice, John Wiley & Sons, New York, NY.

Banks, J., J.S. Carson, B.L. Nelson, and D.M. Nicol (2005), Discrete-Event System Simulation, Fourth Edition, Prentice-Hall, Upper Saddle River, NJ.

Bashir S., Rizwan M., Qureshi J. (2012): Hybrid Software Development Approach For Small To Medium Scale Projects: Rup, XP & Scrum http://www.sci-int.com/pdf/197094943510-Integration%20of%20XP-RUP-Scrum%20Rizwan%20Jamil%20-S381-384.pdf

Beam, W. (1990): Systems Engineering, Architecture and Design. New York: McGraw-Hill Publishing

Beck, K. et al. (2001): Manifesto for Agile Software Development, http://agilemanifesto.org/ 2001

Beck, K.; Andres, C. (2008): Extreme programming explained: Embrace change, 2nd Edition. Reading, MA: Addison-Wesley

Beer, S. (1959): Cybernetics and Management, English Universities Press

Beer, S. (1966): Decision and Control. London, Wiley

Bergsjö, Dag (2009). Product Lifecycle Management – Architectural and Organisational Perspectives. Chalmers University of Technology

Berkun, Scott (2008): Making Things Happen: Mastering Project Management. O'Reilly Media Inc.

Berkun, S. (2007): The Myths of Innovation. O'Reilly Media, Inc.

Blanchard, B. S.; Fabrycky, W.J. (2014): Systems Engineering and Analysis, 5th ed. Pearson

Boehm, Barry (1988): A Spiral Model of Software Development and Enhancement. IEEE Computer, S. 61–72,

Boehm, B.; Turner, R. (2004): Balancing Agility and Discipline. A Guide for the Perplexed. Boston, San Francisco. New York, Addison-Wesley

Bogan, C.E.; English, M.J. (1994): Benchmarking for Best Practices: Winning through Innovative Adaptation. New York: McGraw Hill.

Bohl, Marilyn and Rynn, Maria (2007): Tools for Structured and Object-Oriented Design, Prentice Hall.

Bordegoni, M.; Rizzi, C., eds. (2011): Innovation in Product Design: From CAD to Virtual Prototyping. Springer

Boxwell, R.G. (1994): Benchmarking for Competitive Advantage. New York. McGraw Hill.

Brandenburger, A M.; Nalebuff B. J. (1996): Co-Opetition. Currency Doubleday

Brealey R A., Myers S C. and Allen F. (2013): Principles of Corporate Finance, 8th Edition. McGraw-Hill/Irwin

Brisley, Chester L.: Work Sampling and Group Timing Technique. In: Zandin, Kjell B. (2001): Maynard's Industrial Engineering Handbook. 5. Ed. New York: McGraw-Hill

Brown, S. F. (2004): Toyota's Global Body Shop, Fortune Magazine, 2/9/2004, Vol. 149, Issue 3, pp. 120–123

Brown, T. (2008): Design Thinking. In: Harvard Business Review, June 2008, p. 84–92

Brown, T. (2009): Change by Design: How Design Thinking Transforms Organizations and Inspires Innovation. Harper Business

Browning, T. (2001): Applying the Design Structure Matrix to System Decomposition and Integration Problems: A Review and New Directions. In: IEEE Transactions on Engineering Management. 48(3), 2001, pp. 292–306

Bryman A. (2008): Social Research Methods. Oxford University Press

Büchel, A. (1969): Systems Engineering. Eine Einführung. Industrielle Organisation 38 (1969) Nr. 9, S. 373–385

Buzan, Tony (2010): The Mind Map Book : Unlock your creativity, boost your memory, change your life, Pearson Education

Cadle, J; Paul, D. and Turner, P. (2010): Business Analysis Techniques. 72 Essential Tools for Success. BCS The Chartered Institute for IT

Canetta, L.; Redaelli, CL.; Flores, M. eds. (2011): Digital Factory for Human-oriented Production Systems. The Integration of International Research Projects. London: Springer

Cellier, F.E. and Kofman, E. (2006): Continuous System Simulation, New York, Springer.

Chapman, Ch. S (2005): Controlling Strategy. Management, Accounting, and Performance Measurement. Oxford University Press.

Checkland, P. (1999): Systems Thinking, Systems Practice: Includes a 30-Year Retrospective. Chichester: John Wiley & Sons

Chesbrough, H.W. (2006): Open Business Models: How to Thrive in the New Innovation Landscape. Harvard Business Press.

Chesbrough, H.W. (2005): Open Innovation: The New Imperative for Creating and Profiting from Technology. Harvard Business Press.

Chestnut, H. (1965): System Engineering Tools. Chichester: John Wiley & Sons

Chestnut, H. (1967): System Engineering Methods. Chichester: John Wiley & Sons

Chrissis, M.B.; Konrad, M.; Shrum, S. (2011, 3rd Edition): CMMI for Development: Guidelines for Process Integration and Product Improvement. Pearson Education Inc.

Christensen, C. (2003): The Innovator's Dilemma: The Revolutionary Book that Will Change the Way You Do Business. Harper Paperbacks

Christensen, C.; Raynor, M.E. (2003): The Innovator's Solution: Creating and Sustaining Successful Growth. Harvard Business Press

Christensen, C.B.; Beard. S. (2000): Iridium: Failure & Successes. International Astronautical Federation, Proceedings of the 51st Internat. Astronautical Congress, Rio de Janeiro, Brazil, Oct. 2–6, 2000

Christensen, C.M.; Grossman, J.H., Hwang J. (2009): The Innovator's Prescription: A Disruptive Solution for Health Care. McGraw-Hill.

Christensen, C.M.; Johnson, C.W.; Horn, M.B. (2008): Disrupting Class: How Disruptive Innovation Will Change the Way the World Learns. McGraw-Hill

Churchman, C. W. (1979): The Systems Approach. New York: Delacorte Press

Clausing, D; Fey, V. (2004): Effective Innovation. The Development of Successful Engineering Technologies. London: Professional Engineering Publishers

Clifton, B.C.; Bird, H.M.B.; Albano, R.E.; Townsend, W.P. (2004) Target Costing: Market-Driven Product Design. New York. Marcel Dekker, Inc;

Coad, Peter; Lefebvre Eric; De Luca, Jeff (1999): Java Modeling in Color with UML: Enterprise Components and Process, Prentice Hall

Cockburn, Alistair (1998): Surviving Object-Oriented Projects. Boston: Addison Wesley

Cockburn, A. (2004): Crystal Clear: A Human-Powered Methodology for Small Teams. Addison-Wesley Professional.

Cockburn, Alistair (2005): Crystal Clear. Boston: Addison Wesley

Cockburn, A. (2006): Agile Software Development. Software Development: The Cooperative Game. Pearson Education

Colgan St. (2009): Joined-Up Thinking. Pan Books.

Conklin J. (2006): Dialogue Mapping. Building Shared Understanding of Wicked Problems. Wiley

Copeland, T.; Antikarov, V. (2001): Real Options: A practitioner's guide. New York: Texere Publishing Ltd

Cornelius, P.; Van de Putte, A. and Romani, M. (2005): Three Decades of Scenario Planning in Shell. California Management Review, Nov. 2005

Covey St. R. (2004): The 7 Habits of Highly Effective People. Touchstone

Crawley, E. (2009): Systems Architecting. Course Material 2000–2009. Cambridge, MA: MIT Press

Crawley, E.; Cameron, B.; Selva, D. (2015): System Architecture: Strategy and Product Development for Complex Systems. Pearson.

Csikszentmihalyi, M. (2013): Creativity: Flow and the Psychology of Discovery and Invention. Harper Perennial

Dantzig, G. B. (1963): Linear Programming and Extensions. Princeton University Press, Princeton.

Davenport, T. H. (1993). Process Innovation: Reengineering Work Through Information Technology. Boston MA: Harvard Business School Press

De Bono, E. (1970): The Use of Lateral Thinking. Harper & Row, New York

De Luca, Jeff; Coad, Peter; Lefebvre, Eric (1999): Java Modeling in Color with UML: Enterprise Components and Process. Upper Saddle River, NJ: Prentice Hall Ltd

de Weck, O. L., de Neufville R.; Chaize M. (2004): Staged Deployment of Communications Satellite Constellations. In: Low Earth Orbit, Journal of Aerospace Computing, Information, and Communication, Vol. 1, No.3, pp. 119–136

Delligatti, Lenny (2013): SysML Distilled: A Brief Guide to the Systems Modeling Language. Addison-Wesley Professional

DeMarco, T.(1997): The Deadline. A Novel About Project Management. Dorset House

DeMarco, T.; Lister, T.R. (1987): Peopleware: Productive Projects and Teams. Dorset House

DeMarco, T.; Lister, T.R. (2003): Waltzing With Bears: Managing Risk on Software Projects. Dorset House.

Deming, W.E. (1997): *Out of the Crisis*, Cambridge, Mass.: The MIT Press

Department of Defense (2008). Office of the Deputy Under Secretary of Defense for Acquisition and Technology: Systems Engineering Guide for Systems of Systems. Version 1.0. August 2008

Dhillon Balbir, S.; Reiche, H. (1985): Reliability and maintainability management. Van Nostrand Reinhold.

DIN 55350-11:2008-05: Concepts for quality management – Part 11: Supplement to DIN EN ISO 9000:2005

Dixit, A.K.; Pindyck, R.S. (1994): Investment under Uncertainty. Princeton University Press

Dori, D. (2002): Object-Process Methodology. Berlin: Springer

Dori, Dov (2016): Model-Based Systems Engineering with OPM and SysML. Springer

Dörner, D. (1980): Heuristics and Cognition in Complex Systems. In: Groner, R., Groner, M. & Bischof, W.F. (Eds.): Methods of Heuristics. Hillsdale New York: Erlbaum.

Dörner, D. (1980): On the Difficulties People have in Dealing with Complexity. Simulation & Games, 11, 87–106.

Dörner, D. (1997): The Logic Of Failure: Recognizing And Avoiding Error In Complex Situations. Basic Books

Drevdahl, J. E. (1956): Factors of importance for creativity. In: Journal of Clinical Psychology. Nr. 12, 1956, S. 21–26.

Drucker P. (1955): The practice of management, London: Heinemann

Durward K. Sobek II, Allen C. Ward and Jeffrey K. Liker (1999): Toyota's Principles of Set-Based Concurrent Engineering. MIT Sloan, Management Review, Winter 1999. (http://sloanreview.mit.edu/article/toyotas-principles-of-setbased-concurrent-engineering/)

Edivandro C. Conforto; Fabian Salum; Daniel C. Amaral; Sérgio Luis da Silva and Luís Fernando Magnanini de Almeida (2014): Can Agile Project Management Be Adopted by Industries Other than Software Development? https://www.researchgate.net/profile/Luis_Almeida23/publication/262809231_Can_Agile_Project_Management_Be_Adopted_by_Industries_Other_than_Software_Development/links/569a630208aeeea9859c4df1.pdf

Eilam, Eldad (2005). Reversing: secrets of reverse engineering. John Wiley & Sons

Eiselt, H.A. (2012): Operations Research. A Model Based Approach. Springer.

Eppinger, St., Ulrich, K. (2003): Product Design and Development. Columbus: McGrawHill Irwin

Eppinger, St.; Browning, T. (2012): Design Structure Matrix Methods and Applications. MIT Press Books

Ericson, Clifton A. (2011): Fault Tree Analysis Primer. CreateSpace Inc.

Evans, J.R., Lindsay, W.M., 1995. The management and control of quality, 3rd edn. West Publishing, New York.

Eversheim, W. (2008): Innovation Management for Technical Products: Systematic and Integrated Product Development and Production Planning. Springer

Farris, P W.; Bendle N T.; Pfeifer Ph E; Reibstein D J. (2010). Marketing Metrics: The Definitive Guide to Measuring Marketing Performance. Upper Saddle River, New Jersey: Pearson Education, Inc

Feibel B J. (2003): Investment Performance Measurement. New York: Wiley

Feigenbaum, A.V. (1991): Total Quality Control. McGraw-Hill

Fey, V., Rivin, E. (2005): Innovation on Demand: New Product Development Using TRIZ. Cambridge University Press

Field, A. (2013): Discovering Statistics using IBM SPSS Statistics. Sage Publ.

Finkelstein, M. (2008): Failure Rate Modelling for Reliability and Risk. London. Springer.

Fisher, R.; Ury, W. and Patton, B. (1991). Getting to Yes: Negotiating Agreement Without Giving In. Second Edition. New York: Penguin Books.

Forbes, P. (2005): The gecko's foot. Bio-inspiration – Engineered from nature. W.W. Norton & Co, London

Forrester, J. W. (1977): Industrial Dynamics (9. ed.). Cambridge MA, The MIT Press

Forsberg, K. and Mooz, H. (1991): The Relationship of Systems Engineering to the Project Cycle. First Annual Symposium of the National Council On Systems Engineering (INCOSE)

Forsyth D.R. (2009, 5.ed.): Group Dynamics. Cengage Learning.

Fossa, C. E., e.a. (1998): An overview of the Iridium low earth orbit satellite system, Proceedings of IEEE 1998. National Aerospace and Electronics Conference; (A99-17228 03-01): pp. 152–159.

Foster, Provost; Fawcett, Tom (2013): Data Science for Business. What You Need to Know about Data Mining and Data-Analytic Thinking. O'Reilly Media.

Fricke, E.; Gebhard, B.; Negele, H.; Igenbergs E. (2000): Coping with changes: Causes, findings, and strategies. Systems Engineering, Vol. 3, Issue 4, 2000, p. 169–179

Fricke, E.; Schulz, A. (2005): Design for Changeability – Principles to Enable Changes in Systems Through Their Entire Lifecycle. Journal on Systems Engineering Vol. 8 Issue 4, 2005, pp. 279–341

Friend, J.; Hickling A. (1987): Planning under pressure. Oxford: Pergamon Press

Gallagher B. P.; Phillips, M.; Richter, K.J; Shrum. S. (2009): CMMI-ACQ. Guidelines for Improving the Acquisition of Products and Services. Addison-Wesley.

Gerardin, L. (1968): Bionics. McGraw-Hill

Giambene, Giovanni (2014): Queuing Theory and Telecommunications Networks and Applications. Springer

Gigerenzer, G. (2008): Gut Feelings. The Intelligence of the Unconscious. Penguin

Gigerenzer, G.; Todd P. M. (2000): Simple Heuristics That Make Us Smart (Evolution and Cognition). Oxford University Press, New York.

Gitlow, Howard Seth; Levine, David M. (2005): Six Sigma for Green Belts and Champions: Foundations, Dmaic, Tools, Cases, and Certification. Pearson Education.

Gloger, Boris (2014): https://borisgloger.com/wp-content/uploads/2014/07/Whitepaper-Hardware. pdf?882268

Goldratt, E. M. (1997): Critical Chain. Great Barrington MA. The North River Press.

Goode, H.; Machol R. E. (1957): Systems Engineering. New York: McGraw-Hill

Goodwin, P.; Wright, G. (2004): Decision Analysis for Management Judgment (3rd ed.). Chichester: Wiley

Gorbea, C., Fricke, E., Lindemann, U. (2008): The Design of Future Cars in a New Age of Architectural Competition. New York: Proc. ASME IDETC/CIE

Gordon, William J.J. (1961): Synectics: The Development of Creative Capacity. New York. Harper & Row Publ.

Gordon N J.; Fleisher, W L. (2011): Effective Interviewing and Interrogation Techniques. Elsevier Ltd

Grosskopf, A.; Decker, G.; Weske, M. (2009): The Process: Business Process Modeling Using BPMN.

Groover, M. P.: Work Systems and Methods, measurement, and Management of Work. Pearson Education International, 2007

Gross, Donald; John F. Shortle, James M. Thompson, Carl M. Harris (2008): Fundamentals of Queueing Theory. John Wiley&Sons

Gryna, F M. (2001): Quality Planning and Analysis: From Product Development Through Use. McGraw-Hill

Guindon, R. (1990): Designing the Design Process: Exploiting Opportunistic Thoughts. Journal Human-Computer Interaction archive, Volume 5 Issue 2, June 1990, p. 305–344

Haberfellner, R.; de Weck, O. (2005): Agile SYSTEMS ENGINEERING versus AGILE SYSTEMS engineering. In: Proceedings of the 15th International Symposium INCOSE (CD-version), Rochester N. Y., July 2005 (PDF, 385kb)

Hale, R.; Whitlam, P. (1995): Target setting and goal achievement, London: Kogan Page

Hall, A.D. (1962): A Methodology for Systems Engineering. Princeton N. J.: Princeton University Press

Hammer, M.; Champy, J.A. (2003): *Reengineering the Corporation: A Manifesto for Business Revolution. Harpers Business Essentials*

Harsanyi, J C. and Selten R A. (1988) General Theory of Equilibrium Selection in Games, vol 1, The MIT Press

Haugan, Gregory T. (2003): The Work Breakdown Structure in Government Contracting. Management Concepts Inc.

Henderson, R., Clark, K. (1990): Architectural Innovation. The Reconfiguration Of Existing Product Technologies and the Failure of Established Firms. Administrative Science Quarterly. 35/1. p. 9–30

Hester, R E. and Harrison, R M., eds. (1998): Risk Assessment and Risk Management. Cambridge. The Royal Society of Chemistry, Thomas Graham House.

Hillier, F.S; Lieberman, G.J. (2010): Introduction to Operations.. (9th ed.): New York: McGraw-Hill

von Hippel, E. (1994): The Sources of Innovation. Oxford University Press

von Hippel, E. (2006): Democratizing Innovation. The MIT Press.

Hirano, Hiroyuki and Makota, Furuya (2006): JIT Is Flow: Practice and Principles of Lean Manufacturing. PCS Press, Inc

Hitoshi, Takeda (2006): The Synchronized Production System: Going Beyond Just-in-time Through Kaizen. Kogan Page Publishers

Hoda, R.; Noble, J.; Marshall, S. (2008): Agile Project Management. https://nzcsrsc08.canterbury. ac.nz/site/proceedings/Individual_Papers/pg218_Agile_Project_Management.pdf

Hommes, Bart-Jan; van Reijswoud Victor (2000):The evaluation of business process modeling techniques. Proceedings of the 33rd Hawaii International Conference on System Sciences – 2000

Hopkin, P. (2012): Fundamentals of Risk Management: Understanding, Evaluating and Implementing. Kogan Page Ltd.

Horn, R. E. (2001): Knowledge Mapping for Complex Social Messes, a Stanford University presentation to the "Foundations in the Knowledge Economy" conference at the David and Lucile Packard Foundation, July 16, 2001

Howell David; Windahl Charlotta; Seidel Rainer (2010): A project contingency framework based on uncertainty and its consequences. International Journal of Project Management, Volume 28, Issue 3, April 2010, Pages 256–264 https://www.econbiz.de/Record/a-project-contingency-framework-based-on-uncertainty-and-its-consequences-howell-david/10009508980

Hsu, Chia-Chien & Sandford, Brian A. (2007). The Delphi Technique: Making Sense of Consensus. Practical Assessment Research & Evaluation, 12(10). Human-Computer Interaction, Vol. 5, pp. 305–344. Available online: http://pareonline.net/getvn.asp?v=12&n=10,

IMA: Implementing Target Costing (1994): Published by Institute of Management Accountants Montvale, NJ 07645, IMA Publication Number 98377

Imai, Masaaki (1986): Kaizen: The Key To Japan's Competitive Success.

Imai, Masaaki (2012): Gemba Kaizen: A Commonsense Approach to a Continuous Improvement Strategy. 2. Ed. McGraw Hill

INCOSE Systems Engineering Handbook (2015): A Guide for System Life Cycle Processes and Activities. 4. ed. Wiley

Innes, J. (2009): The Interview Book: Your Definitive Guide to the Perfect Interview Technique. Pearson Education Inc.

IPMA Symposium on Project management (ed. 1997): Managing Risks in Projects. Helsinki

Ishikawa, K. (1990): Introduction to Quality Control. Ed.: 3A Corporation

ISO 9001 in Plain English, (2015) by Craig Cochran

ISO 9001 (n.d.) – What does it mean in the supply chain? Available from: http://www.iso.org/iso/pub100304.pdf

ISO 10007 (2003): Quality management systems – Guidelines for configuration management.

Jänsch, J. and Birkhofer, H.(1993): The Development of the Guideline VDI 2221 -The Change of Direction. International Design Conference – Design 2006 Dubrovnik – Croatia, May 15–18, 2006.

Jensen, Randall W.; Tonies, Charles C. (1979): Software Engineering. Prentice-Hall.

Jiambalvo J., Johnson Frazier, J. (2001): Managerial accounting. Wiley.

Jones, James V. (2006): Integrated Logistics Support Handbook. Mcgraw-Hill Education

Juran, J. M. (2004): *Architect of Quality*. McGraw-Hill.

Juran, J. M. (1995): A history of managing for quality: the evolution, trends, and future directions of managing for quality. ASQC Quality Press

Kahn, H. (1967): The Year 2000. Calman-Levy (1967)

Kahneman, K.; Tversky, A. (2000). Choice, Values, Frames. The Cambridge University Press.

Karlesky, M.; Van der Noord, M, (2008): Agile Project Management (or: Burning Your Gantt Charts). https://www.researchgate.net/profile/Michael_Karlesky/publication/229042037_Agile_Project_Management/links/5512b1c70cf270fd7e3332b1/Agile-Project-Management.pdf

Keeney, R.L.; Raiffa, H. (1976): Decisions with Multiple Objectives; Preferences and Value Tradeoffs. John Wiley & Sons

Keeney, R.L.; Gregory, R.S. (2005). Selecting attributes to measure the achievement of objectives. Operations Research, 53, (pp.1–11).

Kenneth B. Kahn, ed. (2013): The PDMA Handbook of New Product Development. John Wiley&Sons.

Kent, Beck u. a. (2001): Manifesto for Agile Software Development http://agilemanifesto.org/

Keplinger, W. (1991): Merkmale erfolgreichen Projektmanagements. Diss. TU Graz

Kepner, Ch. H.; Tregoe, B. B. (1965). The Rational Manager: A Systematic Approach to Problem Solving and Decision-Making. McGraw-Hill.

Kerzner, Harold R. (2013): Project Management. A Systems Approach to Planning, Scheduling, and Controlling. John Wiley & Sons Inc

Kloppenborg, T.; Petrick, J. (2002): Managing Project Quality. Washington D.C.: Management Concepts Inc.

Koch, R. (2004): Living the 80/20 Way: Work Less, Worry Less, Succeed More, Enjoy More, London: Nicholas Brealey Publishing

Komus, A.; Kuberg, M. (n.d.): Studie "Status Quo Agile" – wie werden agile Methoden in der Praxis eingesetzt? https://www.projektmagazin.de/artikel/studie-status-quo-agile-wie-werden-agile-methoden-der-praxis-eingesetzt_1101303

Kotter, J.P.; Cohen, D.S. (2002). The Heart of Change. Boston: Harvard Business School Publishing.

Krapohl, D. (2013): A Structured Methodology for Group Decision Making. AugmentedIntel.com. AugmentedIntel. Retrieved 26 April 2013.

Kreuter, F.; Presser, St.and Tourangeau, R. (2008): Social Desirability Bias in CATI, IVR, and Web Surveys: The Effects of Mode and Question Sensitivity. *Public Opinion Quarterly*, 72(5): 847–865 first published online January 26, 2009

Kruchten, Ph. (1995): Architectural Blueprints – The "4+1" view model of software architecture. in: IEEE Software 12 (6), pp. 42–50

Kruchten, Ph. (2011): Agile Teenager. https://www.infoq.com/articles/agile-teenage-crisis

Kusiak, A. (1993): Concurrent engineering: automation, tools, and techniques. John Wiley & Sons

Lemieux, Ch. (2009): Monte Carlo and Quasi-Monte Carlo Sampling. Springer

Leveson, Nancy (2013): An STPA Primer http://psas.scripts.mit.edu/home/wp-content/uploads/2015/06/STPA-Primer-v1.pdf

Leveson, Nancy (2012): Engineering a Safer World. Systems Thinking Applied to Safety. MIT Press

Levin H M (1983): Cost-Effectiveness: A Primer. Sage Beverly Hills, California

Lewis, James P. (2004): Project Planning, Scheduling & Control. McGraw-Hill Education (India).

Levitin, Gregory, G. (2005): The Universal Generating Function in Reliability Analysis and Optimization. Springer. London.

Lidwell, W.; Holden, K.; Butler, J. (2010): Universal Principles of Design, Revised and Updated. Rockport Publishers, UK

Lindblom, Ch.E. (1959): The Science of Muddling Through. Public Administration Review, Vol. 19 (1959), No. 2

Lindemann U.; Maurer, M.; Braun, T. (2009): Structural Complexity Management. Berlin: Springer

Linsey J S. and Becker B. (2011): Effectiveness of Brainwriting Techniques: Comparing Nominal Groups to Real Terms. In: Design Creativity 2010. London: Springer London, 2011. 166.

Lyon, D.D. (2000): Practical Configuration Management. Oxford: Butterworth Architecture

Magee, J.F. (1964): Decision Trees for Decision Making. Harvard Business Review. 12 pages. Jul 01, 1964

Maier, M. (1998): Architecting Principles for Systems-of-Systems. in: SystEng 1(4) (1998), 267–284

Mann, D. (2002): Hands on Systematic Innovation: For Technical Systems. CREAX Press

Mann, D. (2004): Hands on Systematic Innovation: For Business and Management. Edward Gaskell Publishers

Mann, P.S. (2012): Introductory Statistics. Wiley

Markman, A.B.; Kristin L.; Wood, K.L. (2009): Tools for Innovation. Oxford University Press.

Marteka, V. (1965): Bionics. JB Lippincott.

Martin, R.C. (2002): Agile Software Development, Principles, Patterns, and Practices. Prentice Hall.

McCain, Roger A. (2014): Game Theory: A Nontechnical Introduction to the Analysis of Strategy. World Scientific Publishing Co

McDermid, J. A. and Rook, P. (1991): Software Development Process Models. Software Engineer's Reference Book, ed. by McDermid, J. A. London, Butterworth-Heinemann

McKeown, M. (2016): The Strategy Book. How to Think and Act Strategically to Deliver Outstanding Results. FT Press.

McKinnon, Ronald C. (2012): Safety Management: Near Miss Identification, Recognition, and Investigation. Taylor & Francis Group

Meadows, D.; Meadows, D.; Randers, J. (1972): The Limits to Growth. Chelsea Green Pub Co

Meadows, D.; Meadows D.L.; Randers J. (2004): Limits to Growth: The 30-Year Update. Chelsea Green

Mendling, J.; Reijers, H. A.; van der Aalst, W. M. P. (2010). Seven process modeling guidelines (7PMG). Information and Software Technology 52 (2), 127–136.

Michalewicz, Z.; Fogel, D.B. (2004): How To Solve It: Modern Heuristics. Springer.

Miles, Lawrence D. (1972): Techniques of Value Analysis and Engineering, *McGraw-Hill Book Company*

Milosevic, D. Z. (2003): Project Management ToolBox: Tools and Techniques for the Practicing Project Manager. Wiley.

Mintzberg, Henry and Quinn, J.B. (1988): The Strategy Process, Prentice-Hall, Harlow.

Moeller, M.; Stolla, C.; Doujak, A. (2008): Strategic Innovation: Building New Growth Business. Wien: Goldegg Verlag

Morgan, J. M.; Liker, J. K. (2006): The Toyota Product Development System: Integrating People, Process and Technology. Taylor & Francis

Morris, Peter W. G.; Pinto, Jeffrey K.; a.o. (2012): The Oxford Handbook of Project Management, OXFORD University Press

Nas, T. F. (1996): Cost-benefit analysis: Theory and application. Sage Publications.

N. N. Authors: Harvard Business School Press (2009): Innovator's Toolkit: 10 Practical Strategies to Help You Develop and Implement Innovation. Harvard Business School Press.

N.N. (2009): Set Based Design. http://silkandspinach.net/2007/01/14/agile-set-based-design/

N.N. Hybrid project management: the best of both worlds. (n.d.). https://www.microtool.de/en/what-is-hybrid-project-management/

Nadler, G. (1967) Work Systems design. The Ideals Concept. Homewood, Ill. Richard D. Irwin Inc.

Nakagawa, Toshio (2005): Maintenance Theory of Reliability. London. Springer.

Nambisan, Satish ed. (2010): Information Technology and Product Development. Springer

Nash, J. (1950): Equilibrium points in n-person games. Proceedings of the National Academy of Sciences. 36(1):48–49.

Negele, H., Fricke E., Igenbergs E.. (1997): ZOPH. A Systemic Approach to the Modeling of Product Development Systems. Proc. of the 7th Int. Sympos. of INCOSE

Newell, A. (1972): Human Problem Solving. Prentice-Hall.

Oak A. (2011) What can talk tell us about design? Analyzing conversation to understand practice. Design Studies Online publication date: 13-Jan-2011.

Oak, A. (2012): You can argue it two ways: The collaborative management of a design dilemma. Design Studies: The International Journal for Design Research in Engineering, Architecture, Products and Systems. 33(6) pp. 630–648.

O'Brien, J.J.; Plotnick, F.L. (2010). CPM in Construction Management, Seventh Edition. McGraw Hill.

Omachonu, Vincent K.; Ross, Joel E. (2004): Principles of Total Quality, Third Edition. Taylor & Francis,

Ortmeier, P. J. (2001): Security Management: An Introduction. Prentice Hall.

Osborn, A. F. (1957): Applied Imagination. New York: Charles Scriber's Sons

Pahl, G.; Beitz, W.; Feldhusen, J.; Grote, K.-H. (2007): Engineering Design: A Systematic Approach. Springer.

Palmer, S.R.; Felsing, J.M. (2002). A Practical Guide to Feature-Driven Development. Prentice Hall.

Pearl, J. (2000): Causality. Cambridge University Press.

Peterson, M. (2009): An Introduction to Decision Theory. Cambridge University Press.

Pfletschinger, Th. (2008): Risiko-Management – Ein Beitrag zur methodischen Berücksichtigung von Risikofaktoren bei der Projektabwicklung und zum Nachweis des Nutzens eines Risiko-Managements. Dissertation. TU-Graz, 2008

Pitman Jim (1999): Probability. Springer Texts in Statistics edition

PMI (2013): A Guide to the Project Management Body of Knowledge (Pmbok Guide). Project Management Institute

Popper, K. (1959): The Logic of Scientific Discovery. London, Routledge

Popper, K. (1957): The Poverty of Historicism. London, Routledge.

Popper, K.; Peterson A.F.; Mejer, J. (1998): The World of Parmenides, Essays on the Presocratic Enlightenment. London, Routledge

Probst G. J. B. and Gomez P. (1992): Thinking in Networks to Avoid Pitfalls. Springer

Project Management Institute (2006): Practice Standard for Work Breakdown Structures. Project Management Institute.

Quinn, R.E.; Faerman, S.E. (2007): Becoming a Master Manager: A Competing Values Approach. John Wiley & Sons

Raftery, J. (1994): Risk Analysis in Project Management. E & F N Spon, London.

Raja, V.; Fernandes, K. J. (2008): Reverse Engineering. An Industrial Perspective. London: Springer

Rantanen, K.; Domb, E. (2007): Simplified TRIZ: New Problem Solving Applications for Engineers and Manufacturing Professionals. AUERBACH; 2 edition

Rechtin, E. (1991): Systems Architecting, Creating and Building Complex Systems. Upper Saddle River N. J.: Prentice-Hall

Rechtin, E.; Maier, M. (1997): The Art of Systems Architecting. Boca Raton: CRC Press Inc

Reichert, F., Kunz, A., Moryson, R, 2008, MAE-P3 – A System to Gain Transparency of Production Structure as a Basis for Production Relocation Planning, icseng 19th International Conference on Systems Engineering, Proceedings: 458–463.

Rendgen, S.; et al. (2012): Information Graphics, Taschen edition

Rescher, N. (1998): Predicting the future: An introduction to the theory of forecasting. State University of New York Press.

Ribbens, J.A. (2000): Simultaneous Engineering. John Wiley & Sons

Rittel, H. (1972): On the Planning Crisis: Systems Analysis of the First an Second Generation. Reprint No. 107, The Institute of Urban and Regional Development, University of California, Berkeley, California,

Rittel, H. W., Webber, M. M. (1973): Dilemmas in a General Theory of Planning. In: Policy Sciences 4, 155–169

Roberts, N. H.; W.E. Vesely (1987): Fault Tree Handbook. Government Printing Office

Robins, D. (2016): Is the Hybrid Methodology the Future of Project Management? https://www.projectmanagement.com/articles/356356/Is-the-Hybrid-Methodology-the-Future-of-Project-Management-

Rodov, A. & Teixidó, J. (2016). Blending agile and waterfall: the keys to a successful implementation. Paper presented at PMI® Global Congress 2016—EMEA, Barcelona, Spain. Newtown Square, PA: Project Management Institute. https://www.pmi.org/learning/library/blending-agile-waterfall-successful-integration-10213

Robinson, St. (2004): Simulation: The Practice of Model Development and Use. John Wiley & Sons.

Rodov, A. and Teixidó, J. (2016). Blending agile and waterfall: the keys to a successful implementation. Paper presented at PMI® Global Congress 2016. https://www.pmi.org/learning/library/blending-agile-waterfall-successful-integration-10213

Rouillard, L.A. (2002): Goals and goal setting: Achieving Measured Objectives, Crisp Learning

Rose, Kenneth H. (July 2005). Project Quality Management: Why, What and How. J. Ross Pub., 2005

Ross, A.; Rhodes D.; Hastings, D. (2008): Defining changeability: Reconciling flexibility, adaptability, scalability, modifiability, and robustness for maintaining system lifecycle value. http://web.mit.edu/adamross/www/Ross_JSE07_preprint.pdf, http://onlinelibrary.wiley.com/doi/10.1002/sys.20098/full

Rossmann T.; Tropea C.; Vincent, J. (2007): Bionics – Natural Technologies and Biomimetics. Springer, Berlin.

Royce, W. (1970): Managing the Development of Large Software Systems. In: Technical Papers of Western Electronic Show and Convention (WesCon), Aug. 25–28, 1970, Los Angeles, USA

Rubinstein, R. Y.; Kroese, D. P. (2007). Simulation and the Monte Carlo Method (2nd ed.). New York: John Wiley & Sons

Rummler, G.A.; Brache, A.P. (1995): Improving Performance. How to Manage the White Space in the Organization Chart. (Kepner-Tregoe-Method). Hoboken: Jossey Bass Business and Management Series

Saaksvuori, Antti (2008). Product Lifecycle Management. Springer.

Saaty, Th. L. (2001): Decision Making for Leaders – The Analytic Hierarchy Process for Decisions in a Complex World (3.ed.). Pittsburgh: RWS Publishing

SAE (2009): Potential Failure Mode and Effects Analysis in Design (Design FMEA), Potential Failure Mode and Effects Analysis in Manufacturing and Assembly Processes (Process FMEA). Product Code: J1739

Sassone, P.G., and Schaffer, W.A. (1978): Cost Benefit Analysis. A Handbook. London: Academic Press.

Sato, Yoshihiko; Kaufman, J. Jerry (2005): Value Analysis Tear-Down: A New Process for Product Development and Innovation. Industrial Press Inc.

Schantin, D. (2004): Makromodellierung von Geschäftsprozessen. Kundenorientierte Prozessgestaltung durch Segmentierung und Kaskadierung. Wiesbaden: DUV Gabler Edition Wissenschaft

Schrijver, A. (1998): Theory of Linear and Integer Programming. John Wiley and Sons.

Schröer, B.; Kain A., and Lindemann U. (2010): Supporting Creativity In Conceptual Design: Method 635-Extended. DS 60: Proceedings of DESIGN 2010, the 11th International Design Conference, Dubrovnik, Croatia

Schulz, A.; Clausing, D.; Fricke, E. and Negele, H. (2000): Development and Integration of Winning Technologies as Key Competetive Advantage. Systems Engineering, Vol. 3, No. 4. (2000), pp. 180–211

Schuyler, J. R. (2001): Risk and Decision Analysis in Projects. Newtown Square, PA: Project Management Institute PMI

Schwaber, K. (2004): Agile Project Management with Scrum. Microsoft Press

Schwaber, K. (2007): The Enterprise and Scrum (Best Practices). Microsoft Press

Schwaber, K.; Beedle, M. (2001): Agile Software Development with Scrum. Upper Saddle River N. J.: Prentice Hall

Scrambler, J. (2017): 10 Frameworks for Mobile Hybrid Apps. https://blog.jscrambler.com/10-frameworks-for-mobile-hybrid-apps/

Scrum.org (n.d.): What is Scrum? A Better Way Of Building Products https://www.scrum.org/resources/what-is-scrum

Senge, P.M. (1990): The Fifth Discipline: The Art & Practice of The Learning Organization. New York, Doubleday

Senge, P.M. (1994): The Fifth Discipline Fieldbook: Strategies and Tools for Building a Learning Organization. New York, Doubleday

Sherwood, D. (1998): Unlock Your Mind: A Practical Guide to Deliberate and Systematic Innovation. Ashgate Publishing

Silver Bruce (2011): BPMN Method and Style with BPMN Implementer's Guide. Cody-Cassidy Press

Silver, N. (2013): The Signal and the Noise: Why So Many Predictions Fail — but Some Don't. Penguin Press

Silverstein, D.; Samuel, Ph.; DeCarlo, N. (2008): The Innovator's Toolkit: 50+ Techniques for Predictable and Sustainable Organic Growth. John Wiley & Sons.

Simister, S. J. (1994): Usage and benefits of project risk analysis and management. Internat. Journal Of Project Management, 1994, 12, (1), 5–8.

Simon, Herbert A. (1969) The Sciences of the Artificial. MIT Press, Cambridge Mass.

Snijders, Paul; Wuttke, Thomas; Zandhuis Anton (2009): PMBOK Guide. Van Haren Publ.

Spear, St.; Bowen, H. K. (1999): Decoding the DNA of the Toyota Production System. In: Harvard Business Review HBR. Sept-Oct 1999, pp. 97–106

Stelzmann, E. (2011): Agiles System Engineering. Eine Methodik zum besseren Umgang mit Veränderungen bei der Entwicklung komplexer Systeme. PhD thesis. TU Graz.

Sterman, J. D. (2006): Business Dynamics – Systems Thinking and Modeling for a Complex World. McGraw-Hill Higher Education.

Steward, D.V. (1981): The Design Structure System. A Method for Managing the Design of Complex Systems. In: IEEE Transactions on Engineering Management. 28(3), 1981, pp. 71–74

Stjepandić Josip; Wognum Nel; Verhagen Wim J.C. (eds.) (2016): Concurrent Engineering in the 21st Century: Foundations, Developments and Challenges. Springer

Suh, N. (1990): The Principles of Design. Oxford University Press

Sundarapandian, V. (2009): Queueing Theory, 7.ed. *Probability, Statistics and Queueing Theory*. PHI Learning.

Sutherland J., Schwaber, K (1995). Business object design and implementation. Springer, London

Tague N R. (2005): The Quality Toolbox. ASQ Quality Press

Taha, H.A. (1992): Operations research: An introduction. MacMillan

Takeda, H. (2006): The Synchronized Production System: Going Beyond Just-In-Time Through Kaizen. London, Kogan Page

Tennant, Geoff (2001). SIX SIGMA: SPC and TQM in Manufacturing and Services. Gower Publishing, Ltd.

Teramata, T.; Nijstad, B.A. (Eds.) (2003): Group Creativity: Innovation Through Collaboration. Oxford University Press.

Terano, T., H. Kita, T. Kaneda, K. Arai, and H. Deguchi, Eds. (2005), Agent-Based Simulation: From Modeling Methodologies to Real-World Applications, Springer, Berlin, Germany.

Tidd, J.; Bessant, J. (2009): Managing Innovation: Integrating Technological, Market and Organizational Change. Wiley

Torgerson, W. S. (1968): Theory and Methods of Scaling. New York. Wiley

Triantaphyllou, E. (2000): Multi-Criteria Decision Making: A Comparative Study. Dordrecht, The Netherlands: Kluwer Academic Publishers.

Tufte, Edward R. (2001): The Visual Display of Quantitative Information, Graphics Press

Ulrich, K. (1995): The role of product architecture in the manufacturing firm. In: Research Policy 24 (3), pp. 419–440

Ulrich, P.H.; Probst, G.J. (1995): Anleitung zum ganzheitlichen Denken und Handeln. Bern: Haupt

Ulwick, A. (2005): What Customers Want: Using Outcome-Driven Innovation to Create Breakthrough Products and Services. McGraw-Hill

VanGundy, A.B. (2007): Getting to Innovation: How Asking the Right Questions Generates the Great Ideas Your Company Needs. AMACOM

VDI-2221 (1993): Methodik zum Entwickeln und Konstruieren technischer Systeme und Produkte. Berlin: Beuth

Verganti, R. (2009): Design Driven Innovation: Changing the Rules of Competition by Radically Innovating What Things Mean. Harvard Business Press

Vester, F. (2007): The Art of Interconnected Thinking. Ideas and Tools for tackling complexity. MCB-Verlag, München

de Ville, B.; Padraic, N. (2013): Decision Trees for Analytics Using SAS® Enterprise

Walleck, A.S., O'Halloran, J.D. and Leader, C.A. (1991): Benchmarking world-class performance. The McKinsey Quarterly, No. 1, pp. 3–24.

Walpole, R.E. (2007, 9.ed.): Probability & Statistics For Engineers & Scientists. Pearson Education

Weihrich, H. (1985): Management excellence: productivity through MBO, NY, McGraw Hill

Weilkiens, T. (2008): Systems Engineering with SysML/UML: Modeling, Analysis, Design. Elsevier

Weisberg, D. (2010): The Engineering Design Revolution, E-Book, www.cadhistory.net, May 2010

Westland, J. (2016): What is Hybrid Methodology? https://www.projectmanager.com/blog/what-is-hybrid-methodology

Wenzel, St.; Bauch, Th.; Fricke, E., Negele, H. (1997): Concurrent Engineering and more. A Systematic Approach to Successful Product Development. In: Proceedings of the 7th Symposium of INCOSE. Los Angeles.

Wenzel, St.; Igenbergs, E.; Michl, Th.; Megerle, F. (2000): Coupling Changes to Product-, Process-, and Agent-System Architectures – A Holistic Framework for Change in Product Development Organizations. In: Proceedings of the "Second European Conference on Systems Engineering" EuSEC 2000. München, Utz-Verlag

Wenzel, St.; Negele, H.; Schulz, A. (2000): Reducing Time to Market. An Empirical Case Study about Strategies, Methods, Tools, and Actions. In: Proceedings of the Regional Conference of INCOSE. Denver, USA.

West, M. (2013): Return On Process (ROP): Getting Real Performance Results from Process Improvement. CRC Press. Taylor & Francis Group

Wheelan, Ch. (2013) Naked Statistics: Stripping the Dread from the data. New York. Norton &Company

White, Stephen A.; Bock, Conrad (2011). BPMN 2.0 Handbook: Methods, Concepts, Case Studies and Standards in Business Process Management Notation.

Wideman, R.M. (1992): Project and Program Risk Management. Newtown Square, PA: Project Management Institute.

Wilson, B. (1990): Systems: Concepts, Methodologies and Applications (2nd ed.). Chichester NY: John Wiley & Sons

Winston, W.L. (2008): Operations Research: Applications and Algorithms. Thomson Business Press.

Womack, James P. and Jones, Daniel T. (2003): Lean Thinking: Banish Waste and Create Wealth in Your Corporation. Harper Business

Wright, G.; Cairns, G. (2011): Scenario thinking: practical approaches to the future, Palgrave Macmillan

Wysocki, R.K., Beck Jr, R., Crane, D.B. (2000): Effective Project Management. John Wiley

Wysocki, R. (2014): Effective Project Management: Traditional, agile, extreme. John Wiley

Zelazny, Gene (2001): Say It With Charts: The Executive's Guide to Visual Communication, McGraw-Hill Education

Zingel Chr. (2018): AVL List, MBSE/SysML-Workshop, Institute of Automotive Engineering, TUG, V1.0

Zwicky, F. (1969): Discovery, Invention, Research – Through the Morphological Approach. Toronto: The Macmillan Company

# Index

© Springer Nature Switzerland AG 2019
R. Haberfellner et al., *Systems Engineering*,
https://doi.org/10.1007/978-3-030-13431-0

Printed in the United States
by Baker & Taylor Publisher Services